T0211623

"In many introductory methods classes, instructors cobble together numerous sources to design a robust class. This volume has it all in one place—an explanation of research questions, a strategy to research the literature and annotate it, clear explanations of the major research methods of the field of technical communication, instructions on writing up the research report—plus examples of award-winning, full-length, peer-reviewed studies pulled from our own major journals, which are much more relatable to new researchers than trying to see how psychology or education studies apply them. Throughout, its tone is reassuring, its explanations crystal. I strongly recommend anyone teaching or becoming a new researcher use this volume as their introduction."
– *Angela Eaton, PhD, Associate Professor of Technical Communication and Rhetoric,*
Texas Tech University

"This revised edition offers thorough, practical, and applied information for anyone interested in technical communication research. The rich content and supplementary materials give readers diverse techniques for improving their research practices. This is a must-read primer that is as effective for the classroom as it is the workplace."
– *Ryan Boettger, PhD, Associate Professor of Technical Communication,*
University of North Texas

"As the field of technical communication has grown in the last decades, well-designed research has become especially crucial to the advancement of knowledge. Written in an accessible style, the second edition of *A Research Primer for Technical Communication: Methods, Exemplars, and Analyses* will prove useful to advanced undergraduates and graduate students studying technical communication research approaches."
– *Marjorie Rush Hovde, PhD, Associate Professor of Technical*
Communication, Indiana University-Purdue University Indianapolis

"*A Research Primer for Technical Communication: Methods, Exemplars, and Analyses* (2nd ed.) delivers what its title promises. This edition retains the most important characteristic of the first one: clear, straightforward explanations that should be accessible and engaging to all readers. A must-have for any student or professional technical communicator."
– *Saul Carliner, PhD, Professor of Educational Technology,*
Concordia University (Montréal)

A Research Primer for Technical Communication

This fully revised edition provides a practical introduction to research methods for anyone conducting and critically reading technical communication research.

The first section discusses the role of research in technical communication and explains in plain language how to conduct and report such research. It covers both quantitative and qualitative methods, as well as surveys, usability studies, and literature reviews. The second section presents a collection of research articles that serve as exemplars of these major types of research projects, each followed by commentary breaking down how it corresponds to the information on that research type. In addition to five new chapters of exemplars and commentaries, this second edition contains a new chapter on usability studies.

. This book is an essential introduction to research methods for students of technical communication and for industry professionals who need to conduct and engage with research on the job.

George F. Hayhoe is Professor Emeritus of technical communication at Mercer University. He has served as editor-in-chief of the *IEEE Transactions on Professional Communication* since 2016, and was editor of the Society for Technical Communication's journal, *Technical Communication*, from 1996 to 2008. He is a Life Fellow of IEEE and a Fellow of the Society for Technical Communication.

Pam Estes Brewer is Professor and Director of the MS in Technical Communication Management program at Mercer University. She focuses on the communication of international virtual teams, online communication strategy, and usability research. Dr. Brewer is a Fellow in the Society for Technical Communication and authored the 2015 book *International Virtual Teams: Engineering Global Success*.

A Research Primer for Technical Communication

Methods, Exemplars, and Analyses

2nd edition

George F. Hayhoe and Pam Estes Brewer

Routledge
Taylor & Francis Group

NEW YORK AND LONDON

This edition published 2021
by Routledge
52 Vanderbilt Avenue, New York, NY 10017

and by Routledge
2 Park Square, Milton Park, Abingdon, Oxon, OX14 4RN

Routledge is an imprint of the Taylor & Francis Group, an informa business

© 2021 Taylor & Francis

The right of George F. Hayhoe and Pam Estes Brewer to be identified as authors of this work has been asserted by them in accordance with sections 77 and 78 of the Copyright, Designs and Patents Act 1988.

All rights reserved. No part of this book may be reprinted or reproduced or utilized in any form or by any electronic, mechanical, or other means, now known or hereafter invented, including photocopying and recording, or in any information storage or retrieval system, without permission in writing from the publishers.

Trademark notice: Product or corporate names may be trademarks or registered trademarks, and are used only for identification and explanation without intent to infringe.

First edition published by Routledge (2007)

Library of Congress Cataloging-in-Publication Data
Names: Hayhoe, George F., author. | Brewer, Pam Estes, author. | Hughes, Michael A. Research primer for technical communication.
Title: A research primer for technical communication : methods, exemplars, and analyses / by George F. Hayhoe, Pam Estes Brewer.
Description: Second edition. | New York, NY : Routledge, 2020. |
Revised edition of: A research primer for technical communication : methods, exemplars, and analyses / Michael A. Hughes, George F. Hayhoe, 2008. |
Includes bibliographical references and index.
Identifiers: LCCN 2020014567 (print) | LCCN 2020014568 (ebook) |
Subjects: LCSH: Communication of technical information. | Technology–
Documentation. | Technical manuals. | Technical writing.
Classification: LCC T10.5 .H4 2020 (print) | LCC T10.5 (ebook) |
DDC 808.06/66–dc23
LC record available at https://lccn.loc.gov/2020014567
LC ebook record available at https://lccn.loc.gov/2020014568

ISBN: 978-0-367-53147-8 (hbk)
ISBN: 978-0-367-53148-5 (pbk)
ISBN: 978-1-003-08068-8 (ebk)

Typeset in Sabon
by Swales & Willis, Exeter, Devon, UK

Contents

Preface

We are pleased to present the second edition of *A Research Primer for Technical Communication: Methods, Exemplars, and Analyses*. Pam Estes Brewer of Mercer University has succeeded Michael A. Hughes as co-author of the book.

This second edition has been significantly revised, and those revisions of the original book have been guided by the suggestions of six reviewers who had used the 2008 text and responded to the publisher's request to offer feedback and suggestions. Chapters 1–6 have been significantly rewritten, and a new Chapter 7 on conducting usability studies has been added. All of the exemplar chapters (Chapters 8–12) are new and feature sample articles from four of the major journals in the field. Information on citation styles has been moved from Chapter 3 to the Appendix.

We believe that, whether you are a student or a practitioner, this book contains essential information that you need to know to perform—or to be an informed consumer of—research in the field of technical communication. First, it is a primer for how to conduct and critically read research about technical communication. Secondly, it reprints four research articles with commentary that analyzes how each article exemplifies a particular type of research report. Lastly, the content of the articles themselves provides grounding in important research topics in the field.

The book is presented in two sections. The first section discusses the role of research in technical communication and explains in plain language how to conduct and report such research. This section covers both qualitative and quantitative methods, and presents the required statistical concepts and methods in ways that can be understood and applied by non-mathematicians. This section is not intended to be an exhaustive discussion of qualitative, quantitative, and mixed methods, but it is sufficient to enable you to conduct research projects and write reports of your findings. The chapters are structured and sequenced to help you identify a research topic, review the appropriate literature, construct a test plan, gather data, analyze the data, and write the research report. Each chapter contains specific activities that will help student readers complete a research project. For practitioner readers, this first section also stresses how to apply its principles as a critical consumer of research articles within the field.

The second section is a collection of articles from *Technical Communication*, the *IEEE Transactions on Professional Communication*, *Technical Communication Quarterly*, and the *Journal of Usability Studies*. The articles have been selected both for their contribution to the body of research in this field as well as for their ability to illuminate the principles of research explained in the first section of the book. Each chapter in this section begins with an introduction that places the subject article in the context of its research methodology and its contribution to scholarship in the field.

Following the full text of each article is a commentary that explains how it exemplifies a particular type of research.

People learn best by doing, and this book will help you learn about research by tutoring you through your own research projects and letting you read actual research articles in this field. Activities and checklists at the end of each chapter will help you apply the principles of that chapter to your actual projects. We think that you will find this book an essential *how-to* if you are a student doing research projects, or as background reading if you are a technical communication professional who *reads the literature* as a way of staying abreast of your ever-changing field.

George F. Hayhoe and Pam Estes Brewer

March 2020

Acknowledgments

First, we are most grateful to Michael Hughes for his contributions to the first edition of this book. You laid an excellent foundation.

In addition to the authors and publishers of the excellent exemplars presented in this book, we would like to thank the following individuals. Many thanks to Angela Eaton and David Nelson for their review of quantitative concepts. Lee Olson, Head of Research Services at Mercer's Tarver Library, provided much appreciated advice on research databases. Thanks also to Mercer graduate students Hannah Nabi and Jodie O'Driscoll for sharing examples from their own research. And finally, thank you to Evan Gambill, a Mercer undergraduate student, for his good eye in editing.

Part I
Methods

1 Research
Why We Do It and Why We Read It

Introduction

If you were asked to envision someone "doing research," you might imagine a student in a library sitting amidst stacks of books. Or you might envision an internet search starting with Google. Or you might picture a scientist in a white coat sitting in a laboratory surrounded by flasks of colorful, bubbling liquids, peering through a microscope. Certainly, these are all examples of research, but they illustrate a narrow, although popular, view about research: namely, that it is academic and somewhat removed from our everyday world.

This book, however, is about research that shapes our professional practice— research that informs practical decisions that technical communicators make. The purpose of this chapter is to introduce you to the role that research plays in technical communication and the types of research that can be done. Its goal is to encourage you to think about your own research potential as a student or practitioner of technical communication, or the ways that you could apply the research of others to your own practice as a technical communicator. It also tries to help you understand that the way a researcher thinks about and approaches a research project can influence what he or she finds.

Learning Objectives

After you have read this chapter, you should be able to:

- Classify research based on its goals
- Classify research based on its methods
- Describe the role of industry, academia, professional communities, and government in regulating the admission of new knowledge into a field of practice
- Describe the different hierarchies of research publications

What is Research?

Peter Senge (1990) uses the term *abstraction wars* to describe a kind of debate typified by a free-for-all of opinions and personal beliefs. You encounter these types of arguments quite frequently among technical communicators (as you would among practitioners of any profession). For example, a group of writers trying to collaborate on a set of documents might argue about the media used to deliver the content so as to protect intellectual property. They might also argue about the mechanics of documents, such as what kind of font should be used as the body text, how many steps procedures should be limited to, whether important notes should be printed in boldface, and so forth.

But in a professional field of practice such as technical communication (like other fields of practice), abstraction wars should not dictate the tenets of the practice. What makes up good technical communication should not rest on arbitrary whims of the individual writer or the personal persuasiveness of those advocating a particular standard or technique. There needs to be a way that the best practices of a profession can emerge as a recognized and reliable consensus among the practitioners of that profession. Well-conducted research can be such a way.

When someone in a meeting says, "Users don't want to go to an online reference to get this information," a reasonable counter is to ask, "What makes you say that?" Typically, what the questioner is seeking is evidence or *data*, and beyond that, an indication of how the data support what the speaker is advocating. What differentiates reasonable arguments from abstraction wars is the use of verifiable observations to support the point being advocated, what the field of action science calls building a *ladder of inference* back to directly observable data (Argyris, Putnam, & Smith, 1985).

A Definition

In essence, the linking of actions, decisions, or advocacy to observable data is what research is all about. In this book, the term *research* is used to mean *the systematic collection and analysis of observations for the purpose of creating new knowledge that can inform actions and decisions*. Let's look at what this definition is saying.

- Research is systematic—The purpose of this book is to describe a repeatable process for conducting research, one that has protocols and safeguards meant to deliver reliable outcomes.
- Research involves the collection and analysis of data—The importance here is that these are two separate activities. The mindset of the researcher is first to focus on gathering good data and *then* to determine what they mean. Data gathered to "prove a point" will almost invariably prove that point, meaning that researcher preconceptions and biases will influence the research design and the data analysis. For example, news sources sometimes select information that supports one point of view and ignore information that may support another. Thus, information can be removed from its context and prove virtually any point. Good researchers must always be willing to go where the data take them.
- Research creates new knowledge—Don't be misled by misconceptions from high school where a "research paper" was intended to show the teacher what you had learned reading from the internet. In a field of practice, such as technical communication, research should advance our collective knowledge about our field.
- Research should inform actions and decisions—Because technical communication is a field of practice, the outcome of research should enable us to do our jobs better. Technical communication, like engineering, is an applied discipline. That is, we apply theories to solving problems. Research in our field takes on a pragmatic aspect associated with the kind of research often called *action*

research. "In action research, a researcher works with a group to define a problem, collects data to determine whether or not this is indeed the problem, and then experiments with potential solutions to the problem" (Watkins & Brooks, 1994, p. 8).

The concept of research also carries with it the assumption that the knowledge created is applicable at a generalized level and is repeatable over multiple instances, producing the same results. For example, although we may sometimes wish to describe the characteristics of one person or a small group to further a goal within an organization, we most often wish to add value by describing how people in general or a class of people behaves. Similarly, it would not be the role of research merely to describe how readers used a specific document; its value would come from describing how readers use documentation or a genre of documentation. These last examples illustrate one of the primary challenges that researchers face: *Research must ultimately articulate generalized truths from specific instances.* For example, if a researcher wants to know how readers in general process information from web pages, that researcher cannot look at all readers nor analyze all web pages. The researcher has a narrow access to *some* readers of *some* web pages, and must optimize this limited opportunity to learn as much as possible about readers and web pages in general.

Research in Technical Communication

Research in technical communication is not an activity conducted in a vacuum; it is generally initiated by a problem or a need to understand a phenomenon related to technical communication. Nor is it an activity conducted for its own sake; its conclusions should move the field of technical communication forward, improve technical communicators' decisions, and make their actions based on those decisions more effective than if they had acted without that knowledge. Just as individuals in a meeting want inferences based on data to support someone's assertions, the field of technical communication relies on research to inform best practices within that field.

Classifying Types of Research

As much as researchers and readers would like to believe that research is a totally objective undertaking, it is not. Ultimately, it is shaped by the goals of the researchers and the methods that they choose to use. Those who do not acknowledge these influences will not be able to manage their own biases or perspectives as researchers, nor be able to critically evaluate potential bias or the effects of researcher perspective in the research that they read. The purpose of this section is to examine the different goals that researchers have, the methods that they employ, and the ways that those goals and methods can direct and affect the outcome of research.

The Scientific Method

Starting with Francis Bacon in the 17th century, research has been associated with a methodology called the scientific method, characterized by the following process (Bright, 1952):

1. Observing a phenomenon or aspect of the physical universe
2. Stating a tentative description, called a hypothesis, that explains the observations
3. Using the hypothesis to make a prediction about what effect would follow a certain action
4. Conducting an experiment to see whether the predicted effect actually does result from the stated action

This model evolved primarily within the physical sciences and is still widely used. However, with the advent of social sciences such as psychology, anthropology, and sociology, research has developed a broader set of methods. Although all of these methods rely on the observation of data and on rigorous techniques for validating the conclusions drawn from those data, research is no longer bound to a strict reliance on hypotheses and experiments as defined in the scientific method.

The following discussions describe the more complex landscape of modern research by classifying research genres by their goals and their methods.

Goals of Research

Research goals act as lenses that affect how the researcher filters and interprets data. Just as camera lenses can help a photographer focus on certain details of a landscape, they can also de-emphasize other details. Therefore, part of the photographer's science and skill is being able to select the correct lens for a specific objective or type of subject. Similarly, part of the science and skill of a researcher or critical reader of research is to select or recognize what goals are driving the research being conducted or studied.

This concept of researcher-as-filter is an important one and is often overlooked by researchers and readers of research alike. For example, suppose that your teacher is leading the class on a tour of your university's usability lab, and the lab director points out the logging station. The director explains that the role of the logger is to keep a running narrative of the user's actions. Your teacher takes the opportunity to interject and point out to the class the importance of the logger's role since this is the first filtration of the data—that is, the logger makes decisions about what data to log and what not to log.

The lab director disagrees with this comment and claims that their loggers record all of the data. Your teacher asks whether loggers typically note details such as when they scratch their arms, shift in their chairs, or brush back their hair. The director almost snorts and replies, "Of course not." And certainly they shouldn't, but the point is that every researcher applies filters and constantly makes decisions about which data are important and which are not. When researchers do not acknowledge these filters, and when readers of research are not sensitive to the fact that some data have been filtered out before even being shared with the reader, they lose the ability to analyze the results critically.

Categories of Research Goals

To form your own research agenda or to understand better the research done by others, it is helpful to understand the various goals that drive research. Reeves (1998), as the editor of the *Journal of Interactive Learning Research*, identified six categories with which to classify educational research by its goals. Technical communicators face many of the same questions and situations faced by educators, and these classifications can help researchers in technical communication as well. Table 1.1 describes Reeves's classifications in the context of technical communication.

Table 1.1 Classifications of research based on goals

Research Goal	Description
Theoretical	Focuses on explaining phenomena without necessarily providing an immediate application of the findings.
Empirical	Focuses on testing hypotheses related to theories of communication and information processing in the context of their application to technical communication.
Interpretivist	Focuses on understanding phenomena related to technology's impact on the ways that humans interact with technology, users interact with the products that technical communicators produce, or technical communicators interact with people in other roles within an organization.
Postmodern	Focuses on examining the assumptions that underlie applications of technology or technical communication with the ultimate goal of revealing hidden agendas and empowering disenfranchised groups.
Developmental	Focuses on the invention and improvement of creative approaches to enhancing technical communication through the use of technology and theory.
Evaluative	Focuses on a particular product, tool, or method for the purpose of improving it or estimating its effectiveness and worth.

THEORETICAL RESEARCH

Technical communication is a field of practice more than a field of study. As such, not many technical communicators engage in theoretical research, but they often rely on the theoretical research conducted by other fields, such as cognitive psychology or human factors.

For example, a classic theoretical research article that has had a profound impact on technical communication is George Miller's "The magical number seven, plus or minus two" (1956). In that article, Miller defines limitations in human capacity to hold information. His observations have influenced many practices in technical communication, such as the optimal way to chunk data so that it is easily processed and remembered.

The point is that although Miller's article has heavily influenced technical communication, he does not specify concrete practices within it. The purpose of theoretical research is to understand the phenomenon, not necessarily to point to its ultimate application. This seeming lack of an action-based resolution does not belittle the practicality of this kind of research; it merely points out that theoretical research is foundational, often driving the other kinds of research described in the following discussions.

EMPIRICAL RESEARCH

Empirical research is the type that most people are familiar with. It has the following characteristics.

- It is based on observation or experience.
- It is based on quantitative and/or qualitative data that are collected systematically.
- It is verifiable.

This type of research is typified by the scientific method. It often involves the testing of hypotheses that come out of theoretical research. For example, empirical research often tests a hypothesis by comparing the results of a control group against the results of a group that has had a particular intervention applied. A technical communicator might read Miller's study "The magical number seven, plus or minus two" and form the hypothesis that the error frequency when users enter long software identification keys needed in an installation procedure might decrease if the numbers were chunked into groups of five digits. This hypothesis that includes a prediction could then be tested by comparing the performance of users who have the long, unchunked software identification key against the performance of users who are presented the same key that has been chunked in accordance with Miller's findings.

The advantage of empirical research is its relative objectivity. Its conclusions are based on observations and well-accepted analysis techniques. On the other hand, one of the disadvantages of empirical research is its narrow focus: *What you learn is tightly constrained by the question that you ask.* In the previous example, the researcher certainly might learn which way of presenting the software key was better (chunked versus unchunked), but it is highly unlikely that while using this approach, the researcher would discover an entirely novel approach to secure software that did not involve user-entered numbers.

INTERPRETIVIST RESEARCH

Interpretivist research is relatively new (compared to theoretical and empirical) and comes from the social sciences, where it was originally termed *naturalistic inquiry* (Lincoln & Guba, 1985). The main focus of interpretivist research is to *understand* rather than to test. Hendrick (1983) states that the purpose of this kind of research is to illustrate rather than provide a truth test. Whereas the example about testing the hypothesis of chunked text strings tested a very specific question ("Is chunked better than unchunked?") an interpretive approach would take a more open-ended strategy. Instead of asking, "Is this way better than that way," the interpretive researcher might ask, "How do people make sense of strings of text? How do they use them in the course of trying to accomplish their own tasks?" The observations might not be as quantifiable in this case as in an empirical study, consisting more of descriptions of user behavior or transcripts of interviews.

The advantage of this more open-ended approach is that it allows for the discovery of unexpected knowledge. The disadvantage, however, is that its non-experimental approach (that is, its lack of hypothesis testing) can allow researchers and readers alike to apply their own subjectivity and not reach agreement on the validity and reliability of the conclusions. (The concepts of *validity* and *reliability* are discussed in later chapters; for now, think of them as describing the quality of the research.)

POSTMODERN RESEARCH

Postmodern research is typified by a general cynicism about technology and an interest in social or political implications, especially where technology might disenfranchise certain groups. This type of research has gained a greater foothold in the field of education than in technical communication, but there are issues in technical communication that could be attractive to a postmodernist researcher. For example, where empiricists and interpretivists might try to apply information design principles to developing better approaches to electronic voting machines, postmodernists might want to research the impact that high technology has on discouraging older or less-educated voters. Their research might advocate that technology favors the *haves* and the status quo over the *have-nots* or disenfranchised constituents. Technical communicators might also be interested in researching how disinformation propagated through social media is used to sway consumers in particularly emotional or complex contexts.

Koetting (1996) summarizes the differences among empirical (which he calls positive science), interpretivist, and postmodern (which he calls critical science) research: "Positive science has an interest in mechanical control; interpretive science has an interest in understanding; and critical science has an interest in emancipation" (p. 1141).

DEVELOPMENTAL RESEARCH

Developmental research is targeted at producing a new approach or product as its outcome. Accordingly, much developmental research is conducted by companies. In a strong sense, professional associations play an important role in this regard. For example, some articles published in *Technical Communication*, the official journal of the Society for Technical Communication, are written by practitioners who are sharing discoveries made while working on developmental projects for their companies.

Continuing with the example of different kinds of research that could be spawned by Miller's original theoretical research, a technical communicator might conduct usability tests on different approaches to the interface of online help to discover how different chunking schemes could make user searches easier and more successful. The advantage of developmental research is its emphasis on practical application. The disadvantage is that its conclusions might not be generalized as easily or as broadly as other models.

EVALUATIVE RESEARCH

The main difference between developmental and evaluative research is that evaluative research starts with a completed product, whereas developmental research is conducted during the design phase of a product.

Michael Hughes interviewed Dr. John Carroll (personal communication, December 28, 1997) at the beginning of his own doctoral studies. In the interview, Hughes asked Carroll, the founder of Minimalism in information science, for advice on directing his own doctoral research. Carroll commented that one of the misconceptions people have is the notion that academics conduct all research to discover new knowledge and then business goes about applying it.

Carroll believed that the more exciting discoveries are being made by businesses, but that they often lack the time or expertise to fully understand or document why their advances work. He thought that academic research can add value by helping in this area. The type of research Carroll was suggesting would be a good application for evaluative research. The advantage of evaluative research, as Carroll points out, is that it draws on what is actually being done by the true thought and technology leaders. The disadvantage is that it is more difficult to generalize the lessons learned when a specific product is the focus of the study.

Exercise 1.1 Classifying Research by Goals

This exercise gives you some idea how varied research can be in the field of technical communication while letting you practice classifying research by goals.

Label the following descriptions of research projects by the type of research goal each seems to be pursuing. Use the following codes:

T   Theoretical
Em   Empirical
I   Interpretivist
P   Postmodern
D   Developmental
Ev   Evaluative

See the answer key at the end of the chapter for the recommended responses.

1. _____ Tests the hypothesis that boldfacing terms in a definition increases how well the term is remembered by comparing two groups: one reading definitions with boldfaced terms and the other reading definitions without boldfaced terms.
2. _____ Exposes the disparity in the treatment of women technical communicators versus male technical communicators in terms of salary and technical credibility, and advocates that women are not treated fairly in technology cultures.
3. _____ Assesses the success of a new online product database in improving customer service levels in a technical support call center.
4. _____ Describes the experience of seven first-time software application users, trying to understand what emotions or impressions that experience evokes in novice users.
5. _____ Analyzes the effect of animation on a computer screen on a viewer's field of focus.
6. _____ Compares and contrasts five leading master's programs in technical communication.
7. _____ Tries to understand what strategies users employ when looking for information in online help.
8. _____ Tries to determine the most effective layout of navigation links on a specific web site to optimize its usability.

Methods of Research

In addition to being categorized by its goals, research can also be categorized by its methods. The methodology that a researcher chooses to employ can shape the outcome of the research in the same way that tools affect an artisan's output. For example, a wood carver could be given the same raw material (wood) and objective ("carve a pelican"), but a different set of tools—for example, a chainsaw, a knife, or a set of large chisels. The outcome would be significantly different based upon which set of tools the carver used.

At a more technical level, the look and feel of a web site or online help might differ, even if only in subtle ways, based on the authoring tool chosen. The same applies to research. The outcome of the research can be greatly influenced by the methodology chosen to conduct it and analyze the data. Therefore, researchers and critical readers of research need to understand the differences in the methods and the ways that they influence the outcome of the research.

Categories of Research Methods

Table 1.2 lists the five categories described by Reeves (1998) in the guidelines for authors for the *Journal of Interactive Learning Research*.

QUANTITATIVE

Quantitative data and their analysis are typically associated with our traditional view of research. Quantitative research relies on statistics to analyze the data and to let the researcher draw reliable inferences from the findings. For example, a research project that compared the average times to complete a task using two styles of online help would use quantitative methods (the capturing of data in numeric format and the statistical analysis of those data).

QUALITATIVE

Qualitative methods rely on non-numeric data, such as interviews with users or video recordings of users trying to perform tasks. Qualitative data are more difficult to analyze in some respects than quantitative data, since they can be more susceptible to subjective interpretation. An example of a qualitative approach would be a field study where the researcher observes workers using a new software product and takes notes about how they go about using the documentation to learn the new product. The researcher might also supplement the observations with interviews. The resulting data,

Table 1.2 Classifications of research based on methods

Method	Description
Quantitative	Primarily involves the collection of data that are expressed in numbers and their analysis using descriptive and inferential statistics. This is the usual method employed in empirical research involving hypothesis testing and statistical analysis.
Qualitative	Primarily involves the collection of non-numeric data (data represented by text, pictures, video, and so forth) and their analysis using ethnographic approaches. This method is often used in case studies and usability tests where the data are the words and actions of the test participants. It can also be used in empirical studies.
Critical Theory	Relies on the deconstruction of "texts" and the technologies that deliver them, looking for social or political agendas or evidence of class, race, or gender domination. This method is usually employed in postmodern research.
Literature Review	Primarily involves the review and reporting on the research of others, often including the analysis and integration of that research through frequency counts and meta-analyses. This method is usually applied in research that integrates prior research.
Mixed Methods	Combines multiple methods, usually quantitative and qualitative. Mixed methods are often found in usability tests, which are a rich source of quantitative data, such as time to complete tasks or frequency of errors, and qualitative data, such as user comments, facial expressions, or actions.

then, would consist of notes and transcripts that would be analyzed to determine whether meaningful patterns emerge.

CRITICAL THEORY

Critical theory looks closely at texts—that is, formal or informal discourses—to determine what they are "really saying" (or what they deliberately are *not* saying) versus their superficial meanings. For example, a study might examine how technical communicators handle known product deficiencies in user guides by analyzing their language styles to see whether stylistic devices such as passive voice or abstraction obscures the meaning of the text and avoids frank discussion of the deficient features.

LITERATURE REVIEW

Although all research projects should include a literature review, some exclusively review other research projects published on the same subject (usually referred to as "the literature"). The purpose might be to look for trends across the research or to collect in one document a cross-section of the important opinions and findings about a particular topic. Think of such a review as a meta-analysis of the literature. For example, a researcher might do a literature review looking at research about the usefulness of readability measures and try to draw conclusions about general trends or findings. This kind of review can be particularly helpful to practitioners who want to adopt best practices.

MIXED METHODS

It is not unusual to find more than one of the above methods being employed in the same research project. For example, a usability test might use quantitative methods to determine where in a web site the majority of users abandon a task and then employ qualitative methods to understand why. Another example would be an evaluative study that uses interviews to determine which section of a manual is considered the most important by users and then compares how rapidly information from that section can be retrieved versus a similar section in a competitor's manual.

Exercise 1.2 Classifying Research by Method

This exercise asks you to apply the categories just discussed in the context of actual technical communication topics.

Label the following descriptions of research projects by the type of method each seems to be applying. Use the following codes:

QN   Quantitative
QL   Qualitative
C   Critical Theory
L   Literature Review
M   Mixed Methods

See the answer key at the end of the chapter for the recommended responses.

1. _____ Interviews graduates of technical communication degree programs after their first year out of the program to see how they think that their education has affected their professional growth.

2. _____ Records the time that it takes each of 12 users to install a product and calculates the average installation time.

3. _____ Examines the wording of the online ads for technical communication jobs to uncover how age-discrimination messages are being communicated.

4. _____ Counts the number of times users go to the online help and interviews each user to understand why they consulted help at the particular times that they did.

5. _____ Reports on seven different studies of font legibility to see whether there is a consensus among researchers concerning serif versus sans-serif typefaces.

6. _____ Compares average salaries of technical communicators based on their level of education.

7. _____ Conducts focus groups to determine the critical issues facing technical communicators today.

Association of Goals and Methods

There is a strong association between the goal of the research and the methods that a researcher employs. Table 1.3 shows associations often encountered in research.

If you are a student researcher or an experienced practitioner starting your first formal research project, the choices can seem overwhelming. But your personal interests or work requirements will dictate the goal, and most technical communication research relies on quantitative or qualitative methods, both of which are covered in this book.

As a critical reader of research, one question that you should ask is whether the researcher has applied the appropriate method for the stated or implied research goal. A mistake that researchers sometimes make is to choose the method of their research before they have clarified their goal. Often this problem results from a preference for qualitative over quantitative (by those afraid of "number crunching") or vice versa (by those who feel that only measurable phenomena can be trusted). The better approach is to *classify the goal first and then to choose the appropriate method(s)*. Classifying the goal is a direct outcome of choosing the research topic and research question, which is discussed in the next chapter.

Table 1.3 Common associations of goals and methods

Goals	Methods
Theoretical	Quantitative
Empirical	Quantitative, qualitative, or mixed
Interpretivist	Qualitative
Postmodern	Critical theory
Developmental	Quantitative, qualitative, or mixed
Evaluative	Quantitative, qualitative, or mixed

Exercise 1.3 Brainstorming Research Topics

The purpose of this exercise is to get you thinking about all the possibilities for a research project that you might conduct.

1. For each type of research goal presented in Chapter 1, identify one to three examples of possible research topics within the field of technical communication. Draw on topics from your studies or real life problems that you deal with as a technical communicator or user of technical documents.
2. For each topic identified in step 1, discuss which method(s) could be appropriate, and give a general example of what kinds of data might be gathered.

Research Sources

Earlier in this chapter, we stated that research in technical communication helped inform best practices within the field. But how does that happen? How does that research get sponsored and disseminated? In effect, there are four sources of research:

*   Industry
*   Academia
*   Professional societies
*   Government

The Role of Industry

One of the great contributions that industry makes is that it keeps research in a field relevant and practical. New ideas are scrutinized against two criteria.

1. Is it worth the cost? (Would users pay for it or would it reduce the price of products or services?)
2. Can you really do it? (That is, can you obtain the results that you promise?)

Both of these criteria drive the need for the collection and analysis of data: research! The disadvantage is that industry generally lacks both the motivation and the mechanisms for distributing the new knowledge that it creates. Businesses also need to protect any proprietary information developed through the research that they conduct.

The Role of Academia

Academic programs in technical communication are another source of research in this field. They have the advantage of having professors who provide both continuity and expertise combined with students who can provide the mental (and sometimes physical) labor. If you are a student getting ready to embark on a required research project as part of your studies, you are part of this mechanism that keeps the field of technical communication vibrant and relevant. Faculty and students also have easy access to the conference proceedings, journals, and books that publish research results. Faculty can also incorporate results into courses and influence students who then carry those results into the workplace.

The Role of Professional Societies

Another source of support for research is the professional societies associated with technical communication. These organizations often sponsor research directly, but more importantly, they publish journals that disseminate the results of relevant research to their members. The exemplars in this book are taken from four such journals:

- *Technical Communication*, the journal of the Society for Technical Communication
- *IEEE Transactions on Professional Communication*, the journal of the IEEE Professional Communication Society
- *Technical Communication Quarterly*, the journal of the Association of Teachers of Technical Writing
- *Journal of Usability Studies*, the journal of the User Experience Professionals Association

The Role of Government

Local, state, and federal governments are yet another source of support for research in technical communication. For example, governments might want to gauge the effectiveness of a communication campaign promoting safe sex or evaluate the usability of a web site that disseminates information about health issues. Government funding of technical communication research is often embedded in grants to natural or social scientists, information technologists, or engineers. Government funding for technical communication can also be found in unexpected places. For example, the Georgia Department of Transportation has funded grants for technical communication research; topics such as the usability of databases, readability of signage, usability of drivers' apps, and data visualization are just a few potential research areas. Funding from such government agencies can often help advance knowledge for the field.

Hierarchies of Publications

Anyone who wishes to have his or her ideas or research published or who wishes to be a critical reader of ideas and research in a professional field needs to understand the different hierarchies of publications and their requirements for publication. There are three levels:

- Open publications
- Editor-regulated publications
- Refereed journals

An *open publication* is one in which anyone can publish without the scrutiny of anyone else as to the validity or reliability of their assertions. For example, anyone can create and post a web site without anyone's permission or review. Another example of an open communication is called the "vanity press." In these venues, an author or a company pays all the publication expenses. Another category of open publication is the "white paper," a document written and published by a company as a marketing tool to promote its technologies or processes, or to share the results of its research in the hope of attracting future business or investment. Read open publications with a high degree of skepticism, since no filtering for adherence to standards of research has been applied by a third party.

An *editor-regulated publication* is one in which the editor or an editorial staff decides which submissions to publish. Standards such as interest of the subject matter to readership, reputation of the author, and quality of the writing are often applied. Examples of editor-regulated publications are newsletters and magazines. These publications may have an informal style and may not reference other research to support the author's assertions.

Refereed journals represent the most rigorous screening of submissions for publication. Submissions are initially evaluated by the editor and then referred to an independent review committee recruited for the purpose of evaluating the manuscript. These reviews are called "peer reviews" because they are conducted by practitioners or academics who share the author's level of expertise (or an even higher level of expertise). These reviewers critique the manuscript, often requiring that the author elaborate on or re-evaluate assertions made in it. Sometimes enthusiastic researchers make unwarranted assertions or claims about their results, and peer reviewers often help them make their claims more conservative and reliable. Often these peer reviewers will reject the submission for not meeting the standards of reliable research. For these reasons, research articles that come from refereed journals have higher credibility and reliability than those appearing in open or editor-regulated publications.

Some book publishers—university presses, for example—combine the methods of editor-regulated publications and peer-reviewed journals. In these cases, an editor makes an initial judgment about the quality of the work, and then all or portions of the manuscript are subject to peer review.

Summary

Research is the systematic collection and analysis of observations for the purpose of creating new knowledge that can inform action. Research in technical communication informs practical decisions that technical communicators make. Its value is that it shapes the best practices of the field.

Research can be categorized by its goals:

- Theoretical—Focuses on explaining phenomena without necessarily providing an immediate application of the findings.
- Empirical—Focuses on testing hypotheses related to theories of communication.
- Interpretivist—Focuses on understanding phenomena related to technology's impact on the ways that humans interact with technology, users interact with the products, or technical communicators interact with people in other roles within an organization.
- Postmodern—Focuses on examining the assumptions that underlie applications of technology or technical communication with the ultimate goal of revealing hidden agendas and empowering disenfranchised groups.

- Developmental—Focuses on the invention and improvement of creative approaches to enhancing technical communication through the use of technology and theory.
- Evaluative—Focuses on a particular product, tool, or method for the purpose of improving it or estimating its effectiveness and worth.

Research can also be categorized by its methods:

- Quantitative—Primarily involves the collection of data that are expressed in numbers and their analysis using descriptive and inferential statistics.
- Qualitative—Primarily involves the collection of non-numeric data (data represented by text, pictures, video, and so forth) and their analysis using ethnographic approaches.
- Critical theory—Relies on the deconstruction of "texts" and the technologies that deliver them, looking for social or political agendas or evidence of class, race, or gender domination.
- Literature review—Primarily involves the review and reporting on the research of others, often including the analysis and integration of that research through frequency counts and meta-analyses.
- Mixed methods—Combines multiple methods, usually quantitative and qualitative.

The publications that disseminate research can be sorted into three hierarchies:

- Open publications—No filtering or selection process is applied.
- Editor-regulated publications—Submissions are selected by the publication's staff.
- Refereed journals—Submissions are peer reviewed not only for the quality of the writing, but for the rigor of the research methods employed.

Answer Key

Exercise 1.1

1. Empirical. The key words *tests* and *hypothesis* point right away to a scientific method approach, placing this study squarely in the empirical category.
2. Postmodern. The emphasis on social disparity (focusing on gender inequity) coupled with a definite point of advocacy (women are not being treated fairly) makes this a postmodernist study. In effect, it is pointing out how technology is being applied to a group's disadvantage—a common theme in postmodernist research.
3. Evaluative or empirical. The results of a specific product are being evaluated. Although the study would look at one instance, it could lead to the generalization that online databases might improve customer service levels in other instances.
4. Interpretivist. This study is trying to understand a general phenomenon, looking to learn how people experience or make sense of a technology experience. It has no prior assumptions, hypotheses, or political agenda. Interpretivist studies can be recognized by this open-ended approach of "What happens when" or "How do people make sense of" to technical communication-related experiences.
5. Theoretical. The clue here is that the root of the question is in human psychology or physiology. The findings are likely to be broad descriptions of how human

sensing or mental processing works. Its findings would not lead to a specific out-come as much as they would lead to additional research that applied its findings to specific designs or processes.

6. Evaluative or empirical. In this case specific programs are being compared and contrasted. The outcome of this study could be applied to making a decision about which program to attend, or it could give insight to someone contemplating starting such a program within his or her own university.

7. Interpretivist. The emphasis is on trying to understand how users make sense of a technical communication task, in this case using online help. It is open-ended and does not purport to test a theory.

8. Developmental. The researcher is trying to optimize a specific product or design. Even so, its findings may point to a best practice that can be applied to other designs; thus, it is included in the field of research.

Exercise 1.2

1. Qualitative. The data being taken and analyzed consist of interviewees' words.
2. Quantitative. The data consist of numeric measurements.
3. Critical Theory. Even though the data consist of words, the analysis is looking for nuances in the text.
4. Mixed methods. The study uses quantitative (number of times users go to help) and qualitative data (interviews to understand why).
5. Literature Review. The study is a *study of studies* and not an original research project.
6. Quantitative. The data are all numeric (that is, salaries and levels of education).
7. Qualitative. The data are the words and other forms of expression gathered or observed in the focus group.

Exercise 1.3

The answer to this exercise will be unique for each person who prepares it, so there is no key to this exercise.

References

Argyris, C., Putnam, R., & Smith, M. M. (1985). *Action science: Concepts, methods, and skills for research and intervention.* Jossey-Bass.

Bright, W. E. (1952). *An introduction to scientific research.* McGraw-Hill.

Hendrick, C. (1983). A middle-way metatheory. *Contemporary Psychology, 287,* 504–507.

Koetting, J. R. (1996). Philosophy, research, and education. In D. H. Jonassen (Ed.), *Handbook of research for educational communications and technology* (pp. 1137–1147). Simon & Schuster Macmillan.

Lincoln, Y. S., & Guba, E. G. (1985). *Naturalistic inquiry.* Sage.

Miller, G. (1956). The magical number seven, plus or minus two: Some limits on our capacity for processing information. *The Psychological Review, 63,* 81–87.

Reeves, T. (1998). The scope and standards of the *Journal of Interactive Learning Research.* www.aace.org/pubs/jilr/scope.html

Senge, P. (1990) *The fifth discipline: The art and practice of the learning organization.* Doubleday.

Watkins, K. E., & Brooks, A. (1994). A framework for using action technologies. In A. Brooks & K. E. Watkins (Eds.), *The emerging power of action inquiry technologies* (pp. 99–111). Jossey-Bass.

2   The Research Phases and Getting Started

Introduction

This chapter introduces the components of a formal research report so that as a researcher, you understand what you are trying to produce, and as a reader of research, you have some landmarks to help navigate through your reading. It also provides a sample project plan to help you plan and track a research project.

This chapter also explains an important requirement for conducting research: gaining the informed consent of the participants.

Finally, the chapter helps the student or first-time researcher tackle the beginning of a project: defining the research goal and articulating the guiding research questions.

Learning Objectives

After you have read this chapter, you should be able to:

- Describe the major sections of a formal research report
- Plan a research project that is replicable, aggregable, and data-driven (RAD)
- Describe how to protect the rights of human participants
- Write a research goal statement
- Write appropriate research questions

The Structure of a Formal Research Report

Before undertaking any project, it is useful to have a picture of what the finished product should look like. Not only is this statement true for assembling bicycles and home entertainment centers, but it applies to research projects as well. The outcome of an academic research project is a research report, a formal document that describes how the research was conducted, summarizes the results, and draws meaningful conclusions. It is often the basis of a research article that gets published in a journal.

The following structure serves as a foundation for reporting research, and its elements should be reflected to some degree in every research report and article (Charles, 1998):

- Introduction with statement of the research problem
- Review of the literature
- Description of the methodology
- Results
- Discussion

Especially in the natural sciences and medicine, this structure is sometimes referred to by the acronym IMRAD: Introduction (of which the literature review is understood to be a part), Methodology, Results, and Discussion.

In a large report, such as a doctoral dissertation or master's thesis, each of these sections is a separate chapter. In shorter reports, they can be sections with their own headings. Still, a good research report reflects these elements in this logic flow, and a critical reader of research articles expects and looks for them.

Introduction with Statement of the Research Problem

The section or chapter of a research report that states the research problem is often called the introduction. Its purpose is to provide the context for the research—in essence, why the researcher thought that it was an important topic and what in particular he or she set out to learn. (Note: It is no accident that we say "set out to learn" and not "set out to prove." A good researcher brings an open mind to every project and must be willing to be surprised by the findings if that's where the data lead. Also, the results of research seldom definitively "prove" a conclusion.)

In writing the introduction, the researcher needs to focus on three areas.

- Background—In general, the researcher needs to describe the problems or activities in the field of technical communication that allow the reader to put the research topic into context. In short, why should the reader care about this research?
- The goal of this research—The researcher focuses on a specific aspect of the general problem area that this research project addresses.
- The research question(s)—The researcher identifies one or more specific questions—that is, what he or she will observe or measure to meet the research goal.

For example, a researcher might want to conduct research related to distribution media. The background section could discuss that there are many media available to technical communicators and explain that media selection is often hotly debated during the design phase of a product. This context establishes that the topic is current and has importance for technical communicators.

Next, the researcher narrows the scope of the research by stating a specific goal for the research project. In the media example, it might be: Determine whether media selection has copyright implications. Now the reader has a better idea of what aspect of the general topic the researcher intends to investigate. But the kind of information or types of data that the researcher will pursue in trying to meet that goal remain vague.

The third component, the research questions, helps clarify how the researcher hopes to achieve the goal—that is, what specific aspects or phenomena will the researcher observe or measure? In the distribution media example, the researcher might specify the following research questions.

1. Does the choice between text or multimedia affect schedule because of the need to obtain copyright permissions?
2. Does the choice between text or multimedia affect user comprehension?
3. Does the choice between text or multimedia affect ability to efficiently update content?

Note that the research questions are what actually shape the character and scope of the research. In the example above, the research study will probably go on to be empirical/quantitative because it asks quantitative questions such as how long and how efficiently. However, it could end up being a mixed methods study depending on how the researcher decides to deal with the questions. The researcher may decide to deal with perceived speed *quantitatively* by measuring effect on project schedule or *qualitatively* by using observations and interviews as the source data.

In a quantitative study, the researcher often states the test hypotheses as well. These are derived from the research questions and state the assumptions that will be tested, usually naming the intervention, the control group, and the test group. For example, one of the hypotheses from the earlier sample questions on media might be the following:

> There is a statistically significant difference in the average time to produce and update product information between text files and multimedia files.

We will discuss how to write test hypotheses in Chapter 4 on quantitative research methods.

Review of the Literature

Research reports are filled with references to other articles or books that the researcher has consulted as part of the study. In the introduction, for example, the researcher will probably refer to several sources in making the case that the topic is significant or in giving the reader the necessary background information to understand the study's context. Even so, there is usually a separate section of the report devoted specifically to reviewing certain kinds of literature, and in a formal report this section is often called the literature review, though it doesn't always bear that title. We have an entire chapter in this book dedicated to the literature review, but we include a brief discussion here to put this important section in the context of the overall structure of the research report.

The purpose of the literature review is threefold.

1. Establish the scope of prior research—The purpose of new research is to *advance* a field's understanding of a topic. To accomplish this goal, the researcher must be familiar with what has been done already. A research advisor would not let a student just decide to "see whether media selection has an effect on project schedule" and set off to gather data. The advisor would point out that others have probably already looked into this question and insist that the student start by studying what has been learned so far. The literature review tells the reader what the researcher found in reviewing what others had already done in the area being studied.

2. Educate the reader—The literature review often presents a comprehensive review of the topic being researched so that the reader can more fully understand the context of the findings that the researcher will report. The introduction started this reader-education process by establishing the importance of the topic within the field of technical communication, and the literature review continues the process in more depth. Often, the literature review presents the technical background that the reader would need to understand the research. For example, in a study

about typeface selection, the literature review could contain references from articles that define the key attributes of typeface design, such as serif versus sans serif, x-height, weight, size, and so on.

3. Ground the researcher's premises in data—People carry a lot of assumptions with them into a discussion of any topic. Some of those assumptions are valid, and some are not. Researchers are no different. Part of the purpose of the literature review is to challenge these assumptions by expecting the researcher to support them with data. For example, a statement such as "Users want an interface that is easy to use" could be uttered in a product design meeting without even a hint of resistance from any of the practitioners in the room. That same sentence in a research study could invoke the response "Says who?" from a critical reader of the article. It's not that we do not believe it; it is just good protocol to ask that the researcher support all of the assertions and assumptions that he or she brings into a study. That assertion, for example, could have been supported by citing studies showing that products rated as user-friendly are more commercially successful, or by citing studies showing that online documentation rated as easy to use was referred to more often than documentation rated as difficult to use.

If, in fact, a basic assumption that you want to make cannot be supported by prior research, testing that assumption might be the more appropriate path for your study. This process is similar to going through security at the airport: We X-ray all the carry-on bags to screen out the dangerous ones. In research, we question all the assumptions to screen out the invalid ones. In this regard, the literature review is an important product to help researchers rid themselves of biases and ungrounded assumptions that they might bring into the study.

Description of the Methodology

Once the researcher has introduced the background and reported on the literature in the area of interest, the next task is to tell the reader how the study was conducted.

Name the Method

The researcher should explain what research method was chosen and why:

- Quantitative
- Qualitative
- Critical theory
- Literature review
- Mixed methods

As a critical reader of research, you should ask yourself whether the researcher's method matches the goal and research questions. For example, if the research question was "Why do users go to online help?" and the method chosen was quantitative—count how often users open the help—the critical reader should question the researcher's choice of methodology. A qualitative method that looked to illuminate user motives would have been a more appropriate choice.

As a researcher yourself, you should carefully select the method based on the goal and the question, not on your individual comfort level with a particular method or the ease of applying it.

Describe the Data Collection Instruments

Closely linked to the method type is the actual way that the researcher gathers data. The following list is just a sample of instruments available to you:

- Case studies
- Surveys
- Interviews
- Focus groups
- Usability tests
- Document or content analyses (analysis of texts, such as open-ended responses to survey questions)
- Controlled experiments (for example, bringing people into a lab and measuring performance such as speed to complete a task)
- Analyses of existing data (for example, analyzing salary data reported by the Society for Technical Communication to discover trends or relationships)

Describe the Sample and the Sample Selection Process

Unless the study is collecting data on the entire population of interest (which is unlikely), the data will be derived from and therefore represent only a sample of that population. The methodology section should explain how the researcher went about selecting that sample and why it is a valid representation of the population as a whole. As a researcher, you should articulate any concerns you might have in that regard and the ways that you have tried to manage them. Doing so will add to the credibility of the study.

For example, a student might choose to conduct a survey of fellow students in a master's program because they are easy to contact and perhaps more willing to participate than others. Such a sample is called a "sample of convenience." Although choosing such a sample is acceptable, the good researcher will note that fact and warn the reader that the attitudes or behaviors of graduate students in a field, for example, might not be indicative of the full range of practitioners in the field. No researcher ever has access to a perfect data collection environment; that fact does not stop us from conducting good research. Being honest with yourself and your reader about the limitations or constraints that you have managed, however, will add value and credibility to your findings.

Describe the Data Analysis Techniques

Finally, the methodology section should provide an overview of the methods used to analyze the data. We talk about data analysis techniques in detail in chapters 4–7. But for now, know that in the methodology section, the researcher should describe at a high level how the data were analyzed.

If you are doing an experimental/quantitative study, it is customary to state your test hypotheses and the statistical tests that you applied. If you are doing a qualitative study, you need to describe the general methods that you used to analyze the qualitative data. By clearly describing your analysis techniques, you help to ensure that your research is replicable, aggregable, and data-supported.

Results

The theme of the results section in a research report is "It is what it is." As a researcher, you need to distance yourself as much as possible in this section from judgmental statements, conclusions, or recommendations. For example, the data might say that Version A of an app resulted in users performing a task more quickly, with fewer mistakes, and with higher satisfaction ratings than with Version B. Do these results mean that Version A is better or preferred? Although you may state that conclusion later in the report, you should not do so in the Results section. Remember that research is designed to separate findings from interpretation.

Good research tries to ensure that our conclusions and recommendations are driven by the data and not by whatever preconceptions we brought into the study. For this reason, it is good protocol to first review what the data state before drawing conclusions or making specific recommendations. The fact that the readers will start to draw their own conclusions about the superiority of Version A after reading the results will add credibility to your discussion section when you explicitly draw that conclusion.

Discussion

In the final section of the report, the voice of the researcher is allowed to emerge. You have done your due diligence in the literature review by grounding your study in what has been done and said previously, and you have kept your judgments and opinions out of the results section. Now you may speak your own mind and voice what the data mean to you in the discussion section.

A word of caution, however: Do not get too far out in front of your data. For example, if you tested to see how readers used indexes, and you based your study on a sample of convenience using students from a master's program in technical communication, avoid drawing broad conclusions about how "users" use indexes or the unwarranted claim that the index approach tested is "superior to conventional indexing methods." People in general might not read the way that graduate students in technical communication read.

You should also temper your analysis based on the kinds of documents used in the study. For example, a study that used software reference manuals might not tell us how people would use a book on automobile maintenance. One of the more common researcher errors is overstating the application of a study's findings. Our advice: Don't overreach. Be satisfied to add a little to the general knowledge about a subject.

Exercise 2.1 Identifying Sections of a Research Report

Skim "The image of user instructions: Comparing users' expectations of and experiences with an official and a commercial software manual" in Chapter 9, and identify in that article the sections of a research report that we have discussed in this chapter.

The Phases of a Research Project

The actual phases of conducting a research project parallel very closely the structure of the final report:

- Identify a research goal
- Formulate the research question(s)
- Review the literature
- Design the study
- Acquire approvals for the research
- Collect the data
- Report the data
- Analyze the data and state your conclusions

A common mistake is to think that you write the research report at the end of the project. Actually, each section of the report is meant to help guide your study, and the report should start to emerge from the very outset of the project. It will be an iterative process, for sure. You will start the introduction, read some literature and begin the literature review, then gain a better understanding of the problem and revise the introduction, etc. So at some times the report points you forward, and at other times it documents where you've been. It is a living document during the project.

In a very formal study, such as a master's thesis or doctoral dissertation, or in the proposal for a research project for which you are seeking funding, you must often write a proposal or research prospectus before receiving the go-ahead or funding to proceed with the study. In this case, the first three sections—the statement of the research problem, the review of the literature, and the description of the methodology—serve as the body of the proposal or prospectus.

Even if a formal proposal or prospectus is not required, you should draft these three sections before gathering any data. Often the insight that comes from the literature review or the planning of the data analysis help determine what data need to be gathered and how the data are gathered. A common mistake among beginning researchers is to gather data (for example, creating and distributing a survey) before they know what their goal and research questions are, or how they intend to analyze the results. In short, they go looking for answers before they understand the questions.

In general, if you are writing a research proposal or prospectus, the methodology section should be written in the future tense—telling the reader what you intend to do. If you are writing a research report or article, the methodology section is written in the past tense—telling the reader what you did.

Managing the Research Project

Figure 2.1 shows a 16-week project plan for conducting a research project in the form of a Gantt chart. Some research projects will take less time, some more. A doctoral dissertation, for example, may require several years to complete. A project plan such as this one helps ensure that you keep your documentation up-to-date as you proceed through the project. You could also add task owners and specific deliverable dates as needed.

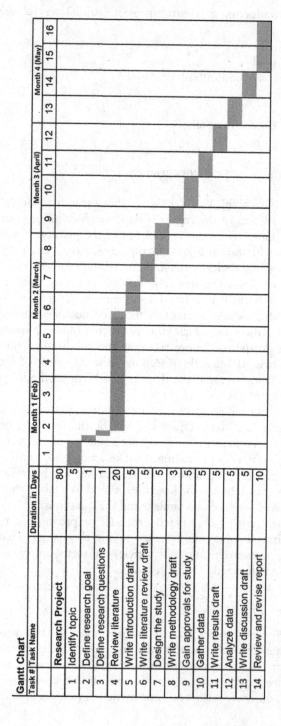

Figure 2.1 Sixteen-week plan for a research project

Bear in mind that real projects tend to circle back on themselves (Smudde, 1991). That is to say, although you will have "completed" a draft of your introduction before you start to collect data, you may gain a better understanding of your research question during the data-gathering phase and want to make some significant revisions or additions to the introduction. This fact could, in turn, lead you to add to or change the focus of your literature review. Technical communicators should be used to this recursive process; we often do not fully understand the product that we are documenting or the users' needs at the beginning of a project, so we often revise early chapters of a manual to accommodate our expanded understanding gained from researching and writing later chapters. Be prepared for the same thing to happen in your research projects.

There is also a final task for revising the entire report. This last task lets the researcher apply his or her final understanding of the research topic across the entire report.

Gaining Permissions

It is critical that a researcher respect the rights of those who are affected by the research. Ensuring those rights often means getting permission or approval to conduct the research. There are three tiers that a researcher should consider and seek permission from if appropriate:

- Institutional review boards (IRB)
- Organizational sponsors
- Individual participants

Institutional Review Boards

Organizations that routinely conduct research involving human participants will have an institutional review board (IRB) that must approve such research. These boards are common in universities, research hospitals, and government health organizations such as the US Centers for Disease Control.

Governmentally funded research in the US is bound by the Common Rule, a set of rules guiding ethical research with human participants. Most university IRBs also follow these rules. The Common Rule was significantly updated in 2017 and went into effect in July 2018 (Phelps, 2020). Each IRB has its own format and procedures that you will have to follow if you are subject to its review, though studies in technical communication rarely require a full review. Instead they are exempted or expedited due to typical lack of risk to human participants in studies in our field.

If you are a student researcher, your professor will tell you whether IRB approval is required.

Sponsors

If you want to do research within an organization, such as a company, you must have the company's knowledge and approval. Sometimes the approval of the manager of the department or division within which the research will take place will suffice, but you should be certain to determine in advance what level of authorization your company requires. For example, some companies require that research projects be approved by the corporate legal department. It is important that researcher and sponsor agree upon any conditions—for example, whether employee time will be taken away from productive tasks (such as asking them to participate in interviews or surveys) and whether the sponsor will be identified in the report.

Sponsor permission is important for consultants who wish to publish the results of a particularly interesting project as a case study. If you are a consultant, it is wise to include such a provision in your contract, make sure that the client is comfortable with that intention, and protect any proprietary knowledge assets that rightfully belong to the client. The same caveats apply to employees doing research within the companies that employ them.

Individual Participants

Many research projects involve working with people who are the source of the data, and it is important that the rights of those people are respected during a research project. Biomedical researchers have developed stringent guidelines and rules of ethics to protect human subjects who are participating in research that can directly impact their health and welfare. For them, protecting human rights includes protecting the subjects from physical harm, ensuring that the benefits of the research justify its risks, advocating that those groups or populations chosen to participate in the risks of the research will be among those who realize its benefits, and ensuring that those who participate in the research do so willingly and knowingly (Office of Human Subjects Research, 1979).

Technical communication researchers should hold themselves to the same standards. Unlike biomedical research, however, research in technical communication seldom exposes participants to risk. However, the concept of *informed consent* is still an important ethical requirement for research in technical communication.

Informed Consent

The principles of informed consent are ingrained in our modern concepts of ethical research. The Nuremberg Code (1949), the first internationally accepted regulation of ethical research, made it the foremost principle: "1. The voluntary consent of the human subject is absolutely essential" (p. 181). The Belmont Report (Office of Human Subjects Research, 1979), the official policy paper of the Office of Human Subjects Research, says that respect for persons requires that participants be given the opportunity to choose what shall or shall not happen to them. The Belmont Report stipulates three components for informed consent:

- Information
- Comprehension
- Voluntariness

INFORMATION

Let participants know what the research is about, what they will be asked to do (or what will be done to them), what risks they might incur, and what will happen with their data. For example, in a research project involving usability testing, the participants should be told that they are part of a research project to understand how people use a product, that they will be asked to perform some typical tasks with that product, that they will be observed, and that their words and interactions with the product will be recorded. They should also be told how any video, audio, or transcribed words will be used if their interactions with the product will be recorded. For example, if the recording might be used in a highlight video shown at conferences or training workshops, the participants should be told that fact. Above all, participants should be told whether their identities will remain *anonymous* (not known to the researcher as in an online survey), *confidential* (known to the researcher but not revealed), or *public*.

Confidentiality is generally maintained by not using the participants' real names or identifying characteristics in the published report. You need to be particularly careful about a participant's *identifying characteristics*—that is, descriptions that could reveal the participant's identity. For example, a survey that asks for company name and job title will not be anonymous if the participant is the only person in the company with that title, nor will it be confidential if those characteristics are linked to the reported findings. For example, if Jane Smith is the only technical communication manager at ABC Inc., specific results or comments attributed to "a technical communication manager at ABC Inc." are hardly confidential. However, you could refer to her as "a technical communication manager at a software development company" as long as the company name is not revealed elsewhere in the report.

COMPREHENSION

It is not enough that you inform the participants. You must also make sure that they understand what you are telling them. Understanding could be an issue when you undertake research involving minors or where there could be language barriers between the researcher and the participant.

VOLUNTARINESS

Voluntariness is essentially the "consent" component in "informed consent." The principle of voluntariness again goes back to the Nuremberg Code: "During the course of the experiment the human subject should be at liberty to bring the experiment to an end if he has reached the physical or mental state where continuation of the experiment seems to him to be impossible" (1949, p. 182). In short, the participant must be empowered to say "No" at any time during the research, including not participating at all. Where the researcher has authority or power over potential participants—for example, as a teacher over students or a manager over employees—great care must be taken to ensure that coercion does not come into play.

For example, a teacher who wants to use certain class assignments in a research project might ask students to hold onto informed consent forms until the course is over and grades have been issued. That way, the teacher does not know who has agreed to have their work included in the research until after the course has been completed, and students do not have to worry that their decisions might influence their grades.

When is Informed Consent Required?

What if a research project included tracking the percentage of visitors to an organization's web site from among those who registered for that organization's newsletter? Would the researcher need to get the informed consent from each visitor? No. Although the decision of when informed consent is required can be complicated, here are three guiding criteria to consider.

- Are the participants at risk?
- Will the participants' words or identities be used as data?
- Is it research or just routine evaluation of what would have occurred normally?

ARE THE PARTICIPANTS AT RISK?

If your research causes risk to participants above what would ordinarily be incurred in their routine environment, you must get the participants' informed consent. Normally, this is not an issue with the kinds of research that technical communicators do. True, users can get frustrated during usability tests, but users routinely get frustrated by software. But if a test is likely to cause unusual stress or anxiety, participants would have to be informed of that probability and would have to give consent.

WILL THE PARTICIPANTS' WORDS OR IDENTITIES BE USED AS DATA?

People own their words.

If you wish to use people's words or equivalent personal expression—such as a video recording of their reactions, descriptions of their interactions, and so forth— or if you identify the participants' identities, you need to get their informed consent. The only exception occurs when the words or expressions already exist in the public domain in some way. For example, you do not need permission to reproduce a brief quotation from a published document such as a magazine interview. But words spoken aloud in the workplace and words written for limited circulation (such as internal memos) are **not** within the public domain.

This principle also applies when research is done after the fact. For example, a researcher might want to analyze email content between subject matter experts and technical communicators. From a researcher's perspective, the employees own the words in their emails. This approach does not contradict the fact that the company owns all intellectual property created by employees, including their words. (As technical communicators, we know this quite well.) But in respecting the rights and privacy of research participants, the researcher respects the right of the individual to control how their words and expressions are used for research, including whether they are used at all. So in this example, the researcher would need to get the informed consent of the email authors.

On the other hand, if all the researcher wanted to do was to analyze the frequency of email exchanges between subject matter experts and technical communicators during different phases of a product's life cycle, informed consent would not be an issue since employees' words and identities would not be used as data.

Separating research from routine evaluation can sometimes be tricky. Although evaluation can be a type of research, not all evaluations are research. Pilot studies of new training programs or beta tests of new products are common examples of routine evaluations that do not, strictly speaking, constitute research. The guiding question should be, "If I were not doing research, would I have created this program or would I be evaluating it?"

For example, a company might introduce a new instant messaging app in a pilot study in one plant while using the existing app in other plants. The company's intent might be to compare productivity of employees who have access to the new app to the productivity of those using the existing program. Would informed consent be required? No. For one thing, it is well within the company's normal scope of operations to deploy new software and to require its use, and the new messaging app surely fits that situation. The fact that the company is evaluating the app's effectiveness before implementing it across the entire enterprise is just sound operating practice.

Getting Informed Consent

Informed consent can be obtained by having participants sign a form that describes the research and asserts their willingness to participate. The form should include a statement to the effect that the participant may choose to stop participating at any time.

Informed consent can also be made implicit in the act of participation. For example, an online survey can provide all the information that the form would have and stipulate that by participating in the survey, the participant consents. This is one way to gain consent while still maintaining participant confidentiality.

Defining the Goal and the Research Questions

Now that you understand the components of the research report and the phases in the process to conduct the research that generates that report, let's focus on getting started. Do not underestimate the importance of defining the research goal and questions; they are the steering force of your project. The earlier you can identify these key elements, the better off you are. In a long research program, such as work toward a master's or Ph.D., having your topic and goal defined early is especially helpful. That way, many papers that you write for individual classes can be geared toward your final research project. In effect, you turn as many classes as possible into an opportunity to add to your literature review and refine your research questions.

As daunting as the task may seem at first, it is quite manageable if broken down into steps. Table 2.1 shows a worksheet for approaching the task. Later pages in this chapter show completed samples of the worksheet.

Table 2.1 Worksheet for developing goals and research questions

Topic	Goal	Questions	Type	Methods

The first step is to identify the topic or topics in which you are interested. An exercise in Chapter 1 got you started thinking about potential research topics in technical communication. Now you must start to identify specific topics that interest you. Actually, doing so is not as hard as it might appear; you are probably sitting in the middle of dozens of research opportunities. To get started, there are two primary sources of research ideas: the literature and real-life problems.

Getting Research Ideas from the Literature

As you read research articles and books about technical communication, it seems that every topic has already been researched and there are no new topics. Do not dismiss a topic just because someone else has already researched it or it has been discussed at length in the literature. Most research is derived from research that has gone before it. Ironically, it is easier to get sponsorship for ideas that have already been researched than for a completely original idea. Academe and professional communities prefer gradual advances that build on and preserve conventional practices and knowledge over paradigm shifts that overturn what has gone before (Kuhn, 1962).

So do not be discouraged if someone else has already done research and written on a topic that you thought that you had discovered. Instead, use it as an opportunity to seed your own research topic or agenda. Use the following checklist of questions to see whether existing research can fuel additional research on a particular topic.

- Can the research be updated? For example, if a study on online help was done 10 years ago, a legitimate question is "Do its results still apply to readers today who might be more sophisticated users of online documentation?"
- Can the research be narrowed? For example, a study of internet applications in general might conclude that the trend in web design is lower contrast between colors. This conclusion could spawn a new study, however, on the effects that low-contrast design have on disabled users.
- Can the research be broadened? For example, a lot of research uses college students as the test sample because this is a convenient population for academic researchers to tap into. Therefore, a good topic for a research project could be to apply research that has been done on this narrow segment of the user population to a broader population, such as working adults.
- Can the research be challenged? Just because a researcher says something is true does not mean that it is. If a researcher's claim seems suspect to you, analyze what flaw in the researcher's assumptions or methodology might have led to the claim and design a research project to examine the potential flaws. For example, a usability study might conclude that users readily go to help when they encounter difficulty. Upon examination of the protocol, you see that the test respondents were told that this was a test of online help. You might challenge that this information predisposed them to use help, and you could do a similar research project where users were not given this information.

Getting Research Ideas from Real-life Problems

If we go back to our definition in Chapter 1 that research should inform decisions about technical communication practices, then any time a technical communicator makes a design or production decision, an opportunity for meaningful research could present itself. For example, if, during a design meeting, someone proposes that users would benefit more from field-level help than screen-level help, this presents an opportunity to do research. Basically, any time that you hear a question or debate that sounds like "Which way is better?" or "Which would users prefer?" or "How do users ...?" is an opportunity to get *real data* about *real users* rather than to speculate or wage abstraction wars in a conference room.

Using the Worksheet

Let's walk through an example showing how a student, Emily, would identify her goal and research questions for a course-based research project. The first step is to identify some topics of interest at a high level (Table 2.2). In our example, Emily might have an interest in online help because that is what she works with in her job, and she might also have an interest in usability because she has been reading about it in a professional newsletter that she gets every month.

The next step is to narrow each topic to a specific goal (Table 2.3). At this point it might be easier to think about the problem instead of the goal. For example, Emily might think, "One of the big problems that I face with help is hearing that our users don't use it." This, in turn, could spawn the goal statement "Understand why users are reluctant to use help."

Similarly, Emily could also be involved in a debate at work over whether help should be presented in a separate help file or as embedded help displayed as a part of the application interface. This debate could lead to the issue of which type of help is more usable. So a goal for her second topic could be "To compare the usability of conventional help and embedded help."

At this point in the process, Emily might start the literature review to understand the goal areas a little better (Table 2.4). For her first topic, she might read

Table 2.2 Worksheet with topics identified

Topic	Goal	Type	Questions	Methods
Help Usability				

Table 2.3 Goal statements defined

Topic	Goal	Questions	Type	Methods
Help Usability	Understand why users are reluctant to use help Compare the usability of conventional help and embedded help			

Table 2.4 Goals and questions

Topic	Goal	Questions		Type	Methods
Help	Understand why users are reluctant to use help	1.	What prior experiences with help do users have?		
		2.	Does the structure of the help match the user model of the task?		
Usability	Compare the usability of conventional help and embedded help	1.	Do users use one type of help more than the other?		
		2.	Do users find answers faster with one type of help than with the other?		

articles that speculate that users avoid help because they have tried to use it in the past and did not find it helpful. She might ask, "I wonder whether that is true." She might also have attended a workshop on writing online help that emphasized the importance of matching the help's content structure to the user's mental model of the task, and she wonders whether this could be an influencer. So now she narrows the scope of her goal by formulating the following research questions.

1. What prior experiences with help do users have?
2. Does the structure of the help match the user's model of the task?

She goes through a slightly different process with the next goal of comparing the usability of conventional help with embedded help. Here she is driven more by the question "What does 'more usable' mean?" She reads some articles on usability and decides that frequency of use and time to obtain information would be good indicators of usability. This decision forms her research questions for that topic.

1. Do users use one type of help more often than the other?
2. Do users find answers faster with one type than with the other?

Before choosing between one topic or the other, Emily should identify the type of research each would be and describe what methods she would use (Table 2.5). This step will help her gauge the scope of the project.

Now Emily can make an informed choice about which project she would rather undertake for her course assignment. Some students spend a lot of time dithering about this decision, and, quite frankly, they lose a lot of valuable time that would be better spent doing research. The advice we have is, "This is your first research project, not your last." In other words, you can do both eventually, but you can't do both right now. Pick one to do now and *get started*.

Table 2.5 Completed worksheet

Topic	Goal	Questions	Type	Methods
Help	Understand why users are reluctant to use help	1. What prior experiences with help do users have? 2. Does the structure of the help match the user model of the task?	Interpretivist	• Interviews • Field observations
Usability	Compare the usability of conventional help and embedded help	1. Do users use one type of help more than the other? 2. Do users find answers faster with one type of help than with the other?	Empirical	• Usability test • Comparisons of average frequencies and average times

Exercise 2.2 Identifying Research Goals and Questions

Use the worksheet introduced in Table 2.1 to do this exercise.

1. Identify at least three topics of interest.
2. For each topic, write a tentative goal statement.
3. For each goal, write two or three research questions.
4. Based on the questions, for each goal, classify the type of research that you would do and list the methods that you would probably employ.
5. Based upon your skills, access to potential data, and time available for the research, pick one of the goals for your research project.

There! The hardest part is over. The remaining sections of the book will help you with the rest.

Is it RAD?

In the first two chapters of this book, we hope that we have helped you understand how to plan a research project from identifying goals, to developing research questions, to selecting appropriate methods. Before you begin to implement your research strategy with the help of the chapters to come, you should become familiar with one more concept: RAD.

RAD research was described by Richard Haswell in a 2005 article in reaction to a distressing trend that he saw in writing studies. That trend was designing or reporting research studies in such a way that each could be approached only in isolation. In brief, he says that such an approach is fatally flawed. Rather, research must be replicable, aggregable, and data-supported (RAD). In other words, the methodology must be sufficiently defined so that others can repeat the study. The results of the study must be reported in sufficient detail that they can be aggregated or combined with the results of other studies to build a body of data. And the conclusions of the study must be supported by that data, not simply the impressions or gut feelings of the researchers (Haswell, 2005).

We enthusiastically agree with Haswell about the value of RAD research. Thus, as you use the information in the chapters to come, you will see us put great emphasis on how you, too, can design, implement, and report on research that is replicable, aggregable, and data-supported.

Summary

A formal research report is commonly constructed of the following elements:

- Introduction—includes background, goal of research, and research questions.
- Review of literature—references other sources that serve as background or foundation for the current research.
- Description of methodology—includes what data collection instruments were used, why they were selected, how the sample was chosen, and which data analysis techniques were used.
- Report of results—details the findings of the study without interpreting those findings.
- Discussion—analyzes the findings and discusses their meaning.

A sample is the part of the population of interest that is selected for study. A sample should be representative of that population.

The process of conducting a research study closely parallels the elements of a formal research report.

Researchers must gain all necessary permissions for their research, which may include approval from an institutional review board, a sponsoring organization, and the participants themselves. Gaining informed consent from participants requires that they understand what they will be asked to do, how the researcher will protect their privacy, and that their participation is voluntary.

Researchers can get ideas for research from the literature on the topic of interest and from real-life problems. Start with a topic and a goal. Then formulate potential research questions, identify the type of research that might help you address the questions, and finally identify the methods that would best collect data that might address your questions.

Answer Key

Exercise 2.1

- Statement of the research problem ("Introduction")
- Review of literature ("Introduction"; "Image, Source Credibility, and User Instructions")
- Description of methodology ("Study 1: Users' Expectations/Method"; "Study 2: Users' Experiences/Methods")
- Results ("Study 1: Users' Expectations/Results"; "Study 2: Users' Experiences/Results")
- Discussion ("Study 1: Users' Expectations/Conclusions"; "Study 2: Users' Experiences/Conclusions"; "Discussion")

Exercise 2.2

The answer to this exercise will be unique for each person who prepares it, so there is no key to this exercise.

References

Charles, C. M. (1998). *Introduction to educational research* (3rd ed.). Addison-Wesley Longman.

Haswell, R. H. (2005). NCTE/CCCC's recent war on scholarship. *Written Communication, 22*(2), 198–223.

Kuhn, T. (1962). *The structure of scientific revolutions.* University of Chicago Press.

The Nuremberg Code. (1949). *Trials of war criminals before the Nuremberg military tribunals under control council law.* No. 10, Vol. 2. U.S. Government Printing Office.

Office of Human Subjects Research. (1979, April 18). *The Belmont report: Ethical principles and guidelines for the protection of human subjects of research.* U.S. Department of Health and Human Services, National Institutes of Health. www.hhs.gov/ohrp/regulations-and-policy/belmont-report/read-the-belmont-report/index.html

Phelps, J. (2020). Modernization updates to the common rule: Recommendations for researchers working with human participants. *IEEE Transactions on Professional Communication, 63*(1), 85–95.

Smudde, P. (1991). A practical model of the document-development process. *Technical Communication, 38*(3), 316–323.

3 Reviewing the Literature

Introduction

We begin this chapter by exploring the differences between primary and secondary research. We then consider the reasons for doing secondary research as part of a primary research project, and explore the best way to conduct a review of existing literature on a topic. We also describe the process for preparing an annotated bibliography of sources. Finally, we discuss the process of writing a literature review.

Learning Objectives

After you have read this chapter, you should be able to:

- Differentiate between primary and secondary research
- Describe the purposes of a literature review:
  - Determining what has already been learned about the area to be researched
  - Identifying gaps in the research
  - Educating the reader
  - Establishing credibility with the audience

- Conduct a review of online and library sources and assemble an initial reading list
- Do research without access to a university library
- Differentiate between a literature review and an annotated bibliography based on content and purpose
- Describe and evaluate the content of your sources, and prepare an annotated bibliography
- Write a literature review

Primary and Secondary Research

The major focus of this book is primary research. Our goal is to help you learn how to perform—and be an informed consumer of—primary research. A literature review, the focus of this chapter, is an example of what is usually called secondary research.

You may have heard your professors distinguish between primary or "original" research, and secondary or "derivative" research. Although the labels *primary* and *secondary* are helpful distinctions, the terms *original* and *derivative* are less helpful and can even be misleading in some respects.

Primary research involves formulating and testing a hypothesis, collecting information through observation, or conducting a survey or usability study. The results are then reported to others in the field. Such research is called *primary* because it adds to the existing body of knowledge in the field. For example, suppose that a company assigns a team of technical communicators from the US to write, design the layout, and prepare illustrations for a set of instructions for a new social media app intended for senior citizen users in the US. The company then has the English instructions (including any text in the illustrations) professionally translated to Chinese. Once the translation has been completed, the company might design and conduct a comparative usability test of both instruction sets to explore whether the design and the rhetorical patterns affected the two groups of users differently in terms of their speed in completing various procedures. This is an example of primary research because it would add to the current state of knowledge about the relationships among cultural expectations, rhetorical patterns, and page design.

Secondary research is conducted by reading reports of previous research, analyzing the results, and then formulating a synthesis of the current state of knowledge on a topic. This research is called *secondary* because it draws on reports of previous primary and secondary research. For example, a researcher interested in intercultural communication would want to know the results of previous primary research that has investigated the effects of the rhetorical patterns and designs of one culture on users from a different culture, as well as any previous secondary work on the topic. The result would be an example of secondary research because it would describe the research that has already been done on the topic.

Although the distinction between primary research and secondary research is an important one, the terms *original* and *derivative* are not entirely accurate in describing these two classifications of research because primary research can be derivative and secondary research can make very significant contributions to a field.

How can primary research be derivative? Well, one of the most important characteristics of primary research is that it should be replicable and that, when repeated, it should produce the same results each time. When we formulate and test a hypothesis experimentally, when we collect information through observation, and when we conduct a survey or usability study, we draw conclusions or make generalizations based on our analysis of the data that we collect. But all primary research is based on a sampling of the total population—a relatively small number of potential users of a smart watch, for example, or every twentieth registered user of a software product. And sometimes the samples that researchers test, observe, or survey are less than ideal. For example, we may draw on students as samples of convenience, rather than people from a wider range of ages and educational backgrounds, because students are easier to recruit and study, especially for researchers on university faculties.

Because the samples used in much primary research are often relatively small and may be less than ideally representative of the total population, the conclusions based on that research are usually stated tentatively. In their conclusions, the authors of such studies typically encourage other researchers to investigate the same questions using different sample populations, and other researchers sometimes duplicate prior studies by using a larger or more representative sample, or by varying the sample or the experimental treatment in some way. Such primary research that attempts to duplicate or test the reliability of research that others have performed is thus, in a sense, derivative although it is quite important because it attempts to verify the generalizability of results obtained by earlier researchers.

Similarly, secondary research can be quite significant despite the fact that it draws on the primary and secondary research performed by others. For example, an annotated bibliography of primary and secondary sources on international virtual team practices could make a very important contribution to the field of technical communication by collecting and commenting on earlier work that has been done on this topic. Similarly, a literature review could be an essential source for other students of international virtual teams because of the synthesis it provides and the insights it provokes.

The Purposes of a Literature Review

In this section, we explore the reasons why researchers include a literature review in the reports of their own research. Such a review is important to discover the work that others have already done, to identify gaps in that research, and to demonstrate your own credibility to your readers.

Discovering What has Already Been Done

One of the most important reasons for reviewing the literature is the heuristic function of this task: Discovering what others have already done in your area of interest will give you insights into aspects that you want to investigate and suggest contributions that you can make to the field.

For example, suppose that you are interested in doing research on intercultural communication. Conducting a thorough review of existing work in the field will reveal that many researchers have drawn on the concept of cultural dimensions described by Geert Hofstede (1994, 2001) to formulate their research studies—for example, the effect of cultural differences in uncertainty avoidance on the willingness to read software documentation, or the effect of cultural differences in power distance on willingness to follow directions precisely. Reading everything you can find on the theoretical contributions of Hofstede and the research studies based on his theory of cultural dimensions could be very influential in helping you zero in on a primary research project that you would like to conduct.

Whether you intend to prepare an extended survey of the existing literature or not, you need to know what research has already been done on your topic so you do not needlessly repeat work that has already been done. For example, suppose that you are interested in the intersection between page design and reading comprehension. By reviewing the existing literature on the topic, you will likely conclude that there is not much point in studying the relationship between printed line length and user comprehension of the information provided because a lot of work has already been done on that subject with quite similar results. Even if you are not working on a Ph.D. dissertation that is supposed to make an original contribution to knowledge on your topic, you probably do not want to go over ground that others have already covered quite adequately.

Identifying Gaps in Existing Research

Knowing what research has already been done on your topic of interest also helps you discover what remains to be studied. What are the gaps in existing studies that still need to be filled in? What are the areas that have not been explored at all? For example, suppose that you have read all the work that has been previously done on the relationship between the length of printed lines and user comprehension. You believe initially that this topic is tapped out—there is nothing useful remaining to be done. And then perhaps you realize that all of the existing studies of this subject have involved texts in Western languages and have used North American and European users. You wonder whether you would get the same results if you studied text in Chinese with a group of Chinese users.

Similarly, reading the research about printed line length and comprehension might cause you to wonder whether there is a relationship between the spatial distance between a procedural step and the graphic that illustrates it, and the user's comprehension of the instruction. This research question might not have occurred to you if you hadn't read this earlier research.

Educating Readers

So far, we've talked about the illuminating quality of a literature review for the researcher, but the researcher's audience will also benefit because most research reports include a summary of related research on the topic. This literature review within a book, article, or paper serves several functions.

First, the literature review situates the current study. It provides background information that explains why you decided to do your research and often describes how the previous work on the topic helped you frame your research question. This background is helpful to your readers because it provides a kind of intellectual history that explains how you became interested in pursuing the topic and how you have approached it.

Second, the literature review explains the context for your research to those who read your report. Many—perhaps most—will not have read all of the previous work that you've reviewed before designing and conducting your experiment, observation, survey, or usability study. Your review of the literature provides your own readers with the essential information that they need to know about your area of inquiry.

Establishing Credibility

One important effect of the review of the literature included in your article or paper is that it establishes your credibility with your audience, both those who are familiar with the work you discuss and those to whom it is new. In both cases, you demonstrate your awareness of the work that others have done. Your subsequent description of your research methodology, presentation and discussion of your data, and conclusions about your research question thus gain prestige in your readers' minds because of the relationship—the intellectual pedigree—that you have established between the previous work and your own.

On the other hand, imagine the effect in a reader's mind if your article or paper does not mention an important prior work on the topic, especially if your research contradicts the findings or observations of the earlier researcher. That reader is unlikely to give your conclusions much intellectual weight because you will appear to have failed to do your job of thoroughly surveying the previous significant work on the subject.

Conducting a Literature Review

Conducting a literature review is actually rather simple. You begin by assembling as comprehensive a list of sources on your general area of study as possible, and then refine that list to include only the works that deal with your topic in some detail. To do so, you need to identify what kind of information you are searching for and then use one or more tools to identify and locate sources.

Search Terms or Keywords

Whether you are searching catalogs, databases, or online journal archives, you will need to assemble a list of words to search for to locate items of interest. These words are called search terms or keywords.

A search term or keyword is a word or brief phrase that is used to identify potential sources for your literature review. For example, say that you want to use a web search engine to locate someone to replace some boards on the wooden deck of your home. You might type *carpenter* or *handyman* into your search engine to find potential matches for that job.

The place to begin assembling your list of search terms is to read carefully through several articles or book chapters on your general subject that you are already familiar with. What are the major focuses of those works? For example, if you are exploring how teams collaborate electronically, some of the keywords that you note will probably be *collaboration*, *virtual teams*, and *electronic collaboration tools*. Begin compiling your list, adding or subtracting search terms as you go. When you actually begin to search databases and other sources, you will continue adding to and refining your list.

Helpfully, some journal articles or book chapters will identify keywords in the sources that you locate. The *IEEE Transactions on Professional Communication*, *Technical Communication*, *Technical Communication Quarterly*, and the *Journal of Usability Studies* from which we drew the articles reprinted in this book all include a short list of index terms of keywords at the end of the abstract for each article that they publish.

You should ensure that your search terms are neither too broad nor too narrow. Overly broad terms will turn up many more match items than you can possibly read, and extremely narrow terms may fail to find items that would be genuinely helpful. The Boolean operators *AND*, *OR*, and *NOT* in database searches combine search terms in useful ways that will restrict the number of matches returned. For example, a search for *collaboration* might return more than 75,000 matches, and a search for *collaboration AND virtual teams* return more than 100 matches. But a search for *collaboration AND virtual teams AND technical communication* finds seven articles that are likely to be extremely relevant as well as being a number that can be evaluated more realistically. On the other hand, combining those three terms with *AND [company name]* would probably fail to locate any matches unless the company has been the subject of prior research on collaboration in technical communication virtual teams.

Similarly, you should use truncated keywords that will help you locate all possible matches for various forms of the truncated word. For example, the truncated keyword *collabora** will locate works in which the keywords *collaborate*, *collaboration*, and *collaborator* appear.

Once you have arrived at what you think is your final list of search terms, be sure that you search every database using all the keywords that you have collected in every combination that is likely to reveal matches.

Search Tools

In this section, we survey the researcher's toolkit, including library catalogs and databases, journal websites, book reviews, and online booksellers' websites.

Library Catalogs and Databases

If you have access to a good college, university, or public library, you can search for books in its online catalog and for periodical article citations in the databases to which the library subscribes. Two databases are particularly helpful because they also provide links to the full text of many of the articles they index.

EBSCO ACADEMIC SEARCH DATABASES

EBSCO Ultimate, Complete, Premier, and Elite are the four versions of this database with increasingly limited numbers of journals indexed. Academic Search Complete, which most university libraries subscribe to, indexes *Technical Communication*, *Technical Communication Quarterly*, and the *Journal of Business and Technical Communication*. The EBSCO databases will also provide you with a citation for each article in a variety of citation formats, including APA (American Psychological Association, 2020), though you should always check any automatically generated citation for correct form. See Figure 3.1 for an example of an EBSCO database-generated citation, and the Appendix for further information.

Note the icon in the left margin to help locate full text of the article available through Mercer University's library. Also note that the citation does not use "sentence case" for the article title, as APA style prescribes. It is always necessary to check the citations you retrieve from sources such as this against the citation style in the Appendix of this book.

PROQUEST RESEARCH LIBRARY

This database indexes *Technical Communication Quarterly*, the *Journal of Business and Technical Communication*, and the *Journal of Technical Writing and Communication*. The ProQuest database will also provide you with a citation for each article in a variety of citation formats, including APA and IEEE (Institute of Electrical and Electronics Engineers, 2018), though you should always check any automatically generated citation for correct form. See Figure 3.2 for an example of a ProQuest database-generated citation in IEEE format, and Appendix I for further information.

Note the icon in the right margin to help locate full text of the article available through Mercer University's library.

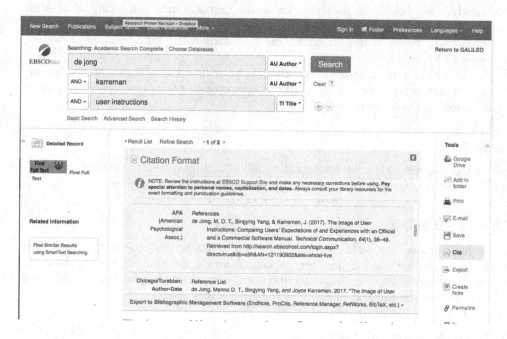

Figure 3.1 EBSCO Academic Search Complete APA citation for an indexed article

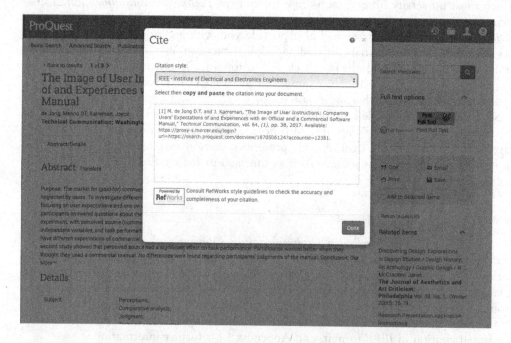

Figure 3.2 ProQuest Research Library IEEE citation for an indexed article

When you find articles in a database, note any terms or keywords different from those that you used in your search, and consider searching for additional articles using those terms. See Figure 3.3 for an example of additional subjects and keywords that you might incorporate into your search.

GOOGLE SCHOLAR

Whether you have access to a good college, university, or public library or not, Google Scholar (scholar.google.com) is a wonderful resource because it is easier to use than the commercial databases and often contains links to full texts of books and periodical articles that your own library cannot access. Although the full-text version of some articles may be a copy of the author's manuscript as submitted to the journal and not the final published article, many others are PDFs of the published articles. Google Scholar lists hundreds of millions of works, and many university libraries indicate which articles they own full-text access to if you use a computer on the university network when you query Google Scholar.

Google Scholar also provides a link to a list of other works that cite a particular work, as well as a link to related articles (see Figure 3.4). This feature is particularly helpful in ensuring that your working list of sources is as complete as possible.

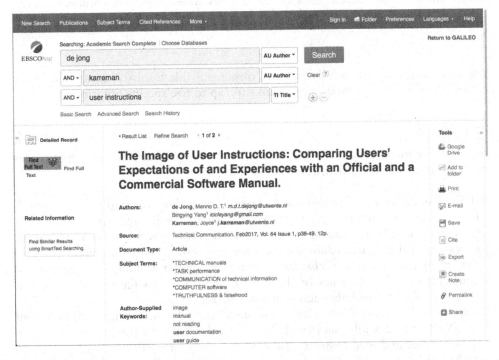

Figure 3.3 EBSCO Academic Search Complete database search detail showing list of subject terms and author-supplied keywords for an article. The ProQuest Research Library offers a similar list of subject terms

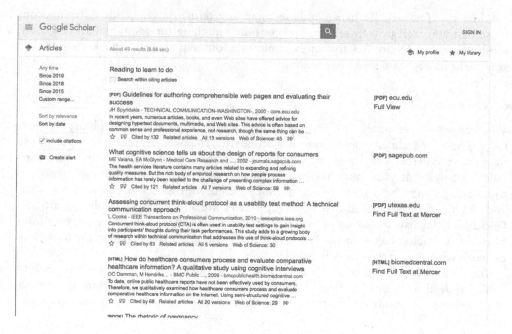

Figure 3.4 Google Scholar includes links to a list of works that cite a particular article, as well as to a list of related articles. Note also the links to find the full texts of articles at your university library

You can use the Google Scholar settings menu available from the menu icon in the upper left corner to show library access links for up to five specific libraries.

Journal Web Sites

Each of the five major journals in the field hosts a web site that includes digital copies of many or all of the issues that it has published.

- *IEEE Transactions on Professional Communication* (https://ieeexplore.ieee.org/xpl/RecentIssue.jsp?punumber=47) includes all issues of the journal from 1972 to the present; the first 16 volumes of the journal are not available online. You can browse issues and read abstracts, editorials, and book reviews without a subscription.
- *Journal of Business and Technical Communication* (https://journals.sagepub.com/home/jbt) includes all issues of the journal from 1987 to the present. You can browse issues and read abstracts without a subscription.
- *Journal of Technical Writing and Communication* (https://journals.sagepub.com/home/jtw) includes all issues of the journal from 1971 to the present. You can browse issues and read abstracts and editorials without a subscription.
- *Technical Communication* (www.ingentaconnect.com/content/stc/tc) includes issues of the journal from 1995 to the present; the first 41 volumes of the journal are not available online. You can browse issues and read abstracts without a subscription.

- *Technical Communication Quarterly* (www.tandfonline.com/toc/htcq20/current) includes all issues of the journal from 1992 to the present. You can browse issues and read abstracts and editorials/issue introductions without a subscription.

"Recent and Relevant"

The "Recent and Relevant" department of *Technical Communication* provides brief abstracts of articles appearing in the other journals in the field as well as in journals in many related fields. The abstracts typically appear about six months following publication, but there are occasionally lags and lack of coverage.

Book Reviews in Journals

The *IEEE Transactions on Professional Communication* and *Technical Communication Quarterly* typically publish reviews of a few titles in the field of technical communication in each issue. However, every issue of *Technical Communication* includes an extended book review section that covers books in technical and professional communication as well as a variety of cognate disciplines. Note that reviews may not appear for several years after publication, so scanning these reviews is not a substitute for other ways to locate recent book-length works in the field, but reviews are typically more timely than many other sources.

Online Booksellers' Websites

Amazon, Barnes and Noble, and other online booksellers allow you to search their databases by topic and provide full bibliographic information about works contained there, as well as the ability to order the books online.

Works Cited by Other Sources

Search through the reference list or bibliography in each of the sources you locate. What additional works do those sources refer to? We have found this method to be one of the most important search tools in our own research.

As you begin collecting potential sources for inclusion in your literature review or annotated bibliography, it may be helpful to use a citation manager such as EndNote or Zotero. See the Appendix for details on how these tools can help you manage your sources.

Doing Research without Access to a University Library

Especially if you work in industry, you may not have ready physical access to a university library, so you may assume that you won't be able to use many of the tools discussed in the previous section. If that is the case, you should not give up on conducting a literature review.

On most university campuses today, the library is no longer the physical place where you go to do research because you can access many of the university's library resources online from your home or anywhere else that has internet access. The library catalogs and databases, the full text of the journals that they index, and of course,

Google Scholar do not "reside" in that building, which these days is far more often used as a quiet place to study than as the center of campus research.

Here are some ideas for gaining access to university or other library collections and resources.

- If you are no longer a student, you may find that your university provides access to alumni of the school after graduation, or a local university may provide short-term guest access to its resources, including remote use of the databases and online collections. Even if the library does not have such a policy, it may be possible to negotiate access for a limited time. Contact your university's alumni office or the university library's administration for information.
- If you are working on a research project for your employer, you should inquire whether the company has (or could make) arrangements for access to the local university library. Community colleges may also provide excellent resources online or through interlibrary loan.
- If your own university library is unable to help and there is no local university or college library, your local public library may be able to help you access materials through interlibrary loan. To do so, you provide them with citations for the books and articles that you would like to borrow (you can obtain citations from Google Scholar), and they will request photocopies or PDFs of articles and copies of books. In addition, very good public libraries may also subscribe to the EBSCO Host and ProQuest databases.
- If all else fails, you can locate citations of articles through a resource like Google Scholar, find the articles on the corresponding journals' websites, and then buy copies of the articles that you need from the journal websites. This is an expensive option, so it should always be your last resort. Your employer may be willing to absorb the cost if the research project will benefit the company.

Describing and Evaluating Your Sources

A literature review is typically a part of a larger research report. It provides the context for the current study by exploring the previous research on the same topic (and often on a particular aspect of it). It usually appears early in the research paper, either as part of the introduction or very soon after it. The literature review is more than a recounting of what other researchers have done, however. It ordinarily identifies common themes or concepts that earlier research has revealed. Although the literature review should be as thorough as necessary in exploring the aspects of interest in the previous work, it need not be exhaustive in covering every earlier article or book that has explored the same aspects, nor comprehensive in including articles or books that are not relevant to the current study. However, it must contain the most important earlier works—those that are considered central to the development of the topic by previous researchers. A literature review that examines earlier work as a standalone study is sometimes called a bibliographical essay.

On the other hand, an annotated bibliography is a much less ambitious work. It consists of a list of citations of earlier books and articles on the topic of interest, with an annotation of each one. The annotations may be purely descriptive, simply summarizing the item's content in a paragraph, or they may also include an evaluative component, explaining why the item is significant in its coverage of the topic. Professors often assign an annotated bibliography to familiarize students with the processes of doing library research, understanding and evaluating sources, and preparing bibliographic citations in a specified format.

Whether you are planning a formal annotated bibliography as a secondary research project or you are surveying the relevant literature as an initial step in a primary research project, you will find that it is helpful to prepare annotations of your sources for several reasons.

- Composing descriptive annotations will help you better understand—and remember—the content of individual books, articles, and other sources about your topic.
- Preparing descriptive annotations will also help you see connections among the various separate works on your research topic.
- Writing evaluative annotations will help you explicitly gauge the relative importance of the existing work in your area of interest.
- Creating evaluative annotations will also help you see the strengths and weaknesses of the existing work, and potentially draw on the strengths and avoid the weaknesses in your own research.

An annotated bibliography consists of a series of entries, each corresponding to a single work that you have selected for inclusion. The entries are typically arranged alphabetically by the last names of the authors (by the last name of the first author in the case of multi-author works). Long annotated bibliographies (those containing more than 25 or 30 entries) are frequently classified by relevant subtopic first, with the entries for each subtopic then arranged alphabetically by the authors' last names.

Each entry consists of three parts:

1. A full bibliographic citation, using the preferred style (APA or IEEE) of the journal, publisher, or course for which you are preparing the bibliography
2. A description of the work's content
3. An evaluation of the work's significance

This book's Appendix contains information about preparing bibliographic citations using APA and IEEE style, so let's examine how to write the descriptive and evaluative annotations that will accompany those citations in your annotated bibliography.

Writing Descriptive Annotations

Your descriptive annotation is essentially a brief abstract of the book, article, or other work. It may take either of two forms.

- *You may reproduce the author's abstract* in full or part if one exists, but you must place it between quotation marks and include the page number(s) or URL (if it is published only online or is not included as a part of the book, article, or other source itself). For example, you might quote published descriptive abstract for an article by Giammona (2004) that appeared in *Technical Communication*:

This article looks at the future of technical communication from the point of view of many of its most seasoned and influential practitioners. And it wraps that point of view around the themes of the [New York Polytechnic University management of technology master's] program—innovation, global concerns, managing technical leaders and practitioners, the impact of new technologies, and the future role of technologists in organizations. It concludes by providing a series of recommendations for the future direction of the profession. http://www.ingentaconnect.com/content/stc/tc/2004/00000051/00000003/ art00002

Although using the published abstract makes preparing the descriptive part of the annotation easier, preparing your own abstract will help you better understand the content of the source. Furthermore, your own abstract may be superior to that provided by the work's author. (You'd be surprised how poor some published abstracts are!)

- *You may prepare your own descriptive abstract.* A good abstract provides readers with all of the essential information that they need about the book, article, or other source, including the research question(s), research method, research results, analysis of the results, and conclusions. Writing your own abstract will help you master the intricacies of the work and better understand its significance, thus helping you write the evaluative part of the annotation. For example:

    Giammona examines the future of the profession based on interviews with and survey responses from experienced practitioners and academics. She organizes her discussion using the following themes: our future role in organizations, management concerns, our contributions to innovation, global concerns, education of future professionals, and relevant technologies. She also provides recommendations of future directions that she believes will help the profession survive and thrive in the future.

Include only essential information in the abstract. For example, if the author reports that the difference between the experimental and control groups' performance is statistically significant, that information is important and should be included in the descriptive annotation. The tests that the author has used to determine the statistical significance are not essential.

Descriptive annotations are typically 100 to 250 words long, depending on the length and complexity of the work you are abstracting.

Once you have located and read the relevant prior research on your topic and have completed your descriptive annotations, you may want to go back and add observations where relevant that connect these sources. For example, you might note in your annotation of the Giammona article that Lanier (2009) agreed that real world skill sets are extraordinarily important to writers of job advertisements for technical communicators, but that the ability to use specific tools is also very significant.

Writing Evaluative Annotations

Preparing evaluative annotations is a very challenging task, especially to those who are new to technical communication research in general or new to the particular area of study. You might ask yourself whether you are qualified to evaluate the work of others, especially if the authors are well-known experts in the field. In fact, as someone new to the area of inquiry or to research in the field generally, you are in a perfect position to gauge the importance of the works you are analyzing.

First, you will be reading the books, articles, and other sources over a relatively short period of time, rather than over a period of years as the works appear. By reading and digesting all the work related to your topic of interest more or less concurrently, you will find it much easier to compare these sources than someone who has read them over a much longer span of time.

Second, if you read the works chronologically in the order that they appeared, you will find it quite easy to determine which works represent truly original research and which are derivative. One way of doing this is to notice which earlier sources other authors cite, especially in identifying their research questions and selecting a research methodology.

Finally, take note of which books, articles, and other sources represent the results of primary research and which are based only on secondary research. How significant are the results of the primary works? How successful are the authors of the secondary works in formulating a helpful synthesis of the primary and secondary work that others have done?

Your evaluative annotations should draw comparisons where appropriate to other works that you are examining. To ensure that your evaluative comments are accurate and helpful, you should revisit and revise each one after you have completed all of your reading and have a first draft of the descriptive and evaluative annotations of all of your sources.

Here is a sample evaluative annotation of Barbara Giammona's article described above:

> This article is significant because it is based on the views of 28 leaders in the field. The many insights in this in-depth qualitative research study will help technical communication practitioners (especially managers) anticipate and respond to trends in the profession, and will help educators prepare students for rewarding, long-lasting careers.

Evaluative annotations are typically 50 to 100 words long, depending on the length and importance of the work you are assessing. It is usually helpful to separate the descriptive and evaluative components of your annotations, devoting a paragraph to each. Begin with the description, and conclude with the evaluation.

Exercise 3.1 Composing an Annotated Bibliography

Prepare an annotated bibliography of at least 12 sources on your research topic. Make sure that you include both books and periodical articles. Prepare the citations using either APA or IEEE style, based on the preference of the journal, publisher, or course for which you are preparing the bibliography. Ensure that your annotations include both a descriptive and an evaluative component.

Writing the Literature Review

In this section, we consider the differences between annotated bibliographies and literature reviews, as well as the reasons for choosing one or the other. We also detail a process for writing a literature review.

Annotated Bibliography or Literature Review?

Now that you have completed an annotated bibliography on your topic, let's consider in more detail the differences between that genre and the literature review, as well as the reasons why you would choose one over the other.

As we have seen, an annotated bibliography is a listing of existing primary and secondary works on a topic. Each entry in the bibliography consists of a full bibliographic citation and both descriptive and evaluative comments about the book, article, or other work. The bibliography is usually arranged alphabetically by the last name of the author (or by the last name of the first author in the case of multi-author works), and bibliographies that include a large number of sources are often classified by subtopic. Many researchers prepare annotated bibliographies for their own use as a preliminary step in conducting their own primary or secondary research on the topic. They are also frequently assigned in courses to help students learn secondary research techniques or prepare them to explore a topic further in a primary research project. Sometimes, annotated bibliographies are published as articles, sections of books, or parts of web sites addressing the topic.

A literature review is always prepared for an audience other than the researcher. It is typically a section of a research article or paper that presents a synthesis of the most important primary and secondary research relevant to the primary research project about which the article or paper is reporting. Rather than being organized by source like an annotated bibliography, the literature review is organized by topic. For example, the literature review for a usability study of an e-commerce web site might include a section on the usability methodology to be used, a section on findings of previous usability studies of e-commerce applications, and another section on the user population(s) to be sampled. The article or paper of which the literature review is a part ends with a list of references that provides a complete citation of each work mentioned in the literature review, as well as any other source mentioned in the article or paper.

You will find that an annotated bibliography is easier and quicker to prepare because it doesn't require you to synthesize the sources as you must do to write a literature review. You must of course read each work that you include in the annotated bibliography, and you must intellectually digest the sources to the extent necessary to describe them and evaluate their relative significance. But an annotated bibliography does not require the same degree of familiarity with the interrelationships of the sources that you must possess to produce a good literature review.

Preparing the Literature Review

To ensure the effectiveness of your literature review, you must define its purpose and audience, determine the most effective organization of the information, and decide the appropriate level of detail for the information.

Purpose and Audience

The audience of your literature review will be the readers of your article or paper. That audience might be the readers of the journal to which you intend submitting a manuscript reporting the results of your research. Or it may be the instructor and other students in a course. Whoever that audience is, analyze those potential readers as you would any other audience. How much are they likely to already know about your topic and previous research about it? What is their level of interest in the topic? How will the information that you report in the literature review section of your paper or manuscript be of interest and use to them in their own research or practice?

In the context of a report on primary research that you have conducted, the purpose of your literature review is to provide a brief overview of work previously done on the topic. For example, if you are conducting primary research on the usability of a web site aimed at a teen audience about alcohol using the plus-minus method, the literature covered in your review would include the findings of previous usability studies about web design, teen audiences, and attitudes toward alcohol, as well as evaluative studies of the advantages and disadvantages of the plus-minus method. Because it is intended to provide your audience with the broad strokes of previous research, your literature review would typically run only a few pages and would not be as detailed as the annotated bibliography that you may have already created for your own use as you were doing your library research. It would also probably include only the most important works, not a comprehensive treatment of every book and article that you have consulted in your secondary research.

Organization

There is no predetermined formula for organizing a literature review, but consider the following questions as you decide how best to arrange the results of your secondary research.

- What audiences or user populations have previous researchers focused on?
- What methods or techniques have they used in conducting their research and analyzing the data that they gathered?
- What are their major findings, and do they more or less coincide or differ?
- What conclusions have they reached?
- Did they expect results that did not materialize or that were not conclusive?
- What work remains to be done?

Level of Detail

Remember that your purpose is to provide your readers with the information that they need to understand what your research area is about, what work has been done already, and what remains to be explored. You may also need to include literature addressing methodology or analysis techniques if they are innovative or unusual. The literature review is not intended as a comprehensive or detailed account of all the work that has been done in an area before your project. Instead, it provides novices in the area with the basics that they need to understand your project and experienced researchers with a reminder of what they have already read themselves.

Determining the appropriate level of detail for your literature review will depend in large part on its purpose, the number of sources you are surveying, and the amount of information that those sources contain about your topic of interest.

If you are surveying the literature as part of a report of primary research, your literature review will likely form a relatively brief part of the larger paper or article. In this type of literature review, you will typically emphasize the major points of the major works, while the less important works receive only a brief mention if they appear at all. If you are writing a standalone literature review or bibliographic essay, you will have a larger scope in which to work. As a result, you will probably provide a much greater amount of detailed information about the various sources that you review.

But even in a literature review that is part of a larger article, note the differences in level of detail in the 1495-word literature review of 29 sources in de Jong, Yang, and Karreman's article on official and commercial user manuals, for an average of 52 words per source (see Chapter 8), and that in the 1382-word literature review of 16 sources in Read and Michaud's report on their survey about the multi-major professional writing course, for an average of 86 words per source (see Chapter 11). And of course, within these literature reviews, the authors devote different amounts of space to various sources, with a few receiving a paragraph of comment while others are batched together with several others in a single sentence.

The number of sources and the depth of coverage that each devotes to your specific topic are usually the key to determining what level of detail you should provide in your review. In general terms, the more important—or the more general—your topic, the more sources you will find about it as you assemble a working bibliography, and the more information that those sources will contain. In such cases, it is usually necessary to devote less space to your consideration of each source and to consider multiple sources that cover similar material together. The narrower your topic, the more reasonable the task of reviewing the literature becomes. For example, a very large number of sources deal with the role of audience in technical communication, but the use of second person in procedures would result in many fewer hits.

Summary

Primary research involves formulating and testing a hypothesis, collecting information through observation, or conducting a survey or usability study, and then reporting the results to others in the field. Such research is called *primary* because it adds to the existing body of knowledge in the field. Secondary research is conducted by reading reports of previous research, analyzing the results, and then formulating a synthesis of the current state of knowledge on a topic. This research is called *secondary* because it draws on reports of previous primary and secondary research.

Reporting on the previous research on your topic helps you master your subject and provides your audience with the basic information that they need to know to understand your research questions and the context for your own research. It also demonstrates the gaps in previous research. Finally, your literature review helps establish your credibility as an expert on your research area.

An annotated bibliography provides a listing of primary and secondary sources, along with descriptive and evaluative notes about each of those sources. Although annotated bibliographies may be published in journals, books, or web sites, they are often assembled by researchers for their own reference as a part of their background investigation for a primary research project.

A literature review is always prepared for an audience other than the researcher. It is typically a section of a research article or paper that presents a synthesis of the most important primary and secondary research relevant to the primary research project about which the article or paper is reporting. Rather than being organized by source like an annotated bibliography, the literature review is organized by topic.

For Further Study

Using the annotated bibliography that you constructed in Exercise 3.1, select at least six sources that deal in some detail with one aspect of your research topic. Then create a draft for a section of what will become the literature review for your research report. Be sure to keep in mind the purpose and audience of the report and the literature review. Decide the appropriate level of detail for each of the sources that you include, using summary, paraphrase, and direct quotation.

Answer Key

Exercise 3.1

Because the annotated bibliography produced for this exercise will be unique for each person who prepares it, there is no key for this exercise.

References

American Psychological Association. (2020). *Publication manual of the American Psychological Association* (7th ed.). American Psychological Association.

Giammona, B. (2004). The future of technical communication: How innovation, technology, information management, and other forces are shaping the future of the profession. *Technical Communication, 51*(3), 349–366.

Hofstede, G. (1994). *Culture and organizations: Software of the mind.* Harper Collins.

Hofstede, G. (2001). *Culture's consequences: Comparing values, behaviours, institutions, and organizations across nations* (2nd ed.). Sage.

Institute of Electrical and Electronics Engineers. (2018). *IEEE reference guide.* http://journals. ieeeauthorcenter.ieee.org/wp-content/uploads/sites/7/IEEE-Reference-Guide.pdf

Lanier, C. (2009). Analysis of the skills called for by technical communication employers in recruitment postings. *Technical Communication, 56*(1), 51–61.

4 Analyzing Quantitative Data

Introduction

If you have chosen to conduct an empirical research project, you may be collecting numeric data, and you will need to analyze those data appropriately—being careful not to draw unwarranted inferences. Or you may be reading research that uses statistics, in which case you need to think critically about whether or not the researchers were justified in making the inferences that they did. The purpose of this chapter is to help you understand the principles that define "good" research when numbers are involved. Without a doubt, this is a huge subject that cannot be covered comprehensively in a single book, let alone a chapter. We'll try to keep the topic manageable in several ways.

First, this chapter limits its detailed discussion of statistical analysis to hypothesis testing involving numeric averages. Hypothesis testing is a common method in research, and most people can easily relate to concepts and calculations involving averages. Secondly, this chapter tries to manage the complexity of statistical calculations by not discussing the mathematical formulas involved. Instead, it discusses the principles that underlie the formulas and shows how to use Microsoft Excel and the open-source tool Jamovi to do all the required calculations.

Overall, we discuss some types of quantitative designs you are likely to encounter in technical communication research articles. You can use this presentation of the basics to support your own replicable, aggregable, and data-supported (RAD) research (Haswell, 2005) and to become a better-informed reader of quantitative research articles.

As with all research that involves human subjects, consider your ethical obligations to your participants by establishing informed consent. Consider as well whether or not your research should be reviewed by an institutional review board (IRB). Refer to Chapter 2 for more information on compliance with ethical standards.

Learning Objectives

After you have read this chapter, you should be able to:

- Identify examples of quantitative data
- Define standards of rigor for quantitative studies
- Differentiate between descriptive and inferential statistics
- Describe the key factors that affect reliability in inferential statistics
- Write a test hypothesis and construct its associated null hypothesis for a difference in means

- Run an analysis of means for two variables using the *t*-test and of means for three or more variables using the Analysis of Variance (ANOVA)
- Calculate a correlation coefficient
- Calculate a simple regression

If you are feeling nervous about running these statistical tests, don't be! If you can follow a set of directions, you can do statistics.

Clarifying Terminology

- Dependent variable—The measurable outcome that the researcher uses to gauge the effect of changing the independent variable.
- Independent variable—The condition that the researcher deliberately manipulates, that is, the intervention.
- Intervention—An action taken to manipulate a variable. "If I do this intervention, will it make a difference?" The intervention is the independent variable in a research study.

Quantitative Data

Quantitative data are data that can be expressed in numbers—that is, where the observed concept of interest can be counted, timed, or otherwise measured in some way. Defining a concept so that it can be measured is called operationalizing that concept. For example, Nielsen (1993) offers a list of items that could be quantified to operationalize the concept of usability:

- Time to complete a task
- Number of tasks that can be completed within a given time limit
- Ratio between successful interactions and errors
- Time spent recovering from errors
- Number of user errors
- Frequency and time using manuals and online help
- Frequency that the manuals or online help solved the problem
- Proportion of positive vs. negative statements
- Number of times the user expressed frustration or joy
- Proportion of users who prefer the test system over an alternative

In other words, any of the above data could help us to measure whether or not a product is usable.

In addition to measuring, quantitative data can also be used in describing. The following are examples of quantifiable data that could be used to describe technical communicators:

- Age
- Salary
- Education level
- Years of experience in the field
- Ratios of writers to developers within companies

The following list identifies quantifiable data that could be used to describe documents:

- Document length
- Word length
- Proportion of graphics to text
- Use of second person
- Percentage of passive voice verbs

And the following list identifies quantifiable data that could be used to describe teaching and training:

- Student evaluations of teaching/training
- Student scores on assignments

Even somewhat abstract concepts such as reader confidence that a procedure has been done correctly can be operationalized by asking test participants to provide a numeric rating in response to a question. For example, the clarity of instructions in a manual could be operationalized by asking study participants to rate their confidence that they have performed a task correctly on a scale of 1 (not very confident) to 5 (very confident). Most data can be interpreted as numeric data.

Standards of Quantitative Rigor

In our earlier discussions, we said that research draws conclusions or inferences about a general subject or group based upon data taken from a smaller sample. Whenever inferences are made based on measurements taken from a sample, two standards of rigor must be addressed:

- The validity of the measurement
- The reliability of the inference

Validity

Validity can be divided into two types: internal validity and external validity. Internal validity addresses the question "Did you measure the concept that you wanted to study?" For example, consider a study that wants to compare reader preference between two different web sites that present similar information in different ways. The researcher might decide to operationalize the concept of "reader preference" by measuring the number of times each site is visited over a fixed period of time, thus equating "more visited" with "more preferred." A critical reader of the research could argue that the test lacked internal validity since the frequency of visits could be affected by many factors other than preference—for example, differences in media promotion of the two sites or a poorly-worded link to one of the sites. In other words, a measure of frequency visited would not be a measure of preference.

In another example, medical researchers conducted a study to see whether the editing process made medical articles easier to read (Roberts, Fletcher, & Fletcher, 1994). They used readability formulas as a measurement, and if they had done research on them, they would have known that those formulas are flawed, particularly when applied to medical jargon. Thus, they did not actually measure whether or not the editing process made medical articles easier to read. They failed to establish internal validity.

External validity addresses the question "Did what you measured in the test environment reflect what would be found in the real world?" For example, a researcher might conduct a test on how highly graphical, interactive design techniques affect users' satisfaction ratings of web sites. The researcher might bring users into a lab and have them use two different versions of the same site, one rendered as plain HTML pages and the other using JavaScript, HTML5, embedded videos, and other highly interactive programmatic techniques. And let's say that the study showed that users rated the highly interactive version more highly than the plainer one. But if the test was conducted on high resolution screens, with super-fast response times, and with the required plug-ins already installed, the test could lack external validity because the test platform might not represent the equipment that a typical user would have. It might be that in real life, with lower screen resolutions, longer response times, and the aggravation of loading plug-ins, the more interactive site would have resulted in much lower satisfaction ratings.

Validity is controlled through test design.

- Internal validity can be managed by taking care when you operationalize a variable to make sure that you are measuring a true indicator of what you want to study. Ask yourself whether other factors produce or affect the measurements that you intend to take but are not related to the attribute or quality that you are studying. If so, your study might lack internal validity. As a researcher, you should also have reviewed the literature to see how other researchers have operationalized the qualities that you seek to measure. If you think that their operationalization is sound, you can use it and then compare your results. If you think that their method is flawed, you should explain why. This is just one way that studies build upon earlier studies.
- External validity can be managed by taking care when you set up the test that the conditions that you create in your test environment match the conditions in the general environment as much as possible. There is an old joke that most of what we know about human psychology comes from college freshmen participating in psych studies. Not exactly representative of the general population, are they? When designing for external validity, ensure that the sample group itself is a fair representation of the general population of interest. (We discuss the effect of sampling on test design later in the chapter.)

Exercise 4.1 Managing Validity

In the following examples, discuss issues that could affect internal and external validity.

1. A company wants to compare ease of installation between one of its consumer products using a quick install guide and that same product without the quick install guide. It uses employees from its help desk as the participants and times how long it takes them to install the product.

2. A study uses a sample of graduate students in a technical communication degree program to estimate how willing users will be to access a proposed online help file in an application to be used by emergency room admissions personnel. The test provides written case studies, and participants are observed as they enter data into the application. The test will measure time to complete task and the number of errors made.

Reliability

Reliability describes the likelihood that the results would be the same if the study were repeated, either with a different sample or by different researchers. Usually, a researcher wants to be able to apply the findings of the study beyond the scope of just the sample observed. The issue of quantitative reliability is one of statistical significance. Statistical significance is a measure of the degree to which chance could have influenced the outcome of a test.

For example, assume that one group of test participants installs one version of a product in an average of 10 minutes, and another group of participants installs a second version in an average of 6 minutes. Assuming that the same environment and setup were used for both tests, then any of the following could be true.

1. The second version installs more quickly.
2. The researcher happened to recruit faster users for the second test.
3. One of the groups was having a really on or off day.

Reliable quantitative research tries to ensure that any differences detected are actually the result of the object of the study and are not caused by differences in the samples or by random variation.

Some Statistics Basics

Statistics can seem confusing and overwhelming. Here are some important things to know about them according to quantitative technical communication researcher, Dr. Angela Eaton (personal communication, December 21, 2019). First, there are statistical calculations that we run before tests, statistical tests that are tests, and statistical tests that are follow-up (called post-hoc) tests. Not all statistical tests perform the same function.

Second, each test can be performed on only a certain type of data. So if your dependent variable is a nominal (named) group such as low, medium, or high, you would apply a different statistical test than if your variable is measured in numbers like seconds.

Third, a test is typically named after a letter that appears in its formula (for example, the *t*-test), the name of the person who created it (for example, the Tukey test), or the name of the person who created it plus a letter in the formula (for example, Cronbach's alpha). Since we aren't showing you the formulas in this chapter, these names may sound like nonsense, but they're not.

Fourth, how do you know which statistical test to choose? You have to determine the form your data are in and then look up that data type in a table that describes appropriate statistical tests. This information can be found in most statistics books.

Finally, you aren't supposed to memorize all of these different statistical tests and formulas. Memorizing the function of the most common ones—the *t*-test and the ANOVA—will allow you to understand most studies. No researcher has memorized them all. And when you run one, simply look it up to make sure that you are doing it correctly.

Descriptive and Inferential Statistics

Researchers talk about two different kinds of statistics: descriptive and inferential. Descriptive statistics describe a specific set of data. For example, you could record the education level of the attendees at a technical communication conference and calculate their average education level at 17 years. That average would be a descriptive statistic because it describes that specific data set.

Inferential statistics, on the other hand, make inferences about a larger population based upon sample data. For example, we might learn that the average education level for attendees at an instructional designer's conference was 16.2 years (another descriptive statistic). It would be tempting to infer from those two descriptive statistics that the average education level of technical communicators in general is higher than the average for instructional designers in general. But you cannot make such leaps unless you apply inferential statistical techniques to help you gauge the reliability of those inferences. The following illustrative example looks at how statistics help manage the uncertainties of making inferences based on data taken from samples.

Example—Registration Time: Jumping to an Unreliable Inference

Bill, an information designer for a health care provider, conducted a usability study of his company's web site. An important feature of the site was its ability to personalize the content based upon the user's personal health profile. This personalization feature, however, depended on getting visitors to register their personal health information on the site, so part of the test looked at problems that users could have with registration. Along with a lot of qualitative data observed and analyzed, the test recorded the amount of time it took each participant to register. Five participants were tested, and the product went through some revisions to eliminate problems that the test had detected. Bill then tested the new design using the same scenarios and five new participants, and once again the test participants were timed. Table 4.1 shows the numeric data (time to register) from the two tests.

Predictably, someone in management asked, "Did the new version work any better than the first one?" Bill was confident that the second version was easier to use based on qualitative data; namely, users better understood the value of registering and getting their page personalized, showed greater willingness to do the registration, and had less difficulty understanding the questions that they were being asked to answer. But because "managers love numbers," Bill decided to look at the registration times from the tests to see whether they would also support the claim that the second version was better—in this case, faster.

From simply looking at the raw data in Table 4.1, it is hard to draw any immediate conclusions. Some scores for version 1 are better than some scores for version 2, and

Table 4.1 Registration times (in minutes)

Participant	Version 1	Version 2
1	7.2	
2	5.6	
3	10.3	
4	6.6	
5	8.9	
6		8.3
7		11.2
8		4.6
9		6.4
10		4.3

vice versa. To help understand the numbers a bit better, Bill calculated the average registration time for the "before" and "after." He found that the average installation time for version 1 was 7.72 minutes and the average for version 2 was 6.96 minutes— a respectable difference of 0.76 minutes. He felt good, then, that the quantitative data also supported that the second version had indeed done better than the first.

But when Bill presented the quantitative results to Ann, the product manager, she seemed a bit skeptical. "How do you know that you didn't just luck out and get some faster participants in the second test?" she asked. She saw Bill's face go blank. She started entering the test times into her computer and said, "Let me ask it another way. Let's say you tested the exact same version twice. What would be the probability of the second test going faster just by luck? You know, you got some smarter users in the second group or a couple of slow-pokes in the first?"

"I don't know," Bill answered, "pretty small, I'd think."

Ann hit the Enter key on her laptop. She turned back to Bill and looked over her reading glasses. "About 32 percent, actually," she said. "The probability is 32 percent that I could get those same results or bigger just by testing the same product twice using two different samples of users."

Bill was crestfallen.

"So how confident are you now that the second version is faster than the first?" Ann asked.

"Not very," Bill conceded. "How did you come up with that number?"

Ann smiled. "Actually, I did it in an Excel spreadsheet using one of its built-in formulas."

"It's just that by looking at the test results, the difference seemed pretty significant," Bill said.

Ann agreed. "The difference does look significant from a practical perspective, a little over 45 seconds. Who wouldn't be happy to knock 45 seconds off of any task? But the computer was looking at statistical significance. In other words, what is the probability that this difference is real versus caused by sampling error or random variation?"

Bill went on, "But you seemed like you had some questions about the results even before you put the numbers in the computer. What were you looking at?"

"The same thing the computer looked at. For one thing, the sample sizes were very small. If you test only five people, the probability of an unusual user throwing off your average is a lot higher than if you test a lot of people. Secondly, the numbers were all over the place; in other words, there was a lot of variability in the data. When data vary a lot, the probability of picking an uncharacteristic group of fast or slow participants is higher than if the data are bunched up more tightly."

Ann smiled again, "Of course, I couldn't come up with 32 percent off the top of my head, but that's what computers are for."

Variance and Sample Size

In the example above, the product manager, Ann, demonstrated how statisticians think about numbers—and how you need to think about numbers when conducting or reading quantitative research. One of her questions strikes to the very core of inferential statistical analysis: "Let's say you tested the exact same version twice. What would be the probability of the test results differing just by luck?" This is essentially the question of reliability that is the cornerstone underlying hypothesis testing. Inferential statistics contain sophisticated formulas and models to answer that question. But two basic principles underlie all of these formulas, two principles that a beginning researcher needs to grasp.

Principle 1: The Smaller the Variance in the Data, the More Reliable the Inference

Formulas that estimate statistical significance (that is, formulas that assess the reliability of an inference) consider how much variance is contained in the sample data. Imagine for a moment that you had to estimate the height of two different populations based on samples taken from each. Let's say that the first population is the city of Atlanta, Georgia, and the second population is all the members of the men's collegiate swimming teams in the United States. Furthermore, let's say that you can take only a very small sample, 10 people randomly drawn from each population. In the case of the Atlanta population, there is a lot more variance in the data than in the case of the men's collegiate swimming teams. Atlanta's population includes infants, young children, big people, small people, and so on. The swimming teams have a much smaller variance—pretty much physically fit males between the ages of 18 and 21. So the probability is much higher that you can get an "unlucky draw" in your Atlanta sample than in your swimming team sample. You could get a lot of children (making your sample average too low) or a greater than normal proportion of tall adults (making your sample average too high). You could get unlucky draws in the swimming team sample as well, but the effects would not be as great. In other words, your tallest swimmer is not going to drag down the average that much, nor would the shortest swimmer have much of an effect.

The point is that you can "trust" the outcome from statistical formulas more when the data in the sample do not vary much. The more the data vary, the lower the reliability of the inference.

How do formulas take variability into account? They calculate a number called the standard deviation (*SD*). You hear and read this term a lot in research reports, and you will probably report it in your own quantitative study. It is an indicator of the variation of the data in the sample. If the population is a typical one, you could take the average of whatever you measure (for example, height or time to complete a task) and calculate that roughly two thirds of the population fall within plus and minus one standard deviation of that average. So, in the case of the men's swimming teams, if the average height was 5 feet, 11 inches and the standard deviation was 2 inches, we would expect to find that two thirds of the members of the men's swimming teams fell between 5 feet, 9 inches and 6 feet, 1 inch tall.

Standard deviation is possibly the most helpful of the statistical calculations. Take the *SD* of a normal data set and add it and subtract it from the mean; 68% of all data points fall into this range. So, if you see a very large *SD*, you know that the data are widely scattered. If the *SD* is small, the data are clustered tightly. But that's not all. If you subtract and add two *SD*s to the mean, 95% of the data points fall into this range. And if you subtract and add three *SD*s to the mean, 99.7% of the data points fall into this range.

SD tells us quite a lot about where our data fall. Standard deviation is helpful for envisioning how widely the data vary from the average, and it is an important component of the formulas that calculate the reliability of an inference.

Principle 2: The Bigger the Sample Size, the More Reliable the Inference

Let's go back to our example of calculating the average height for the population of Atlanta. If our sample size is 10 and we get an unusually tall person, our average will be too high. But if we increase our sample size to 500, two important things happen.

- The probability is greater that our proportion of tall people in the sample more closely matches the proportion of tall people in the population.
- The effect of an unusual data point (called an outlier) is much smaller. For example, one unusually tall person will have a much greater effect on an average that is calculated using 10 data points than on an average that is calculated using 500 data points.

Just as they do with the standard deviation, formulas that assess the reliability of statistical inferences take into account the size of the sample.

It is necessary to understand these principles so that you do not overreach your data and make unreliable inferences. In research reports, standard deviation is represented by *SD*, and sample size is represented by *n*.

Sampling

We have already discussed the importance that sampling can have on the rigor of a study.

- The external validity of the study is dependent on how realistically the profiles of the test participants match the profile of the population of interest.
- The reliability of the study is affected by the size of the sample.

You may be surprised at how small a sample can provide statistical reliability. A common question from new researchers is "How many people should I include in a sample?" Sample size calculators can sometimes help with this decision, but these calculators typically depend on numbers you may not have in technical communication, such as how common a phenomenon is in a population. When you can't use a sample size calculator because of your topic, look at previous studies to see how many participants those researchers used, talk to colleagues, and talk to research methods teachers.

We will now discuss another important aspect of sampling called random selection. In a perfect test, the only factor that would affect the dependent variable would be the manipulation of the independent variable. For example, in the case of the web site usability test, Bill would have preferred that the only thing that could have affected the time to register would be the respective merits of the two designs. But other factors can also introduce differences in performance:

- The innate abilities of a test participant—for example, intelligence, manual dexterity, and so forth
- The experience and education of the test participant
- The condition of the participant—for example, fresh or fatigued, happy or sad, and so on
- Environmental factors—for example, comfortable room conditions for some participants versus uncomfortable for other participants

The researcher can try to control these confounding variables, as they are called, but even the best design will be subject to variances that have nothing to do with the intervention. Rigorous test methods deal with this problem by randomly assigning test participants to the test groups. The theory is that the confounding variables will be evenly distributed across the groups and will, therefore, even themselves out.

Random selection means that the selection and assignment of each test participant is independent and equal. By independent we mean that selecting one participant to be in a particular group does not influence the selection of another. By equal we mean that every member of the population has an equal chance of being selected. In reality, it is challenging and often impossible to meet both of these criteria perfectly. Researchers must often make compromises, and they need to consider (as well as disclose to the reader) how those compromises could affect the validity and reliability of the findings.

The hardest goal to achieve is to randomly select from the total population. Researchers tend to sample from a sampling frame—that is, a list of the members of a group that they want to talk to. According to Eaton (personal communication, December 21, 2019), that approach is great, except that there is no sampling frame for a lot of the studies that technical communicators want to do. For example, there is no list of all the professional technical editors in the country; the closest that might be found is a list of those people who belong to the Society for Technical Communication's Special Interest Group in Technical Editing. As researchers, we have to find the closest list that we can to the population that we are trying to sample.

Many academic research projects, for example, rely on the local student population for their sample pool. Research done by companies often relies on employees or on customer data that have been collected for another purpose. These samples are called samples of convenience and compromise the principle that every member of the population of interest should have an equal chance of being selected. In those cases, the researcher should either demonstrate how the sample still represents the general population or at least qualify the statement of findings with appropriate limitations. For example (and depending on what you are studying), to justify how findings from a single class of students might be generalizable to all undergraduates, you could compare the mean grade point average of your group to the mean grade point average of all students enrolled at the university.

In fact, the use of samples of convenience in an earlier study can be a good springboard for future research—to see whether different sample groups would respond differently. For example, a study might recruit its participants from a rural community because of the location of the university that is sponsoring the study. This test population could prompt another researcher to do a similar study with urban participants to see whether the results would be the same.

Even if you are limited to a geographic area or other constraint, you must be rigorous in assigning participants to one intervention or another. Find a method that eliminates researcher or participant bias in assigning participants to groups. For example, if you are using your classmates to test the effectiveness of one type of search engine versus another, you could use a numbered alphabetical class roster and arbitrarily assign the odd-numbered students to using one type of search engine and the even-numbered students to the alternate type. Alternatively, you could put the names in a hat and draw them out in equal halves and make the assignments that way. In yet another example, if a university office for undergraduate research wants to assess the quality of undergraduate research projects by pulling sample projects from classes across the university, they need to randomly select student names and request those projects from professors. If professors selected the projects, they might choose the best projects. Likewise, avoid letting participants self-select or assign themselves to a group. These methods could compromise the integrity of the sample.

However you select and assign your participants, tell the reader what you've done and discuss any limitations that your method might impose on the reliability of the findings.

Testing for a Difference in Two Independent Means (*t*-Test)

In statistics, the word *mean* denotes average. Throughout the remainder of this chapter, we use the word *mean*. Research in a field of practice such as technical communication is generally practical and often asks the question "If I do this, will it make a difference?" In research, we call this an *intervention*. A common research pattern, and one whose roots are firmly planted in the scientific method, is to test an intervention between two groups: one group that gets the intervention (called the test group) and one group that does not (called the control group). The intervention that you test is called the *independent variable*, and the result that you measure is called the *dependent* variable. You can remember this by thinking that the dependent variable *depends* on what happens in the intervention.

For example, you might want to test what effect using headings in a document has on how long it takes readers to find information. Using headings is the intervention that you want to test (the independent variable). You test one group of participants using a document that does not have headings (the control group) and another group of participants using the same document to which headings have been added (the test group). You give each group the same information-recovery task, and you measure the time that it takes each participant to find the information (the dependent variable). You can absolutely have more than one independent and dependent variable in a study. Since we are introducing them here, however, the examples used will be of one variable each.

At the end, you compare the average time that it took the users of the document without headings to the average time that it took the users of the document with the headings to determine whether headings made a statistically significant difference in successfully performing the task. Once again, a difference is statistically significant if it is not likely to have been caused by sampling error (that is, unlucky selection of exceptionally fast or slow participants in this case) or random variation.

Hypothesis testing can also test such conditions though they do not involve interventions. For example, you might want to test whether there is a difference in time spent editing documents written by contract technical writers and company-employed technical writers. Although employee status is not technically an intervention, you could treat it as the independent variable, and you could treat time-to-edit as the dependent variable. Other pre-existing categories that can be used as independent variables are gender, educational level, and so on.

Example—Registration Time: A Hypothesis Testing Approach Between Two Means

Let's revisit our example where Bill tested the two web site versions and see how he could have done it more reliably as a hypothesis test. Here, we define and illustrate all of the steps that you would need to follow to analyze data from a reliable research project testing two groups and one intervention. The approach we describe has a specific technical name: a *t*-test of two independent means. It is a common research design. However, the principles this case illustrates apply to every quantitative study, regardless of the specific design.

Defining the Hypothesis

Bill should have started by stating his test hypothesis—that is, the assumption that he was trying to test. A test hypothesis should clearly state the following things:

- The independent variable (the thing that the researcher will deliberately manipulate)
- The dependent variable (the measurable outcome that the researcher will use to gauge the effect of changing the independent variable). If it is not clear how the dependent variable will be measured, state how it will be operationalized in the test hypothesis as well. For example, "improve reader satisfaction (as measured by a satisfaction survey) ...."
- The direction (if any) of the hypothesized effect—for example, does the researcher expect the dependent variable to increase or decrease as a result of changing the independent variable?

Here is how Bill could state his test hypothesis.

H_1: The mean registration time will be reduced between the original web site design and a second design that resulted from applying information learned during usability testing.

But there is no way to prove that hypothesis with any certainty. Because humans are all different, and circumstances can vary, it is very likely that the two results would be different, because data always vary, at least a little. Even if Bill tested the same web site design twice, the probability is that the two average registration times would differ by some amount.

The next logical question is, "But would the difference be big enough to say that the difference was significant?" The problem is that there are no general formulas or good models to forecast the probability of finding a difference of a certain size if the two populations are different. More accurately, to make that kind of calculation, you would already have to know so much about the two groups as to make the research unnecessary. Asking every single person in a target group is called a census, and you can't run inferential statistical tests on it because you aren't trying to generalize to any larger group—you've literally asked everyone. (Hence, a national census is a census, not a survey.)

So Bill solves this problem by restating the hypothesis in a form called the null hypothesis (so named because it claims there is no difference):

H_0: The mean registration time will not be reduced between the original web site design and a second design that resulted from applying information learned during usability testing.

Why do researchers formulate and test null hypotheses? Because there are general formulas and models for calculating the probability of finding a difference in the means of two samples drawn from the same or equivalent populations. In addition, logically, it is much more difficult to prove that a thing is true than it is to prove that it is not true. The logic of hypothesis testing is similar to the judicial philosophy of "presumed innocence." Juries must presume that a defendant is innocent unless the prosecutor can introduce enough evidence to make them believe otherwise beyond a reasonable doubt. Similarly, researchers assume that two groups or interventions are equivalent (that there is no difference between them) unless the data introduce enough evidence to make them believe otherwise beyond a reasonable doubt.

The process runs like this.

1. We can't say with any certainty whether an observed difference in registration time is caused by differences in the web sites or just by sampling error.
2. Therefore, let's assume that there will be no difference in mean registration time between the two web sites; the average time to register (if we could test everybody) would be the same for both.
3. We'll test the two sites (with sample users) and see how big a difference we get.
4. We'll ask ourselves, "What is the probability that we would get a difference this big if the two designs really were equivalent?" We will use an Excel spreadsheet to calculate that probability for us using a *t*-test.
5. If the probability is small enough (that is, it's highly unlikely that we'd get a difference this big just by luck of the draw), we will reject the null hypothesis and

be left with the assumption, "The difference in times was caused by differences in the web site designs." (Note that we talk about the observed difference between the two means being large and the probability of random chance being small.)

Analyzing the Data

In this section, we show you how to use an Excel spreadsheet to calculate the descriptive statistics that you will include in your report to support your conclusion. Then we show you how to use that same spreadsheet to calculate the probability that the findings could be the result of sampling error.

First Bill enters the registration time data into a spreadsheet and types labels for the descriptive statistics that he will report (*n*, the size of each sample; the mean time for each version; and *SD*, the standard deviation of the mean time for each sample) (see Figure 4.1).

The spreadsheet can then automatically calculate the three descriptive statistics for each version. Even though it is easy to look at the data and know that *n* equals 5 in this example, we will show you how to automate the calculation in case you work with large amounts of data. Bill follows these steps to calculate *n*.

1. Insert the cursor wherever you wish to display the answer (cell B9 in the example).
2. Select Formulas | Insert Function from the menu. Excel displays the Insert Function dialog box.
3. Select the category "Statistical" (see Figure 4.2).
4. Select the function "COUNT" and click "OK." The Function Arguments dialog box appears as shown in Figure 4.3.

	A	B	C	D	E
1		**Registration Times**			
2					
3		Version 1	Version 2		
4		7.2	8.3		
5		5.6	11.2		
6		10.3	4.6		
7		6.6	6.4		
8		8.9	4.3		
9	n				
10	mean				
11	SD				
12					

Figure 4.1 Data and labels entered into an Excel spreadsheet

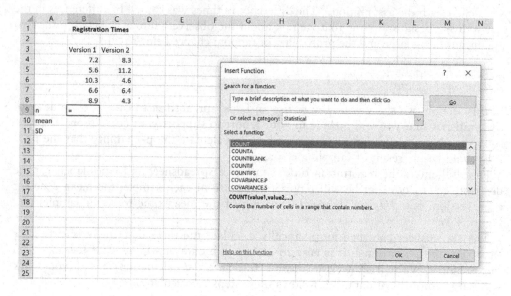

Figure 4.2 Selecting the function to count the *n* of the data

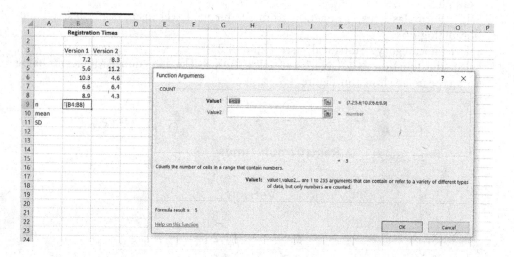

Figure 4.3 Function arguments defined

5. Select the cells that contain the data for one of the versions by clicking and drag-
    ging the mouse cursor.
6. Click "OK."

The calculation of *n* will appear in the cell that you indicated in Step 1.

Bill follows the same procedure to fill in the other descriptive statistics. See Figure 4.4. For the mean he uses the function "AVERAGE," and for the standard deviation he uses the function "STDEV.S." In Excel, choose "STDEV.S" when you are basing

⊿	A	B	C	D	E
1		**Registration Times**			
2					
3		Version 1	Version 2		
4		7.2	8.3		
5		5.6	11.2		
6		10.3	4.6		
7		6.6	6.4		
8		8.9	4.3		
9	n	5	5		
10	mean	7.72	6.96		
11	SD	1.875367	2.860594		
12					

Figure 4.4 Descriptive statistics calculated

the standard deviation on the sample rather than the entire population. (Note: Be careful to select only those cells that contain the test data. For example, do not accidentally include the cells that contain the n and mean values when calculating the *SD*.)

Bill then uses the spreadsheet function "T.TEST" to calculate the probability of random chance being responsible for getting a difference in means this large from two equivalent populations. (Remember, in hypothesis testing we want to determine whether we can reject the null hypothesis.)

1. He creates a space for his probability value (p) on his spreadsheet and inserts the "T.TEST" function just as he did with the "COUNT," "AVERAGE," and "STDEV.S" functions. See Figure 4.5.
2. In the Function Arguments dialog box, Bill selects the Version 1 dataset for Array 1 and the Version 2 dataset for Array 2. See Figure 4.6.

The Tails argument (value can be 1 or 2) refers to a one-tailed or two-tailed test. Use a one-tailed test if the test hypothesis is describing an intervention that you believe will cause a difference in a particular direction (one-tailed tests are also called directional tests). In a two-tailed (or non-directional) test, the test hypothesis would be supported by a difference in either direction. For example, if you were testing a hypothesis that men and women differ in regards to how much time they spend reading help, you would use a two-tailed test because it would not make any difference in your finding if men spent more time or less time than women—a difference in either direction would support your test hypothesis that they spend a different amount of time than women do. On the other hand, if your test hypothesis is that men spend

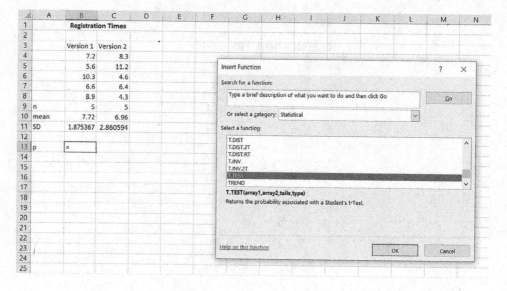

Figure 4.5 Selecting the function for the *t*-test of the data

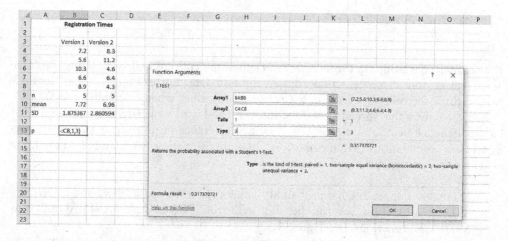

Figure 4.6 Function arguments for *t*-test

less time than women do, then you would use a one-tailed test because the hypothesis would be supported only if the difference were in the direction of less time.

3. In our registration time example, Bill's test hypothesis stated that the changes in the design would reduce the mean time to register, so he selects a one-tailed test.

The Type argument defines what kind of *t*-test you are running (value can be 1, 2, or 3):

- Use "1" if the data points are paired. This type of *t*-test is common in pre-test/post-test studies in which the participant takes a test, then experiences the intervention (such as instruction or training), and then takes the same test again.

- Use "2" on non-paired tests (for example, you use a different group of participants to test the intervention) where the variances around the mean for each group are roughly equal.
- Use "3" on non-paired tests where the variances around the mean for the groups are distinctly not equal.

4. In the registration example, Bill makes the more conservative choice of "3." Selecting "3" makes it slightly harder to reject the null hypothesis (see Figure 4.7).

Now comes the deciding question, "When is p small enough to reject the null hypothesis and accept the test hypothesis?" In other words, when can you say, "The probability of getting these results by chance is so small, I'm going to assume that there is a real difference in the two samples"? This decision depends on how confident you want to be. If you are not conservative enough (that is, if you accept too high a p value), you risk accepting false results. Researchers call this a Type I error. If you are too conservative (that is, if you insist on too low a p value), you risk rejecting true results. Researchers call this a Type II error. Most research in the field of technical communication would be rigorous enough if it accepted results that had a p value of 0.1 or lower. That is, if the p value is 0.1 or lower, you have only one chance in 10 that the results are due to random chance or participant selection. We call this standard the alpha value. In other words, with an alpha of 0.1, the p value must fall below 0.1. In situations where considerable harm could be done by accepting a false claim (for example, scrapping an expensive documentation system already in place), then a more conservative alpha of less than 0.05 or even less than 0.01 might be more appropriate.

▲	A	B	C	D	E
1		**Registration Times**			
2					
3		Version 1	Version 2		
4		7.2	8.3		
5		5.6	11.2		
6		10.3	4.6		
7		6.6	6.4		
8		8.9	4.3		
9	n	5	5		
10	mean	7.72	6.96		
11	SD	1.875367	2.860594		
12					
13	p	0.317371			
14					

Figure 4.7 p calculated

5. Bill decides to use a p value of 0.1. He concludes that there is insufficient evidence to assume that Version 2 is any faster than Version 1.

Don't feel too bad for Bill; he still has a lot of good qualitative data that demonstrated that Version 2 was better; he just can't say it's faster. We discuss how to analyze qualitative data in Chapter 5.

Actually, a lot of research ends up reaching the conclusion that there is no statistically significant difference between the two groups or interventions tested. That finding is not a problem; it does not discount the value of the study. The field of technical communication still benefits from learning that two groups are not different or that a specific intervention does not have the effect that some have thought it would. As Thomas Edison said, "I have not failed. I've just found 10,000 ways that won't work." Sometimes, however, the researcher thinks that the observed trend would be significant if the sample size were increased or a more diverse sample were recruited, and recommends that additional research be conducted with larger samples.

Exercise 4.2 Testing the Difference Between Two Means

Jodie does a baseline study to determine how long it takes surgeons to complete a medical device procedure with documentation that has not undergone a new validation process as compared to surgeons who complete the procedure with documentation that has undergone a new validation process. Here are the times (in minutes) that it took surgeons to complete the procedure with each version of the documentation:

Version 1 (without validation): 56, 51, 41, 55, 56, 60, 47
Version 2 (edited based on validation criteria): 46, 43, 42, 50, 49, 43, 45

Did documentation that had been edited based on the validation process reduce the time it took the surgeons to successfully complete the medical procedure?

- Write the test hypothesis that you would apply.
- Write the corresponding null hypothesis.
- Calculate n, mean, and SD for both samples.
- Calculate the p value for the t-test.
- State what your conclusion would be if you were Jodie.

Testing for a Difference in Three or More Independent Means (ANOVA)

As you conduct research, you may often need to compare more than two variables. Because a t-test can compare only two variables, you need another statistical tool. The ANOVA (Analysis of Variance) tool enables you to compare the means of three or more independent variables. Many types of ANOVAs enable complex comparison of multiple variables and interventions, but in this chapter, we focus on the most common and simplest type: a one-way ANOVA. A one-way ANOVA allows you to identify whether or not there is a statistically significant difference among the means of three or more independent variables.

In a one-way analysis of variance, one independent variable is manipulated as the intervention, and the results of the intervention are analyzed among three or more groups. For example, a researcher might want to examine the effect of type size on reader reaction to a document. But instead of wanting to compare just two type sizes such as 10- and 12-point type for a given typeface, the researcher would like to study the differences at 8, 10, 12, and 14 points. Obviously, this design would not lend itself to the test of two means that we looked at earlier in this chapter because four means are involved. One could use the *t*-test of two means to test all the possible combinations of two type sizes (8 vs. 10, 8 vs. 12, 8 vs. 14, 10 vs. 12, 10 vs. 14, and 12 vs. 14), but this approach would be cumbersome, and the reliability of testing and comparing the results for so many combinations may be low. The ANOVA calculation uses a different analytical technique that gives reliable results for a test like this.

As we mentioned earlier, a number of different ANOVA calculations are available to meet a variety of research needs. A two-way analysis of variance allows even more options, allowing the researcher to examine multiple variables as well as multiple interventions. For example, our researcher in the example above might want to test those four type sizes using different typefaces, such as Times New Roman, Verdana, Arial, and MS Comic Sans. A two-way ANOVA can handle such complex scenarios. It would also enable the researcher to discover interactions such as whether one type size is preferred for one typeface and another type size for a different typeface.

In the remainder of this section, we show you how to conduct a one-way ANOVA of three means with a post-hoc analysis. An ANOVA simply tells us whether there is or isn't a difference in the means; it does not tell us where the difference is. Post-hoc tests of an ANOVA allow you to identify specific relationships within the ANOVA. In contrast, a *t*-test doesn't have any post-hoc tests as it is just testing whether two things differ; there's only one place the difference could be. You can expand on these options of analysis as needed in your own research and study.

Example—Registration Time: A Hypothesis Testing Approach Among Three Means

In this example, we use open-source statistical analysis software: Jamovi. Although Excel will calculate ANOVAs, Jamovi allows you to run post-hoc tests in many fewer steps than does Excel. Thus, using Jamovi in this example requires fewer steps and introduces you to an excellent open-source statistical software option.

Let's return to Bill and his redesign of the web site. Suppose that eventually, because the first redesign resulted in no significant decrease in registration times, Bill and his colleagues decide that they will update the site again based on some of the qualitative data that they collected in their first study. Once they have completed this third version, they again study the performance of users and gather data on registration time. These tests result in three sets of data. The *t*-test won't work for an analysis of variance between three means, so Bill does some research into what type of statistical tests he can run to analyze the three versions. He decides to run a one-way ANOVA.

If Bill were going to use Excel to run this ANOVA, he would first install the Analysis Toolpack add-in for Excel. This process is very easy, and he can use Excel help or a web search to locate instructions. Excel, Jamovi, and other tools such as SPSS have advantages and disadvantages; you can decide on your preference. However, this time Bill is going to use Jamovi because he wants to be able to run a post-hoc analysis in as few steps as possible. To accommodate the new data, Bill recasts his hypothesis.

H_1: The mean registration time will be reduced between the original web site design, a second design, and a third design that resulted from applying information learned during usability testing.

Because this hypothesis cannot be proved with certainty, Bill again develops a null hypothesis.

H_0: The mean registration time will not change between the original web site design, a second design, and a third design that resulted from applying information learned during usability testing.

Bill and his colleagues proceed to test Version 3 resulting in the times you see in Figure 4.8.

Notice that Bill also calculated the mean and standard deviation for the new data. Now Bill begins to calculate the variance in Jamovi.

1. Bill goes to Jamovi.org and downloads this open-source data analysis software to his computer. You can see his untitled Jamovi worksheet in Figure 4.9.
2. Bill enters his data into the Jamovi worksheet a little differently than he would using Excel. First, he clicks on the Data tab and then on the Setup icon.
3. He clicks in the Data Variable field and enters his first variable, in this case the independent variable "Version." At this time he leaves all other fields set to the

E	F	G	H	I	J
		Registration Times			
		Version 1	Version 2	Version 3	
		7.2	8.3	5	
		5.6	11.2	4.8	
		10.3	4.6	4.5	
		6.6	6.4	2	
		8.9	4.3	3.7	
	n	5	5	5	
	mean	7.72	6.96	4	
	SD	1.875367	2.860594	1.222702	
	p				

Figure 4.8 Registration times among the three versions

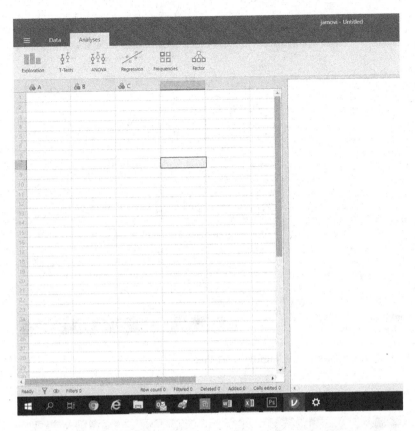

Figure 4.9 Blank Jamovi worksheet

defaults, including the data level as "Nominal." See Figure 4.10. Nominal data are represented by names, such as "version." There is no order to these types of data. For the statistics program to read them, the names are assigned numbers, but they still represent names.

4. He then clicks on the right arrow so that he can define his dependent variable of registration time.

5. To finish, he clicks the up arrow, and his variables become the column labels shown in Figure 4.11.

6. Next he enters his data—his three version numbers as well as the observed registration times. Notice that all version numbers are entered in one column and all registration times are entered in the next column as shown in Figure 4.11.

7. Now that Bill has entered his data, he can run his ANOVA. He clicks on the Analysis tab, and then on "ANOVA."

8. Next, he clicks "ANOVA" on the drop-down menu. Note that Bill could also have clicked on "One-Way ANOVA," but by clicking "ANOVA," he will have the option of a few more descriptors on the upcoming screens.

9. Bill then highlights "Registration Time" and clicks on the right arrow to move it over to the Dependent Variable field.

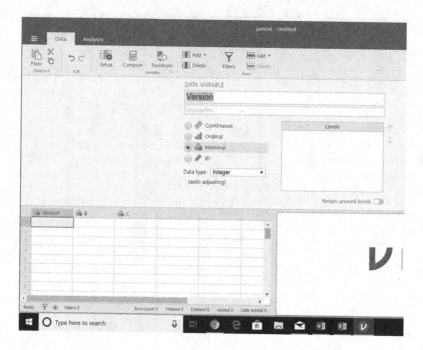

Figure 4.10 Defining variables in a Jamovi worksheet

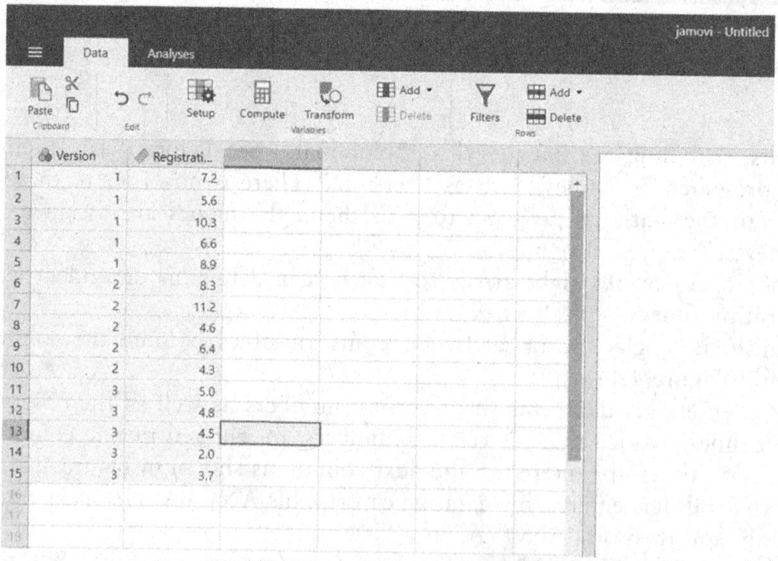

Figure 4.11 Jamovi worksheet with variables and data identified

10. He does the same thing to move "Version" over to the Fixed Factors field, which is the independent variable field. Why isn't this called the Independent Variable field? We don't know. Names aren't always consistent throughout research methods. Just realize that the exact term this chapter uses may not be the exact term used by the statistics program or by another book. We've tried hard to use the most commonly accepted names.

Jamovi immediately calculates the ANOVA as shown in Figure 4.12 and reports that there is a statistically significant difference of means within the array of data. Because the p value is 0.045, which is less than Bill's alpha value of 0.1, the null can be rejected. Version 3 has resulted in significantly different registration times.

To analyze further, Bill knows that there was no significant decrease in registration times between Version 1 and Version 2 of the web site, and he also knows that the difference for Version 3 is significant. But he cannot tell from the ANOVA where the significant difference is. Is it between Versions 2 and 3? Between Versions 1 and 3? To find out, he can run a post-hoc test.

Post-hoc analyses provide specific group-to-group analyses. Where the ANOVA can identify whether or not there is a significant difference in variables overall, a post-hoc test

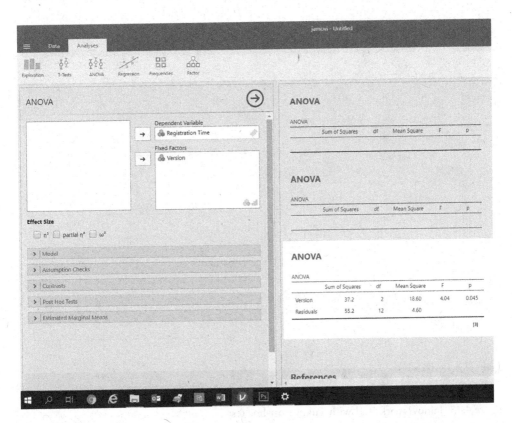

Figure 4.12 Jamovi worksheet with ANOVA calculated

can identify where those differences lie. The Tukey post-hoc test is a common choice in statistical analysis. It controls well for Type I error (that is, it helps us avoid rejecting the null hypothesis when it is true), and it works well on unequal sample sizes. Bill continues to use Jamovi to conduct a post-hoc analysis of his ANOVA and gain more information.

1. He scrolls down to Post-Hoc Tests and clicks on the drop-down menu. The Tukey test is already checked, so he leaves that alone.
2. He moves his factor "Version" from the left to the right box in the post-hoc section.
3. He highlights "Version" and clicks on the right arrow. Jamovi immediately runs the Tukey test as shown in Figure 4.13.

Bill reviews the data and sees that only the difference in registration times between Version 1 and Version 3 is significant. That is, only the comparison of means between Version 1 and Version 3 shows a difference with a p value below his requirement of 0.1. Therefore, Bill can conclude that the redesign for Version 3 resulted in a statistically significant improvement in registration times compared to Version 1.

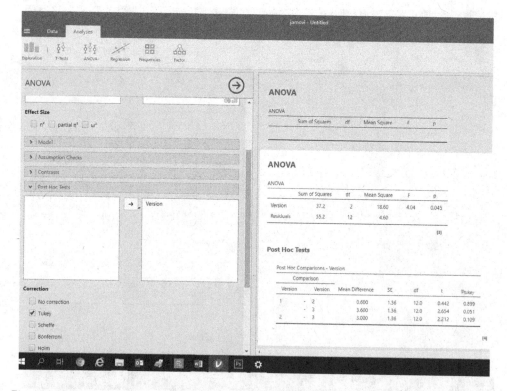

Figure 4.13 Jamovi worksheet with Tukey post-hoc test

Process for Hypothesis Testing

Use this process to design and conduct a hypothesis test involving a difference of means.

1. State the test hypothesis(es). Be sure to identify the independent variable(s), the dependent variable(s), and the expected direction (if any).
2. Recast the test hypothesis as a null hypothesis. This is a formal step but it keeps everyone focused on what the test is truly looking at.
3. Collect the data.
4. Enter the data into a statistical analysis software spreadsheet.
5. Calculate the descriptive statistics of sample size, mean, and standard deviation.
6. For a test of two means, use the *t*-test to calculate the *p* value—that is, the probability that the results could be caused by differences in the samples or random variation rather than the intervention.
   Or
   For a test of three or more means, use the ANOVA to calculate the *p* value and a post-hoc test to calculate group-to-group relationships.
7. Accept or reject the null hypothesis based on the *p* value(s). (Typically, you can reject the null when a *p* value is less than or equal to 0.1.)
8. If you reject the null hypothesis, then accept the test hypothesis.

Exercise 4.3 Testing the Difference Between Three or More Means

Jonathan is studying the perceptions of usability of the engineering web site at his university. Administrators want to better understand how the web site serves some of its most important users: potential students, current students, parents of potential students, and parents of current students. As part of the study, 10 participants from each category are asked to use the web site to accomplish specific tasks. Then each is asked to rate the ease of use (on a scale of 1 to 5, with 1 being very difficult and 5 being very easy) based on their experience completing the tasks. He collects the following data:

> Potential students: 2, 3, 4, 3, 5, 4, 4, 2, 2, 1
> Current students: 2, 2, 1, 2, 3, 3, 2, 1, 1, 3
> Parents of potential students: 3, 4, 4, 4, 4, 2, 5, 4, 3, 4
> Parents of current students: 4, 3, 2, 4, 5, 5, 4, 3, 5, 3

Is there a statistically significant difference among any of these four categories? That is, is there a difference in how user groups perceive ease of use? Follow these steps to answer the questions.

1. Write the test hypothesis that you would apply.
2. Write the corresponding null hypothesis.
3. Calculate the ANOVA.
4. Is there a statistically significant difference among groups within the ANOVA?

5. If there is a statistically significant difference among groups within the ANOVA, use a Tukey post-hoc test to identify where those differences lie.
6. State what your conclusion would be if you were Jonathan.

Correlation Analysis

Some studies do not make interventions, strictly speaking, but just look to see whether or not a relationship exists between two variables. For example, a researcher might want to see whether there is a correlation between time to install a product and the perceived ease of installation. This determination would not necessarily involve testing with independent and dependent variables. The researcher could simply have a number of test participants install the same product, time them, and then ask them to rate the ease of installation. However, the researcher might like to know whether the value of one variable can predict (not cause) the value of a second variable. Correlations exist everywhere, but correlation does not mean causation. For example, as you have aged, prices on goods have gone up, but, obviously, your age did not cause prices to rise.

The researcher could calculate the coefficient of correlation to see how strong the relationship was between the time participants took to install the product and their rating of the ease of installation. The result of that calculation is called r, and its value is always equal to or between +1.0 and -1.0. If the number is positive, the relationship is said to be direct; that is, an increase in one variable is associated with an increase in the other. If the number is negative, the relationship is said to be inverse; that is, an increase in one variable is associated with a decrease in the other. We can assign some numbers to the example we just presented.

Example—A Correlation Analysis of Installation Time and Perception of Ease of Use

Let's use Excel again for this example. Heath asks 10 people to install the software product, and their installation times in minutes are: 5, 6, 4, 6.5, 9, 10, 8.75, 6, 9, and 11.5. Once each person finishes the installation, he asks each of them to rate the ease of installation on a scale of 1 (very difficult) to 5 (very easy). See Figure 4.14 for his results.

Heath wants to determine whether there is a correlation between time to install and the ease-of-use rating.

1. In Figure 4.14, Heath has added a label for the r value. He positions the cursor in cell B:13 and clicks Formulas and then Insert Function.
2. On the Insert Function screen, he selects the "Statistical" category, and then scrolls down to select "CORREL." See Figure 4.15.
3. He clicks "OK" and then highlights the data in the Time to Install column for Array1 and the Ease of Use column for Array2.
4. He clicks "OK" again, and the correlation coefficient is displayed in cell B:13. See Figure 4.16.

▲	A	B	C	D	E
1		Subject	Time to Install	Ease of Use Rating	
2		1	5	2	
3		2	6	1	
4		3	4	1	
5		4	6.5	1	
6		5	9	5	
7		6	10	4.5	
8		7	8.75	4	
9		8	6	3	
10		9	9	4	
11		10	11.5	4	
12					
13	r				
14					

Figure 4.14 Excel worksheet with software installation data

The following general guide can be used to interpret correlation coefficients. (Incidentally, this strength is the same no matter whether the coefficient starts with a positive or a negative. Strength is the same in both directions.)

- If r is less than 0.3, the relationship is weak.
- If r is between 0.3 and 0.7, the relationship is moderate.
- If r is greater than 0.7, the relationship is strong.

In this case, the statistical test has identified a strong correlation between the two variables, 0.822715. Thus, Heath can conclude that time to install can predict perception of ease of use. As a reminder, the correlation shows that one variable can be used to predict the other but not to show that one causes the other. As you can see, such a prediction could be used in forecasting user response.

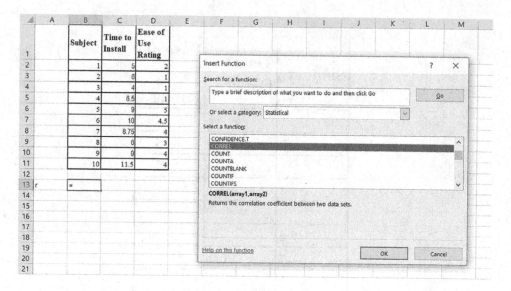

Figure 4.15 Excel worksheet with correlation function selected

	A	B	C	D	E
1		Subject	Time to Install	Ease of Use Rating	
2		1	5	2	
3		2	6	1	
4		3	4	1	
5		4	6.5	1	
6		5	9	5	
7		6	10	4.5	
8		7	8.75	4	
9		8	6	3	
10		9	9	4	
11		10	11.5	4	
12					
13	r	0.822715			
14					

Figure 4.16 Excel worksheet with correlation coefficient (r) calculated

Regression Analysis

Regression analysis is yet another type of statistical analysis that can be very useful to technical communicators, and like the ANOVA and other statistical analysis methods, the calculation can become very complex. Regression analysis is used to add further granularity to interpreting the strength of correlation. In other words, if time to install in our example above can predict perception of ease, how much will perception of ease change when there is a change in time to install? The variable r^2 is used to show the results of such a regression analysis, and how often the independent variable predicts the dependent variable. In this analysis, the closer the r^2 value is to 1, the better, because this indicates a stronger correlation. Once again, we will demonstrate this commonly used method at a basic level, and you can also see it used in the exemplar article in Chapter 9.

Example—Regression Analysis of Installation Time and Perception of Ease of Use

In the previous case, Heath has calculated the correlation coefficient (r), and now wants to continue to calculate the strength of that correlation using a regression analysis. If he has not already done so, he needs to add the Analysis ToolPak to his installation of Excel.

1. He labels an r^2 field in his Excel worksheet and selects the cell next to it to accept the calculation. See Figure 4.17.
2. He selects Data from the tool bar, and then Data Analysis.
3. From the Data Analysis definition box, he selects "Regression" and clicks "OK."
4. Next, he selects the Y data range (dependent variable) by clicking and dragging on the ease-of-use data.
5. He selects the X data range (independent variable) by clicking and dragging on the time-to-install data.
6. He also selects a confidence level of 90% and selects cell B:14 for his output. See his work in Figure 4.17.
7. He clicks "OK" to run the analysis. See Figure 4.18.

Heath sees the r^2 value of 0.67686, which indicates that there is a 67% probability that the independent variable predicts changes in the dependent variable. It is also important to look at the p value, indicated in the regression as the *Significance F*. In this analysis, the p value is much lower than our required 0.1; thus, Heath must reject the null hypothesis.

Chi Square

Thus far in this chapter, we have focused heavily on using means to analyze quantitative data. Here we add one more statistical tool that is often used and focuses on analyzing percentages: chi square. We explain the use of this tool but do not provide a sample calculation. Once you have read this section, however, you will be ready to choose an appropriate tool and calculate it on your own.

Chi-square analyses are often used when the variable of interest has been operationalized as a percentage. For example, a researcher might question whether the writing

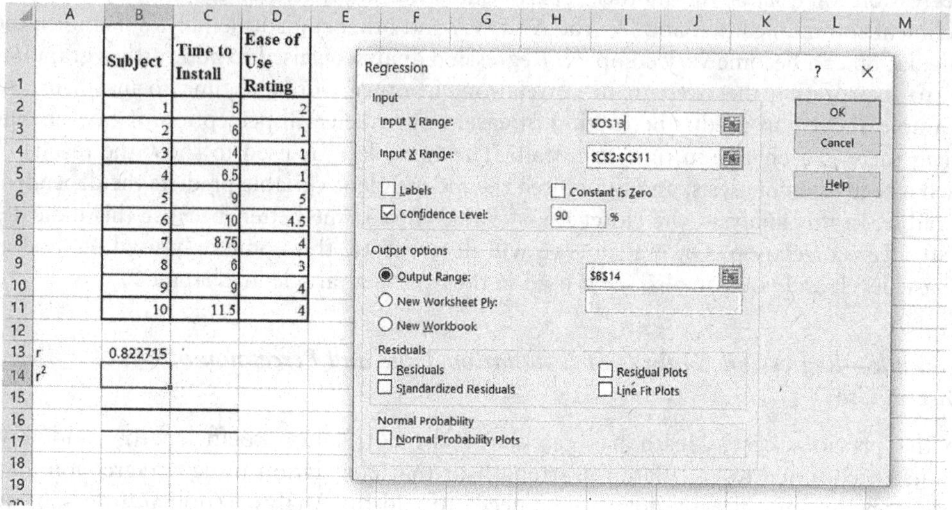

Figure 4.17 Excel worksheet with regression analysis (r^2) defined

	A	B	C	D	E	F	G	H	I	J	K
10		9	9	4							
11		10	11.5	4							
12											
13	r	0.822715									
14	r^2	SUMMARY OUTPUT									
15											
16		*Regression Statistics*									
17		Multiple R	0.822715								
18		R Square	0.67686								
19		Adjusted R	0.636467								
20		Standard E	0.947483								
21		Observatic	10								
22											
23		ANOVA									
24			*df*	*SS*	*MS*	*F*	*ignificance F*				
25		Regressior	1	15.0432	15.0432	16.75704	0.003469				
26		Residual	8	7.181796	0.897725						
27		Total	9	22.225							
28											
29			*Coefficients*	*andard Err*	*t Stat*	*P-value*	*Lower 95%*	*Upper 95%*	*ower 90.0%*	*pper 90.0%*	
30		Intercept	-1.11428	1.037078	-1.07444	0.313959	-3.50579	1.277225	-3.04278	0.814215	
31		X Variable	0.536539	0.13107	4.093536	0.003469	0.234291	0.838786	0.292808	0.780269	
32											
33											

Figure 4.18 Excel worksheet with regression analysis (r^2) calculated

styles of technical communicators differ between journal articles and user manuals. One of the test hypotheses might be that there is a difference between the percentage of passive voice clauses in journal articles and the percentage of passive voice clauses in user manuals. In this case, the appropriate method to test for a statistically significant difference in percentages is the chi-square test.

The chi-square test described in the previous paragraph is an independent samples chi-square test, and it is conducted in a way very similar to the *t*-test of independent means that we explained earlier in this chapter. It tests the null hypothesis: "There is no difference in the percentage of passive voice clauses found in technical communication journal articles and the percentage of passive voice clauses found in user manuals." Like the *t*-test of independent means, it calculates the probability that a difference between two samples could be the result of sampling error or coincidence. The main difference is that the *t*-test of independent means looks at averages, and the independent samples chi-square test looks at percentages.

Another common chi-square test is the goodness-of-fit test. In this chi-square test, a percentage calculated from a single sample is compared to a percentage calculated from a population or some other expected value. For example, a researcher might want to know whether minority students enroll in technical communication degree programs in the same proportion that they enroll in other degree programs. The researcher could compare the percentages of minority students enrolled in a university's technical communication program and compare it to the percentage of minorities enrolled in the university as a whole.

The two main differences between a goodness-of-fit test and the independent samples chi-square tests are that only one sample is taken in a goodness-of-fit test and that the null hypothesis in a goodness-of-fit test states the expected value. For example, if university records show that 16% of the students enrolled in the university are minority students, then the null hypothesis would say that 16% of the students enrolled in the technical communication program would be minority students (in effect, saying that there is no difference between the two populations).

Tools for Quantitative Analysis

As we have noted, a great many digital tools are available to aid you in quantitative data analysis. We have demonstrated two of them in this chapter, one proprietary (Excel) and one open source (Jamovi). We did not demonstrate SPSS Statistics (originally Statistical Package for the Social Sciences) from IBM, but we mention it here because it is a popular choice for technical communication researchers and may be available through your university or company. All of these tools have both advantages and disadvantages, and we suggest that you try several of them when you launch a new research project to identify the one that best meets your needs.

No matter which program you choose, it is often helpful to read a book or watch videos that explain how to use the program and read the output, which can be very hard to interpret. We also recommend that you become familiar with the program of your choice before you are on a deadline!

Is It RAD?

Remember that as you design quantitative methods in your research, you should always seek to produce research that is RAD—replicable, aggregable, and data-supported (Haswell, 2005). Your methodology must be sufficiently defined so that others can repeat the study. Your results must be reported in sufficient detail that the data can be aggregated or combined with the results of other studies to build a body of data. And your conclusions must be supported by the data, not simply be impressions or gut feelings of the researchers.

Summary

Whenever inferences are made based on measurements taken from a sample, two standards of rigor must be addressed:

- The validity of the measurement
- The reliability of the inference

Internal validity addresses the question "Did you measure the concept you wanted to study?"

External validity addresses the question "Did what you measured in the test environment reflect what would be found in the real world?"

Descriptive statistics describe a specific set of data—usually the sample population.

Inferential statistics make inferences about a larger population based upon sample data.

The two underlying principles of inferential statistics are the following.

- The smaller the variance in the data, the more reliable the inference.
- The bigger the sample size, the more reliable the inference.

Hypothesis testing allows you to see whether an intervention makes a difference (the test hypothesis). It tests the assumption that results from two groups (control group and test group) will be alike. If it can reject that assumption with confidence, it accepts the alternate test hypothesis that the intervention made a difference.

Random selection means that the selection and assignment of each test participant is independent and equal.

- Independent means that selecting one participant to be in a particular group does not influence the selection of another.
- Equal means every member of the population has an equal chance of being selected.

The type of data you have collected and what you want to learn from them determine the type of statistical calculations you should perform. See Table 4.2.

Table 4.2 Statistical tests and their uses

Statistical Test	Use
Standard Deviation (SD)	Indicates the variation of the data in the sample
t-test	Calculates the difference between two independent means
Analysis of Variance (ANOVA)	Calculates the differences among three or more independent variables
Correlation Analysis	Calculates the correlation between variables (the ability of one variable to predict, not cause, another)
Regression Analysis	Calculates the strength of a correlation
Chi-Square Test	Determines whether a statistically significant difference exists between independent percentages

Answer Key

Exercise 4.1

1. Regarding internal validity, speed to install could be said to be a reasonable way to operationalize at least an aspect of usability. Depending on how broadly the study wishes to describe usability, one could argue that it should also look at other aspects, such as user satisfaction, confidence that the product was installed correctly, and so on. However, the selection of the product's help desk employees as the test participants seriously compromises the test's external validity. These participants bring skills and product knowledge to the experience that would not be available in the general population.

2. A serious internal validity issue in this case is that the study purports to examine "willingness to access help files" but it measures completely unrelated variables (time to complete task and number of errors). Its external validity is seriously flawed by the use of technical communication students, who might well have an entirely different attitude about using help files than hospital admissions personnel. Also, the external factors associated with admitting someone into a hospital (such as time pressures because of life-threatening conditions) would be difficult to reproduce in a lab environment. Not accounting for such real-world factors could compromise the external validity of the study.

Exercise 4.2

H_1: It takes surgeons more time to successfully complete a medical device procedure with documentation that has not gone through the new validation process than with documentation that has gone through the validation process.

H_0: There is no difference in the amount of time that it takes surgeons to successfully complete a medical device procedure with documentation that has not gone through the new validation process as compared with documentation that has gone through the validation process.

Table 4.3 contains the data and calculations for Jodie's study.

Table 4.3 Data and calculations for Jodie's study

Data Analysis

	Time to complete without validated documentation (Version 1). In minutes.	Time to complete with validated documentation (Version 2). In minutes.
	56	46
	51	43
	41	42
	55	50
	56	49
	60	43
	47	45
n	7	7
Mean	52.28571	45.42857
SD	6.473389	3.101459
p	0.016719	

Conclusion: Jodie concludes that the difference between the time it took surgeons to successfully complete the medical device procedure with documentation that had not been validated as opposed to documentation that had been validated is statistically significant ($p < 0.1$). Therefore, she concludes that editing documentation based on the new validation process does result in decreased time for surgeons to complete the procedure successfully.

Exercise 4.3

H_1: There is a difference in perceptions of ease of use of the engineering web site among potential students, current students, parents of potential students, and parents of current students.

H_0: There is no difference in perceptions of ease of use of the engineering web site among potential students, current students, parents of potential students, and parents of current students.

Jonathan reviews the data in the ANOVA and sees that the probability that the difference in ease-of-use rankings could be attributed to chance is less than 0.001. Because this is less than his required alpha of 0.1, Jonathan rejects the null; there is a statistically significant difference between one or more groups in the ANOVA. Therefore, he proceeds to run a Tukey post-hoc test. As he considers the Tukey test (see Figure 4.19), he sees that the only statistically significant differences between groups are between current students (CS) and parents of prospective students (PPS), as well as between CS and parents of current students (PCS).

Conclusion: Among other things, Jonathan might conclude that although there is not a significant difference in ease-of-use perceptions between potential and current students, there is a difference in these perceptions between current students and both

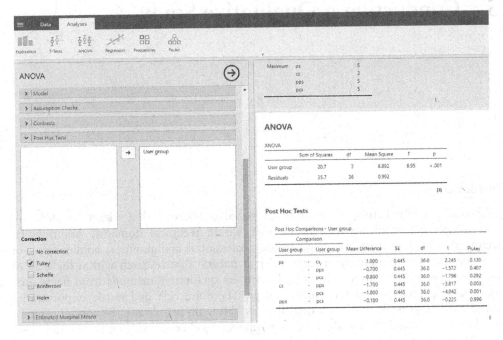

Figure 4.19 Jamovi worksheet with ANOVA and Tukey post-hoc calculated

parent groups. He might also conclude that web designers need to work on better accommodating use by parents.

References

Haswell, R. H. (2005). NCTE/CCCC's recent war on scholarship. *Written Communication*, 22(2), 198–223.

Nielsen, J. (1993). *Usability engineering*. Academic Press, Inc.

Roberts, J. C., Fletcher, R. H., & Fletcher, S. W. (1994). Effects of peer review and editing on the readability of articles published in *Annals of Internal Medicine. JAMA*, 272(2), 119–121. doi:10.1001/jama.1994.03520020045012

5. Conducting a Qualitative Study

Introduction

Technical communication is essentially a social study, one that tries to understand how people communicate with each other within technical domains. Additionally, technical communicators are members of professional communities, communities of practice, and the community at large. So it is natural that research within the field of technical communication should draw heavily on qualitative methods from fields such as psychology, sociology, and anthropology. There are rich and fertile opportunities for research that deals with such questions as "How do people use the communication products we create," "How do we collaborate with others in making these products," and "How do readers make sense of what we write?" Furthermore, there are broad social implications to what we do. For example, do we make technology accessible to those who would otherwise be disenfranchised, or do we systematically exclude classes of potential users by explaining technology using media and vocabulary that favor the "ins" and exclude the "outs"?

In trying to find meaningful answers to questions such as these, we find that the quantitative methods discussed in the previous chapter seem inadequate. For example, although those quantitative methods can tell us with great precision and confidence what percentage of users will not use a help file when in trouble, their precision dulls and our confidence in their accuracy falters when we ask the follow-up question, "Why not?" It is for questions like these that we turn to the disciplines and practices of qualitative research.

As with all research that involves human subjects, consider your ethical obligations to your participants by establishing informed consent. Refer to Chapter 2 for more information on compliance with ethical standards.

Learning Objectives

After you have read this chapter, you should be able to:

- Identify examples of qualitative data
- Define the standards of rigor for qualitative studies:
  - Credibility
  - Transferability
  - Dependability

- Describe ways of ensuring rigor in qualitative studies
- Apply coding and categorization schemes to analyze qualitative data
- Use research memos to improve your research process

The Process of Qualitative Research

Qualitative research can be defined by the type of data it relies on and by the methods it applies in gathering and analyzing that data. Qualitative data is non-numeric—it is words, images, objects, or sounds that convey or contain meaning. Qualitative methods generally involve the following three phases:

1. The observation of behaviors or artifacts in natural surroundings or authentic contexts
2. The recording or noting of those observations in descriptive or narrative formats
3. The systematic analysis of those observations or notes to derive patterns, models, or principles that can be applied beyond the specific events or artifacts studied

The following are common qualitative methods within technical communication research:

- Interviews—what participants say when asked (one-to-one)
- Focus groups—what participants say when asked (one-to-many)
- Usability tests—what users do when observed interacting with a product or process
- Field observations—what users do when observed in their natural environment
- Document analyses—what documents tell us about the author(s) or the intended audience

Research Memos

Research memos (also called in-process memos) provide a valuable tool in the process of qualitative research. They are small, manageable pieces of analysis that support the research process and may become part of the final product. They can help the researcher to focus the research, provide a record and analysis of issues encountered, clarify observations, and aid in collaboration. Research memos accomplish all of this by helping the researcher to make sense of individual parts of the larger research project and commit those thoughts to a written record. These memos are short—often only 500 to 1000 words—and can be used in any phase of the research project from design to data collection to analysis. The researcher might plan to produce these analytical memos at pre-determined points in the project but might also produce them when grappling with intriguing observations or unexpected challenges.

For example, Belinsky and Gogan (2016) used research memos to aid in collaboration during their research into framing. They used one author's field notes to determine the content areas (themes) that they wanted to include in their structured research memos. Each of them then wrote memos based on this structure to extract data from their field notes. Finally, they compared the contents of the memos to identify "connections between the emergent themes and any cultural artifacts" (Belinsky & Gogan, 2016, p. 330). In Table 5.1, you can see the process that Belinsky and Gogan used to transcribe, categorize, and connect field notes, and then write a research memo. The full research memo appears in Figure 5.1.

Table 5.1 Sample process of translating field notes into a research memo. The researcher is observing a marketing pitch. This first page of notes (along with the other pages from the field notes) maps to the research memo shown in Figure 5.1

One-page excerpt from field notes in framing study	Transcription of page 1 of field notes (Belinsky)	Initial list of themes: The "What" Question (Gogan)	Connection with other artifacts
	(1) Pitch Zoo 11/4/2013 Feedback on everything! (My comments)		
	*Bringing resources together in one place – what kinds of resources – be specific!	Resources Logistics Specificity	
	→ Katie brought this up too.		Katie Interview Notes
	*Love the YWCA part (what it is not)	Definition (Negative) Differentiation	Business Model Canvas: Partners
	*Talk about what kinds of meetings you would have	Logistics	
	*Group feedback:		
	– Have you talked to the YWCA about partnership?	Partnership	Business Model Canvas: Partners
	– You don't sound very excited about your idea!	Delivery Audience Evaluation	
	– Convince me you're the person to run this. J.C.'s, MBA, work in your own project management experience	Persuasion Qualifications	
	– Talk somewhat about how you would achieve income	Finances	Business Model Canvas: Revenue

Exercise 5.1 Observing a Situation and Writing a Research Memo

Choose a school or workplace situation of interest. Spend one hour observing the situation and taking notes on what you see. Write a one-page research memo summarizing what you observed and identifying observations that might warrant research and reasons why they may be important.

Memo Date: December 2015

Subject: Analysis of 11/4/2013 Post-Pitch Field Note Themes

I'm recognizing two groups of recurring themes that seem to be emerging from the post-pitch notes taken on November 4, 2013.

First, the post-pitch feedback that Stacy received indicated that the audience members sought a clearer definition of the venture. Audience members pushed for more information about what the venture was—that is, what was the market gap that needed to be filled, what was the vision for the services provided to members, what was the relationship between the proposed venture and other organizations, and what was the tax status (for-profit or non-profit) of the venture. I'm interested to see how subsequent pitches respond to this need for clearer definition and how these definitions work to frame the project.

Relatedly, the notes also suggest that the use of a negative definition in the pitch was memorable to the audience members. While some wanted the strategy to be "soften[ed]" others complimented the approach. I wonder if the reason the strategy was memorable was because the audience was familiar with the YWCA against which the venture was defined. While this rhetorical device influenced the audience, I'm not sure that it can be classified as a framing device. It functions more like the absence of a frame, or the "wall" on which a "frame" is hung.

Second, the notes reveal that a significant amount of the post-pitch feedback addressed the delivery of the pitch and spoke to the way in which the catchers wanted to be affected by the pitcher. These responses were a meta-commentary of sorts, commenting less upon the particulars of the venture and more upon the way in which the pitch was thrown and caught. This kind of discussion would be appropriate for the Pitch Zoo venue. Three of the comments seem interesting given that our project traces the development of Stacy's pitch experience:

- The audience wanted to hear more excitement from the pitcher. I suspect that this kind of comment does not appear after many of Stacy's later pitches, as Stacy will have established more experience pitching the concept.
- The audience wanted to be left with three ideas that could be taken away from the pitch. And, from what I recall, the closing request or "ask" was something that shifted as the pitch approach changed. It would be interesting to see if any connection exists between the frames Stacy used in her pitch and the "ask" that was used to close-out the pitch.
- The audience wanted the personal dimension of the pitch to be rearranged—starting with the personal story and then entering into the venture's services—and also presented more as drive or motivation behind the venture. I'll need to review the pitch transcripts and videos to better chart how this personal element evolved over the pitch development process.
- The audience also wanted to be reassured that the pitcher is the best manager for the venture. This comment surprises me, since the idea for the venture originated with Stacy. Does this comment suggest that Stacy should look to hire a manager for the project? Is this statement suggesting that Stacy lacks the credibility to convince this audience member? Or, does this audience member know more about Stacy's qualifications than those that were shared during the pitch? Who is this audience member?

Figure 5.1 Example of a research memo based on the field notes presented in Table 5.1. This example is based on the complete set of notes from that day, though Table 5.1 shows only the first page of notes and analysis

Standards of Rigor in Qualitative Research

Although qualitative research has different methods for ensuring rigor than does quantitative research, it is no less interested in achieving that same level of rigor. Merriam (1998) states this common concern: "All research is concerned with producing valid and reliable knowledge in an ethical manner" (p. 198). She again reinforces the similar goals of the two methods: "Assessing the validity and reliability of a qualitative study involves examining its component parts, as you might in other types of research" (p. 199).

But Corbin and Strauss (1990) state that many readers of qualitative reports tend to read them with a quantitative interpretation: "Qualitative studies (and research proposals) are often judged by quantitatively-oriented readers; by many, though not all, the judgment is made in terms of quantitative canons" (p. 4).

Recognizing this problem, Lincoln and Guba (1985) held that qualitative research needed a vocabulary that differentiated it from quantitative methods, and they suggested a set of terms shown in Table 5.2 to correlate qualitative concerns with the corresponding concerns in quantitative analysis.

Credibility

In a quantitative study we assess the *internal validity* by asking "Did you measure the concept that you wanted to study?" In a qualitative study, where data are not measurable, we look more to the *credibility* of the data; that is, do the participants truly represent the population or phenomenon of interest, and how typical are their behavior and comments? For example, a researcher might study how help desk personnel use a product's technical documentation, and for this study the researcher interviews help desk supervisors across several companies. A critical reader of the research could challenge the credibility of the study on the grounds that supervisors are not necessarily appropriate spokespersons for how their employees actually use the documentation. Similarly, a study that interviewed only male managers to determine whether female technical communicators were treated equitably would lack credibility because of gender bias.

Table 5.2 Comparison of quantitative terms to qualitative

Quantitative Term	Qualitative Term
Internal validity Did you measure the concept that you wanted to study?	*Credibility* Do the participants truly represent the population or phenomenon of interest?
External validity Does the thing or phenomenon that you are measuring in the test environment reflect what would be found in the real world?	*Transferability* Does the thing or phenomenon you are observing in the test environment reflect what would be found in the real world?
Reliability To what degree can the conclusions reached in your research project be replicated by different researchers?	*Dependability* Does your study employ depth of engagement, diversity of perspectives and methods, and stay grounded in the data?

Credibility can also be affected by the methods employed in the study, especially if the methods affect how freely and honestly participants can participate. For example, a focus group that contains managers and line employees could compromise credibility if subordinates felt pressured to answer questions in a particular way by the presence of their managers in the group.

And finally, *observed behavior* (also called "manifest data" derived from watching what other people do) has higher credibility than *self-reports* (also called "ideal data" derived from asking people to tell you what they do). It is very common in usability tests, for example, to see a conflict between user self-reports during pre- or post-test interviews and their observed behavior during the test. During pre-test interviews participants might say that they rely heavily on product documentation, yet they never actually use the documentation during the test, even in the face of difficulties in accomplishing the task. Brown (2002) cites a study done at Xerox where accounting clerks were interviewed about how they did their jobs. Their individual descriptions were consistent with the written procedures; however, subsequent on-the-job observations showed that their descriptions were very different from how they actually performed the tasks.

The upshot of all this is that if your research relies heavily on self-reported data such as interviews or focus groups *as a way of describing behavior,* you must be concerned about the credibility of your findings. A good guideline is to use interviews and focus groups to discover people's opinions, motives, and reactions—not to learn about their behavior. If you want to know what people *do,* you are better off watching them do it rather than asking them what they would do. But if you want to know *why they do it* or *how they feel about it,* then interviews and focus groups can be credible methods.

Transferability

In quantitative research, *external validity* addresses the question "Does the thing or phenomenon that you're measuring in the test environment reflect what would be found in the real world?" In qualitative research the question is essentially the same, with the substitution of "observed" for "measured." The question could also be stated, "How natural or authentic was the environment in which the study took place?" Sometimes "natural" is difficult to achieve, in which case the emphasis needs to be on "authentic." Authentic means that the context of the behavior being studied is consistent with the context of the real-world situation in which the actual behavior would occur.

For example, in the early days of usability testing, researchers worried a lot about making the lab environment feel like the participant's natural environment. Emphasis has now shifted more to ensuring that the tasks are authentic—that is, that the users are asked to achieve realistic goals with the product. More recent usability research has started to use "interview-based tasks" in which the test starts with an interview to determine what that participant might actually do with the product or web site being tested. Then the test consists of observing the user doing that task, not one made up by the researcher. This approach could be said to have higher transferability than one that used pre-determined or scripted tasks.

Researchers must often balance the ability to observe and record data in a controlled environment versus the increased transferability of using the participant's natural environment. The more the researcher must compromise the naturalness of the environment, the more he or she must attend to the authenticity of the tasks.

In a real sense, much of the burden for transferability falls on the reader of the research report. "Qualitative research passes the responsibility for application from the researcher to the reader" (Firestone, 1993, p. 22). The reader must decide to what degree "these participants and these circumstances" resemble the population and environment of interest to the reader. For example, a manager of web content developers might be interested in problems related to managing creative tasks. That manager might decide that a study of how graphic arts supervisors manage creativity was very relevant to the problems that she was trying to solve. On the other hand, a study of how supervisors manage the productivity of proposal writers, although seemingly more similar to technical communication, might have less relevance to her situation of managing creativity. Researchers help the readers make these decisions about transferability by describing the participants and their circumstances with a greater emphasis on richer descriptions of the participants and their contexts than might be seen in a quantitative study.

Dependability

As with its quantitative counterpart, reliability, *dependability* speaks to the confidence with which the conclusions reached in a research project could be replicated by different researchers. In qualitative studies, more so than with quantitative studies, it is more difficult to turn off the researcher's own subjectivity. Therefore, qualitative researchers must be careful to ensure that their conclusions have emerged from the data and not from the researcher's own preconceptions or biases. Unlike quantitative studies that can rely on procedural, statistical methods to do this, qualitative studies must rely on more humanistic protocols. The literature on qualitative research discusses many techniques that can be applied to verifying the conclusions produced by such research. In this discussion, we summarize them into three categories:

- Depth of engagement
- Diversity of perspectives and methods
- Staying grounded in the data

Depth of Engagement

Generally, the more opportunities that researchers give themselves to be exposed to the environment and to observe the data, the more dependable the findings will be. For example, a case study that looks at writers in a department over a one-day period is not going to be as dependable as one that observes the same department over a period of three months. Longer studies gather more data within a broader array of contexts and are less susceptible to what is called the Hawthorne effect, the phenomenon where behavior might change temporarily just because of the novelty of the new treatment or the attention of the researcher.

But depth of engagement is not necessarily measured in numbers, such as length of time or number of participants. Another gauge is the richness of the context and the study's success in achieving *data saturation*. Data saturation is the point at which staying another day, interviewing another participant, or testing another user ceases to add any new data. Investigations of usability studies reveal that we can reliably predict that this will happen after testing four to seven users, but for many other kinds of studies, you won't know where data saturation occurs until you reach it—in essence, when you start observing the same themes or behaviors occurring with no new insight or data being added. Figure 5.2 shows a graph from a research project that included data from a usability study (Hughes, 2000). The author used the graph to support his assertion that data saturation had occurred in the usability test by noting the sharp drop-off in new data after the third session. (He supported this conclusion with literature references that reported similar findings in other studies.)

Diversity of Perspectives

Qualitative research often relies on *triangulation* to demonstrate its dependability. The term comes from a technique for using a map and a compass to determine location. Essentially it relies on comparing compass readings of a point taken from two other points and plotting their intersection on a map (you are here!). The term is also used in qualitative research to describe methods that look at data from different perspectives. Seeing data from multiple perspectives—for example, using multiple researchers or multiple data collection techniques—increases rigor.

For example, let's say that a researcher does a study that relies on field observations in the workplace, one-on-one interviews with the writers, and analyses of the documentation produced, and then the researcher draws conclusions based on those three sources of data. That study would be more dependable than one that used just one of those data collection methods. Similarly, a usability study in which multiple observers watch the user will be more dependable than one that has just a single observer.

Diversity of perspectives can also be applied to the diversity of the participants that the researcher involves in the study. Here, qualitative research can differ markedly from quantitative. In quantitative studies, the researcher often tries to reduce diversity among subjects as a way to minimize variance. (Remember from our chapter on quantitative methods that reducing variance can increase reliability.) For example,

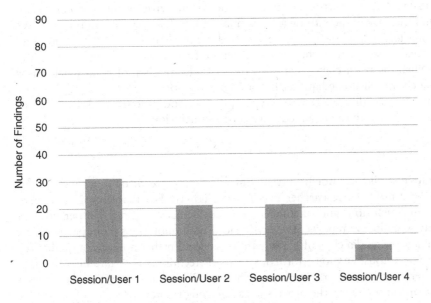

Figure 5.2 Chart of cumulative findings showing data saturation

a researcher who wanted to test which of two user interfaces was quicker to use might recruit users with similar computer expertise and background. This strategy would reduce variance in the data not related to the actual treatments being tested and would make the results more reliable. But a qualitative researcher wishing to understand how users would search for information on a company intranet might purposely recruit participants of varying experience and expertise to get a richer perspective of how the intranet would be used in general.

Similar to triangulation is the technique of *peer review*, which gathers multiple perspectives after the fact. In peer review, the researcher allows others to examine the data and the findings to determine whether they interpret them the same way as the researcher does. One of the vulnerabilities of qualitative research is that it can be subjective. Kidder speaks of researchers experiencing a "click of recognition" or a "Yes, of course" reaction (1982, p. 56) during qualitative research. This is a great strength of qualitative research, the clarification that comes from seeing something through the participant's perspective, but you must guard against the subjectivity that might have led to that "of course" reaction—one person's "Of course!" could be another person's "What the heck was that?"

Peer review helps eliminate some of that subjectivity by viewing the data through multiple subjective lenses. Consensus strengthens the dependability of the conclusion. When consensus cannot be reached, the researcher may choose to say, "Since I'm the only one who interprets this participant statement this way, I'll drop it," or the researcher might choose to note in the report that peer reviewers saw it differently, giving the reader both perspectives.

Another way to bring validating perspectives into a research project is to use *member checking*. In member checking, the researcher solicits the participants' views of the credibility of the findings and interpretations. This can be done at the end of the study by asking participants to read the report and verify the observations and conclusions that the researcher has made, or it can be done within the study by the researcher, validating his or her perceptions at the time with the participants. In the case of the Belinsky and Gogan example in this chapter, we borrowed the data from them, wrote it into the context of this chapter, and then asked them to review the final version for accuracy. In this way, we ensured that our understanding of the example did not depart from what they had intended. This real-time member checking is common in interviews, focus groups, and usability tests when the researcher restates his or her understanding of what participants have said or done for their immediate verification or correction.

When writing up a qualitative research report, it is a good practice to describe all of the methods used to ensure the dependability of the conclusions.

Staying Grounded in the Data

Because qualitative studies do not have the controlled structure of quantitative studies, they are more susceptible to researcher bias or subjectivity. For that reason, it is very important that all conclusions and statements be traceable back to directly observed data within the study. Qualitative reports are heavily laced with quotations from participants, samples of artifacts, or media clips of participant behavior. Furthermore, for a qualitative study to have rigor, it must employ a formal, systematic technique for examining the data and finding patterns or common themes across the data. We cannot overstate the importance of a systematic approach to the process in establishing rigor.

The positive aspect of qualitative studies' looser structure is that they can "go where the data takes them" in ways that quantitative studies cannot. For example, in a technique known as *negative case analysis*, the researcher refines and redefines working hypotheses in the light of disconfirming or negative evidence. Here we see the contrast between quantitative and qualitative methods. In quantitative studies, variance is often controlled by keeping the conditions of the test as constant as possible throughout the test. However, in qualitative studies, it is not at all unusual to change tasks, questions, and even participant profiles as you go.

This pattern of evaluating the data during the study and then modifying the course of the study is very common in *grounded theory* studies, where models are built, evaluated, and modified after every interview, and new interview questions and participant profiles are subsequently created to validate or test the emerging model or theory. In this kind of approach, the emphasis is not on proving or disproving a hypothesis; rather it is on refining it at each iteration. Since the final hypothesis or model reflects a broader diversity of participant input, it is considered more dependable than one tested inflexibly over several participants.

In a simpler application of this principle of "go where the data takes you," it is a legitimate practice in usability studies to change scenarios or even modify the prototype being tested to validate a finding or just to move on and collect more data if a particular problem has become obvious. For example, if the first two users in a study fail to use a critical link, and the observers all have the same "click of recognition" that the link has been poorly signaled, it is acceptable to reword the link and test that alternative wording with the subsequent users. Such a change would be an unforgivable breach of protocol in a quantitative study. However, in a qualitative study, the researchers can follow the data but must do so systematically, capturing evolution and justifying it.

Finally, we present an example of a research strategy that has evolved out of qualitative inquiry and been applied to technical communication. In the past several decades, *ethnomethodology* (also an application of the principle "go where the data takes you") has been used with success in researching questions related to technical communication, design, and work. Spinuzzi (2003) provides a notable example with his research published in *Tracing Genres Through Organizations: A Sociocultural Approach to Information Design*. In this book, he advocates for following the workarounds that people use to improve their use of processes and systems to find solutions to design problems. Just as much debate remains in the field of sociology about how to conduct effective ethnographic research, much debate also exists about how to define and use ethnomethodology. Here, we give you some basics.

Ethnomethodology is not a method; rather it is a research strategy closely related to ethnography that can employ multiple qualitative methods. Ethnomethodology seeks to "give careful and detailed attention to how work is done, how problems are construed and solved" (Sharrock & Randall, 2004, p. 193) and is commonly based on observation of the interaction between users, technology, rules affecting the interaction, and the environment.

When using ethnomethodology, a researcher does not begin with pre-determined theories and constructs as one might when employing ethnography in sociological research. Rather, the researcher uses a rigorous application of selected methods to draw conclusions about the complex components of an authentic experience. Patterns found in the social setting itself lead to theory. While ethnomethodology may appear to be an unstructured approach to research, it is not. It must be based in rigorous methods.

For example, a person can study the instructions for how to use a large student database, but that person doesn't know how it is actually used (or not used) until observing its use in an authentic setting. Through observation, the researcher collects manifest data (observed behavior) that aids in analyzing the interaction between users, technology, rules, and environment.

Because all of this can be quite subjective at worst and messy at best, it is important that qualitative studies incorporate systematic methods for examining and analyzing the large quantities of data that they gather. Later in this chapter, we describe some useful qualitative data analysis techniques.

Qualitative Methods

In this section, we discuss guidelines for planning and conducting qualitative studies that use any of the following methods:

- Interviews
- Focus groups
- Usability testing
- Field observations
- Document analysis
- Surveys

Keep in mind that interviews, focus groups, and usability assessments can be conducted remotely, using technology, or face to face.

Interviews

An interview is a structured conversation between a researcher and a participant. The amount of structure is determined by the *protocol* or list of questions the researcher plans ahead of time. A good protocol usually involves a combination of open-ended questions—questions that call for long, descriptive answers from the participant, and closed-ended questions—questions that call for short, validating answers.

You plan for an interview by defining the profile of the participants whom you wish to interview, planning how you will recruit those participants, and writing out the protocol that you will use. The questions that you ask should be derived from the research goal and questions that you have defined for your study. Some initial questions might be added as "ice breakers"—that is, not so much intended to gather data related to the study as to get the participant talking and feeling comfortable.

When conducting an interview, try to maintain a conversational tone. Be familiar with your questions so that you don't have to read them, and try to provide natural transitions between questions, such as "That was an interesting point that you made about how the subject matter expert's time is hard to secure. I'd like to follow up on that by asking . . ."

Also, don't be afraid to deviate from the protocol if the participant makes an interesting but unanticipated comment. In that case, ask him or her to elaborate or pursue the point with a question not necessarily in your protocol. Do this, however, only if the new direction is consistent with your research goal.

A good way to close an interview is to ask the participant whether he or she has anything else to add on the topic that you have been discussing that has not already come out in your conversation. This question can often unveil some rich data or insight that you had not anticipated.

Focus Groups

Focus groups are essentially group interviews and should address the same issues and follow the same guidelines stated for interviews. In the case of focus groups, however, there are some additional considerations due to group dynamics.

Make an effort to draw out diverse perspectives in non-threatening ways. The purpose of a focus group is to get multiple perspectives, and this can sometimes be lost if some people are reluctant to offer opinions that are different from what other, more vocal, participants have offered. The best way to get at these differing opinions is to avoid positioning them as disagreements. Do not ask, "Does anyone disagree with what Mary just said?" Instead, ask, "I'd like to hear other opinions on that. Can someone share a different perspective than Mary just described for us?"

Also, some participants will simply be less talkative than others. Be prepared to call on them directly to get their input.

Usability Testing

Usability testing is a common form of technical communication research. Because of the robust nature of usability studies and the fact that they most often produce qualitative (non-numeric) and quantitative (numeric) data, we introduce them in this chapter and go into more depth about usability testing and related usability assessment methods in Chapter 7. In usability testing, we observe users performing authentic tasks with a product or process of any kind: mobile apps, documentation, or even a mechanism like a wheelchair. To plan a usability test research project, you must address the following questions:

- What is your purpose? What questions do you want to answer?
- What is the profile of the participants whom you wish to observe?
- What tasks will you have them do?

Once you have answered these questions, you develop a test agenda that identifies all activities from the time you begin to work with a participant until you finish. That agenda is likely to include such items as signing an informed consent form, practicing think-aloud protocol, responding to a pre-test questionnaire, testing, and responding to a post-test interview. More complete details on the steps for designing an effective usability assessment are included in Chapter 7.

Field Observations

Field observation can be a very credible technique for qualitative research because it relies on observations made in the participants' natural environment. Of course, you have to be aware that the presence of a researcher can alter the environment and thereby detract from credibility. Essentially, in a field observation, the researcher observes the participants going through their natural routines.

The observations can be very open, for example, a shadow study where the researcher follows the participants and notes what they do within their normal routine. Let's say that a researcher is studying how help desk personnel use a product's technical manuals in solving problems. One method might be to spend a day or half-day sitting with a customer care representative as the representative takes customer calls and noting how he or she researches the answers to the problems.

The observations can also be very focused. For example, another researcher might be interested in how non-accounting managers use a spreadsheet application to put together their department budgets. In this case, the researcher wants to be present to watch only that particular task. In both cases, open and focused, the emphasis is on observing how the participant actually goes about a task or collection of tasks in their real-world setting.

Note-taking in a field observation study can be as detailed as video transcripts of the entire time of interest or as loose as observation notes taken manually in the researcher's notebook. In the case of the latter, emphasis should be placed on noting actions that the participant performs, the sequence in which steps are taken, the length of time that he or she spends performing a task (it's a good idea to keep field observation notes in the form of time-based journal entries), the tools or resources that he or she uses, the people whom he or she talks to or seeks help from, etc.

Perhaps more so than any other form of data gathering, field observations can seem the most intrusive on a participant's or sponsor's time or sense of confidentiality. Care must be taken to gather the proper permissions—telling both the participant and the sponsor how the data will be used and what their rights as participants are. Refer to Chapter 2 for details on permissions.

Document Analysis

A lot of what technical communicators do is centered on the production of various kinds of documents, and much can be learned from examining the documents that they produce. In a quantitative study, standard methods of analysis might include document length, reading grade level, percentage of active versus passive voice, and so forth. In a qualitative study, however, you would be looking for non-quantifiable areas of interest, such as tone, style, and vocabulary selection.

For example, a quantitative study might note that technical white papers used more passive voice constructions than did user guides on the same topic (a simple analysis of frequency of passive constructions), whereas a qualitative study might note that user guides tended to shift into passive voice when potential negative consequences are described—perhaps in an attempt to distance the manufacturer from accountability or association with the negative consequences. Both are observations about how passive voice is used, but the qualitative is more interpretative.

In another example, a researcher might count the number of questions that the editors of three journals ask in the reviews they give to authors. This numeric data might enable the researcher to compare editors' styles or correlate the number of questions asked during review to the rate of acceptance. However, the researcher might also analyze how the style of writing, including the use of questions, makes criticism less harsh. The quantitative data confirm presence while the qualitative data interpret the cause and effects of style.

Surveys

A survey is a list of questions that prompts participants to provide information about specific characteristics, behavior, or attitudes. A survey can be administered using different instruments including questionnaires (a set of questions rendered as text with a choice of answers), interviews, and focus groups.

Surveys are another common form of technical communication research. Like usability studies, they most often produce qualitative (non-numeric) and quantitative (numeric) data; thus, we introduce them in this chapter and go into more depth about how best to develop and administer them in Chapter 6.

Constructing a survey consists of

- Determining what questions to ask
- Asking them in the appropriate format
- Arranging them in an effective order

Additionally, you need to provide instructions for participants on whether or not they are qualified to participate, how their privacy will be protected, why the study is being conducted, how long it will take to participate, and so on.

Once your survey instrument is prepared, whether it is a questionnaire, interview, or focus group, you should pilot the instrument and make any changes needed before you begin collecting data systematically. You will often be amazed at what a pilot survey can reveal about such things as they way that you worded a question, data that you forgot to gather with a question, etc. Feel free to administer multiple pilots until you have an instrument that is working well. This is time well spent.

Data Analysis

Before you can analyze qualitative data, it must exist in some physical form. This form can be transcripts of the interviews that you conducted, detailed logs that you kept, video transcripts, or documents that you are studying. Having these artifacts in electronic format allows you to manipulate and mark them up during the analysis phase. Many professional data analysis tools as well as conventional spreadsheets and word processor applications can make your analysis easier if the data are in electronic format.

Qualitative data analysis consists of the following three phases:

1. Coding
2. Categorizing
3. Modeling

Coding

Coding involves breaking the data into the smallest chunks of interest and describing it at the chunk level. How small a chunk should be depends on what you consider to be your basic *unit of analysis*. Generally, we analyze discourse (text and transcripts of participant comments) at the sentence or phrase level. In some cases, however, it might make more sense to use larger units, such as document sections, or smaller units, such as individual words, as the units of analysis. Field observation notes are typically analyzed at a higher level since researcher notes are not as detailed as a participant interview transcript would be.

The codes applied during the coding phase can be *predefined codes* or *open codes*. Predefined codes are determined before the analysis begins and are usually taken from an existing model or theoretical structure on which the researcher is basing the analysis. For example, a researcher might be studying to what degree virtual team behavior can be mapped by Tuckman's model for group development: forming, storming, norming, and performing (1965). In this case, the researcher would use the stages of development as the codes and would code sections of team transcripts to reflect which stage was represented. As another example of predefined codes, consider that professional usability labs often have a standard set of codes based on established usability heuristics such as codes for "Navigation" or "Terminology." On the other hand, refer to Table 5.1 and the column on initial themes to see an example of emerging open codes as identified by Belinsky and Gogan.

Open codes are more commonly used in qualitative analysis, especially if the purpose of the study is to identify new patterns, taxonomies, or models. In an open-coding approach, the researcher creates the codes based upon the data. Many times, the open code is taken from the words in the transcript itself, a technique known as *in vivo* coding. For example, a researcher might code the participant phrase "I get so frustrated when the instructions assume that I know where I'm supposed to be in the application" with the word *frustrated*.

When coding data, it is sometimes useful just to scan the transcript first, highlighting statements of interest, and then to go back and apply codes to comments or incidents that seem significant. This approach is especially useful for the first transcript or document that you are analyzing, before you have a sense of what themes might emerge.

Perhaps the greatest challenge for new researchers is deciding how to code; that is, what should the codes look like, how long or short should they be, and how much should be coded? Welcome to qualitative research!

The good news is that coding is flexible and fluid. By that we mean that it can be done correctly in a number of ways and that you can always change your coding scheme as patterns or models start to emerge. In fact, this is an important strategy in qualitative analysis, so if you find yourself changing codes, this just means that your insight is emerging from the data and not from your preconceptions. However, remember that a change in coding scheme requires a systematic recoding of all related data based on those changes.

To add rigor to your coding and to support dependability (reliability), use a process called inter-rater checking. As noted by Frey, Botan, & Kreps (2000), "If the observations recorded by two or more individuals who are not aware of (are 'blind' to) the purposes of the study and each other's codings are highly related (showing 70% agreement or more), their ratings are considered reliable [dependable]" (p. 115). For example, after your codes have been established, you and additional individuals should code a portion (perhaps 10%) of the data using codes that you have established. If your codes and those of your inter-rater(s) reach agreement at a rate 70% or more, those codes can be viewed as dependable. If the coding does not agree at this level, edit the codes themselves. Then recode a fresh portion of the data with the edited codes and have your inter-rater(s) do the same. Continue this process until you have reached an agreement level of 70% or more. Once agreement has been reached, use the dependable coding scheme to code the complete set of data.

Reaching the 70% threshold can be more difficult than you might think. The more categories you have, the more difficult it is to reach 70% agreement, but fewer categories may be less useful to your reader. Though we do not cover it in this book, Cronbach's alpha is a coefficient that can be calculated to check internal consistency such as degree of agreement among raters.

Not all statements need to be coded, only those that seem to have importance to the study, and some data can have multiple codes. In the example above, the sentence "I get so frustrated when the instructions assume that I know where I'm supposed to be in the application" could have been coded *frustration* and *location in the application* and even *instructions* if the researcher wanted to. The important thing is to dive into the data and start coding versus staying outside the data making broad assertions.

Electronic transcripts can be coded using special software applications designed for qualitative research, or you can use the indexing feature in a word processor. We discuss some of these options at the end of the chapter.

Categorizing

The next phase in the analysis of the data is to start to look for patterns or groupings in the codes. In this phase, codes are grouped into categories, and categories can even be grouped under higher-level categories. Where the coding is an analysis of the directly observed data, categorizing is an analysis of the codes themselves. This is an important step in making the findings of the research transferable to events or situations other than just the one being observed.

During this phase, the researcher often rewords codes or even breaks them down into lower levels of coding, making the original code a category. For example, a researcher might have been coding certain types of participant comments as *emotional reaction*, but then decides in the categorizing phase to break that down into *frustration, anger, relief, thrill*, and so forth. Doing this means going back into the data, finding all the occurrences of the code *emotional reaction*, and recoding them. Although this might seem to be inefficient, this bouncing back and forth is actually part of the rigor of good qualitative analysis. Once again, it is an indication that the conclusions are emerging from the data and not from the researcher's preconceptions. A helpful tip, however, is to avoid starting with abstract codes—for example, starting with the code "emotion" rather than "frustration." Your *initial codes should stay close to the data*, and abstractions should emerge primarily from the categorization phase of your analysis.

Categorizing can be done manually with index cards or post-it notes, or you can use mind mapping software products to move and manipulate the codes you have created. For example, you can use an old standard like Microsoft Visio; the brainstorming stencil allows you to enter codes, create categories, and then manipulate them graphically. Or you can use a newer tool like Canva (see Figure 5.3), which handles tasks like mind mapping as well as presentations.

We also favor using paper or online sticky notes from a free sticky-note generator as part of a process of categorizing data as this method allows for easy and iterative categorization. For the categorizing phase to be productive, it should be iterative and

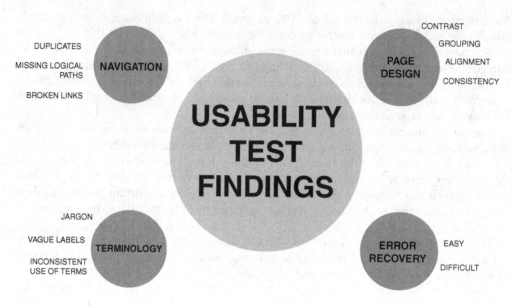

Figure 5.3 Categorizing in Canva

experimental. That is, keep moving things around, renaming them, making more codes, and deleting codes that don't seem to work. In short, keep working with the data until they start to fall into place. Remember that in qualitative research, the aim is *not to eliminate* the subjective insight of the researcher *but to manage* it so that it facilitates transferable, dependable conclusions.

Exercise 5.2 Coding Qualitative Data

[This exercise was contributed by Mercer University graduate student Hannah Nabi.]

As a researcher, you are interested in the service-learning opportunities provided by third-party providers of study-abroad programs. You decide to study the public-facing content provided on the web sites of these organizations. One of your research questions is "What kinds of information are third-party providers sharing about service-learning programs, projects, and partners on their web sites?" Below is a sample of content (data) gathered from several of these web sites.

- The service-learning study abroad program is exciting, engaging, and innovative in its approach to meaningful social impact overseas. This program is centered on you and the community, with the goal of cultivating change makers through civic engagement and structured reflection.
- Our courses are designed to integrate academic learning with experiences that foster cultural immersion, identity exploration, and global engagement. Each project works with and for the community. Here are some selected examples: planting and tending a community garden and orchard; upkeep on visitor and community facilities; engaging with heritage programs designed for children or

elderly locals, and helping to maintain the outstanding natural and cultural beauty of a local town; volunteering with minors.

- The course focuses on developing attitudes and values such as opening up to social reality and sensitivity towards global injustice, social exclusion, and the situation of the victims of such reality. To participate in this project, you will need to bring a state police criminal record with you from your home country. This is a requirement for anyone volunteering with minors.

Your task is to code the sample using the following process.

1. Read the content sample to familiarize yourself with the information.
2. Read the sample again and highlight words or phrases that indicate a fit into the broad categories of the research question: Is the information describing a program, a project, or a partner? You may not find information that fits all of these categories.
3. After your first round of coding, revisit the content that you coded to develop more specific codes. For example, look at the information coded as "project." Is there a type of service project described?
4. After you are satisfied with the level of detail in your coding, group your coded information into categories and review the categories to identify themes.

A sample coding scheme is provided in the answer key at the end of this chapter.

Modeling

The modeling phase is the natural outcome of the categorizing phase, and oftentimes one blends seamlessly into the other. Figure 5.3, where a graphical mapping tool was used to create categories, is a good example. What emerged from the categorizing phase was an affinity map of data produced during a usability test. The map provides a useful model for the researchers redesigning a web site.

In other cases, the emergence of patterns or models might not flow as naturally. What the researcher is looking for in this phase are generalizations that can be applied beyond the boundaries of the particular study. These could take many forms:

- List of principles or axioms
- Table that summarizes roles or relationships
- Set of design heuristics
- Process diagram
- Affinity/mind map

The exact form that a model should take is up to the researcher. The criterion should be what form best communicates the type of findings. For example, Figure 5.3 shows a model of the information that participants thought was important when they used a web site, expressed as an affinity map showing the complex information relationships. Table 5.3 shows a model of the various roles that a facilitator plays during a usability test. The table helps to better contrast the roles. Figure 5.4 shows a model of how team learning occurs during a collaborative usability test shown as a process diagram to illustrate inputs and outputs of that process and the events in between.

Table 5.3 Example of a model shown as a table (Hughes, 2000)

	Facilitator Roles		
	Moderator	*Expert*	*Coach*
Focus	Meeting/event efficiency	Fixing the product	Team and individual learning
Techniques	• Communicates time/procedural status • Directs	• Teaches principles • Puts findings into context • Recommends solutions	• Encourages participation • Encourages reflections • Encourages scrutiny
Knowledge Base	Procedures, facilities, and equipment	• Design guidelines • Heuristics	• Group dynamics • Team learning • Action science
Data	User profiles/scenarios	Historical (other users in other tests)	Directly observable data within the user sessions

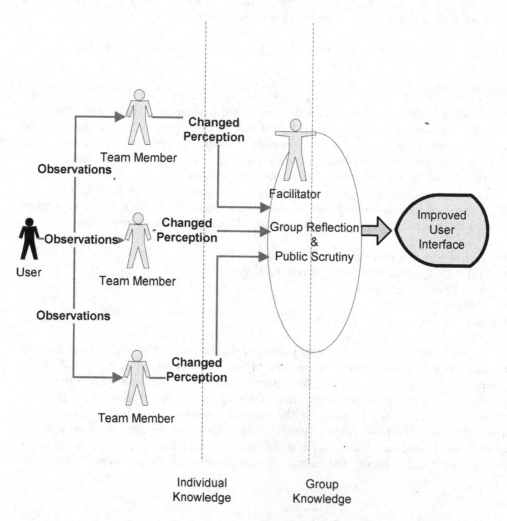

Figure 5.4 Example of a model shown as a process diagram (Hughes, 2006)

Qualitative research takes its rigor from the fact that whatever output your research creates can be traced back to the original data, that is, the model is derived from the categories, the categories were derived from the codes, and each code is tied to one or more instances in the data. In fact, in writing up the research report, it is common to see direct quotes taken from the coded data to support principles or assertions made in the discussion section of the report. Not only does this approach help illuminate the conclusion, but it adds credibility.

Tools and Techniques

Depending on the size of your research project, you might do manual coding such as what you see in the short example in Table 5.1, you might use a basic tool such as the index feature in a word processor as shown in this section, or you might use sophisticated qualitative data analysis tools. For large research projects, you want one or more tools that can handle a large corpus of data (such as text, graphics, audio, and video) and also do analysis and coding. Such tools help you to make sense of large bodies of data more efficiently than you can do manually.

Using a Word Processor for Data Analysis

For small to mid-size bodies of data, you can take advantage of tools that are already available to you, such as word processors like Microsoft Word and graphing applications like Microsoft Visio. Spreadsheets such as Excel can also be very useful for coding, categorizing, sorting, and filtering log sheets such as those created during usability tests. For example, you can import all of your data into a Microsoft Word file and then tag styles, create an index, or do global searches to identify category names and members. Editing and re-editing your categories is then as simple as cutting and pasting. See Figure 5.5 for an example showing how the indexing feature in Word presents categories resulting from your analysis. The resultant output is the findings of your analysis. The categories represent the themes that emerged from the data, and the index entries themselves are links back to the directly observed data (with the page numbers of where in the original file the entries can be found).

More specialized tools make it easier to analyze multiple documents and create visual models, but even for a master's level research project, this simple word processor protocol can give robust results. The creative part of the process is what conclusions you draw from your results, what model you infer from the themes that emerge.

Using Qualitative Data Analysis Software

If you are contemplating a large research project such as a doctoral dissertation, purchasing specialized software for coding and categorizing qualitative data can be a useful investment. Most products can be evaluated using free downloads. (Many of these free demonstration packages are the full application with limited ability to save results.) An internet search for "qualitative research software" or "qualitative data analysis" tools will yield a good list.

Atlas.ti is an example of a popular qualitative data analysis (QDA) tool that works on many platforms including Windows, macOS, Android, and iOS. It offers cloud support and can analyze both qualitative and quantitative data. NVivo is another

Standards

Credibility

credibility, 4
Credibility, 5
internal validity, 4

Dependability

dependability, 8
external validity, 6

Transferability

transferability, 7
Transferability, 6

Methods

Qualitative

Qualitative methods, 2
Document analyses, 3
Field observations, 3
Focus groups, 3
Interviews, 3
Usability tests, 3
analysis, 3
observation, 2
qualitative, 2, 4

Quantitative

quantitative, 1, 4

Society

Community

communities of practice, 1
community at large, 1
professional communities, 1

Social Studies

social implications, 1
social study, 1
sociology and anthropology,1
Technical communication,1

Figure 5.5 Index entries sorted into categories

popular QDA. Once you identify codes, QDA tools can conduct analysis in mere minutes for large bodies of data that can include hundreds of text, audio, video, and graphics files. These tools can also help you identify the original codes and manipulate them as your coding evolves.

For example, you might want to see whether there is a correlation between passive structures and the delivery of negative information. You can set up the software to locate the passive structures, negative information, and the number and type of correlations between the two. Or perhaps you would like to discover how often in a body of data (audio and text) participants from a certain global region use virtual teams in the workplace. The software can identify these connections amidst hundreds of documents and hours of interviews. Within such software, the researcher can set up to perform bottom-up analysis based on open coding as well as top-down analysis from existing structures.

Is it RAD?

Remember that as you design qualitative methods in your research, you should always seek to produce research that is RAD—replicable, aggregable, and data-supported (Haswell 2005). Your study methodology must be sufficiently defined so that others can repeat the study. Your results must be reported in sufficient detail that the data can be aggregated or combined with the results of other studies to build a body of data. Your conclusions must be supported by the data, not simply be impressions or gut feelings of the researchers.

Summary

Qualitative data are non-numeric—words, images, objects, or sounds that convey or contain meaning.

Qualitative methods generally involve the following three phases:

1. The observation of behaviors or artifacts in natural surroundings or authentic contexts
2. The recording or noting of those observations in descriptive or narrative formats
3. The systematic analysis of those observations or notes to derive patterns, models, or principles that can be applied beyond the specific events or artifacts studied

Qualitative research has its own standards of rigor:

- Credibility—Do the participants truly represent the population or phenomenon of interest and how typical are their behaviors and comments?
- Transferability—Does the thing or phenomenon that you are observing in the test environment reflect what would be found in the real world?
- Dependability—Does your study employ depth of engagement and diversity of perspectives and methods, and stay grounded in the data?

The following are common qualitative methods within technical communication research:

- Interviews—What participants say when asked (one-on-one)
- Focus groups—What participants say when asked (one-on-many)
- Usability tests—What users do when observed in a controlled environment
- Field observations—What users do when observed in their natural environment

- Document analyses—What documents tell us about the author(s) or the intended audience

Analyzing qualitative data consists of the following:

- Coding the data
- Categorizing the codes
- Drawing models out of the emergent categories and their relationships

Answer Key

Exercise 5.1

The answer to this exercise will be unique for each person who prepares it, so there is no key to this exercise.

Exercise 5.2

The following is a sample of one way that the data for this exercise can be coded.

Code	Content	Category
Program	• This program is centered on you and the community with the goal of cultivating change makers through civic engagement and structured reflection	Inspiring social responsibility
Project	• Planting and tending a community garden and orchard • Upkeeping visitor and community facilities • Helping to maintain the outstanding natural and cultural beauty of a local town	Conservation and beautification
Project	• Engaging with heritage programs designed for children • Volunteering with minors	Working with children
Project	• Engaging with heritage programs designed for elderly locals	Working with seniors
Partner	• You will need to bring a state police criminal record with you from your home country. This is a requirement for anyone volunteering with minors.	Community partner-defined qualifications

References

Belinsky, S. J., & Gogan, B. (2016). Throwing a change-up, pitching a strike: An autoethnography of frame acquisition, application, and fit in a pitch development and delivery experience. *IEEE Transactions on Professional Communication, 59*(4), 323–341.

Brown, J. S. (2002). Research that reinvents the corporation. *The Harvard Business Review, 80*(8), 105–115.

Corbin, J., & Strauss, A. (1990). Grounded theory research: Procedures, canons, and evaluative criteria. *Qualitative Sociology, 13*(1), 3–21.

Firestone, W. A. (1993). Alternative arguments for generalizing from data as applied to qualitative research. *Educational Researcher, 22*(4), 16–23.

Frey, L. R., Botan, C. H., & Kreps, G. L. (2000). *Investigating communication: An introduction to research methods*. Allyn and Bacon.

Haswell, R. H. (2005). NCTE/CCCC's recent war on scholarship. *Written Communication*, 22(2), 198–223.

Hughes, M. (2000). *Team usability testing of web-based training: The interplay of team learning and personal expertise*. [Unpublished doctoral dissertation]. University of Georgia.

Hughes, M. (2006). A pattern language approach to usability knowledge management. *Journal of Usability Studies*, 2(1), 76–90.

Kidder, L. (1982). Face validity from multiple perspectives. In D. Brinberg & L. Kidder (Eds.), *New directions for methodology of social and behavioral science: Forms of validity in research* (pp. 41–57). Jossey-Bass.

Lincoln, Y. S., & Guba, E. G. (1985). *Naturalistic inquiry*. Sage.

Merriam, S. B. (1998). *Qualitative research and case study applications in education*. San Francisco: Jossey-Bass.

Sharrock, W., & Randall, D. (2004). Ethnography, ethnomethodology and the problem of generalisation in design. *European Journal of Information Systems*, 13, 186–194.

Spinuzzi, C. (2003). *Tracing genres through organizations: A sociocultural approach to information design*. MIT Press.

Tuckman, B. W. (1965). Developmental sequence in small groups. *Psychological Bulletin*, 63(6), 384–399.

6 Conducting Surveys

Introduction

Surveys are a popular way for student and experienced researchers alike to gather data for their studies. With the emergence of free internet surveying software and the proliferation of academic and professional list serves, surveying has become an efficient way to reach a large sample and collect a relatively large body of data. However, the very ease with which surveys can be launched has resulted in a proliferation of surveys arriving in email inboxes; therefore, people tend to be more selective about the surveys they respond to, resulting in potentially lower response rates. In addition, unless correctly designed, implemented, and analyzed, a survey can result in a botched opportunity at best or a misrepresentation of a population at worst. In this chapter, we focus on surveys delivered as questionnaires.

   Given the popularity of surveys as a research tool and their ability to support both quantitative and qualitative studies, we deal with them in their own chapter. As with the chapters on quantitative and qualitative methodologies, no single chapter can deal with a topic as rich as surveying with the depth and scope needed to fully master the topic. Our objective instead is to give the student researcher enough of an introduction to conduct a valid survey and to understand what limitations might constrain its reliability. We also try to give the critical reader of research an understanding of the criteria and considerations that should be applied when reading research articles based on survey data.

Learning Objectives

After you have read this chapter, you should be able to:

- Design an effective survey
- Write clear, valid survey questions
- Collect data successfully
- Calculate response rate, margin of error, and confidence intervals

What Can Surveys Measure?

A survey is a list of questions that asks respondents to give information about specific characteristics, behaviors, or attitudes. Surveys can help us to understand the demographic profile of a sample or to estimate a statistic within a population. For example, surveys can collect data about the respondents' age, gender, salary, years of experience, level of education, and so forth. These data points could be directly pertinent to the research question or could be used to assess whether a sample is representative of the larger population from which it was drawn. For example, if a researcher were examining gender differences in salary levels among technical communicators, then asking the respondents' gender and salary would be directly related to the study's research question. If, on the other hand, the study was about college students' use of the internet as a study aid, questions about age and gender might be used merely to determine whether the demographics of the sample are representative of the college's demographics in general.

Surveys can also measure behavior (or more accurately, self-reported behavior). For example, a survey can ask how frequently students conduct internet searches as part of a homework assignment.

Lastly, surveys are often used to assess attitude, that is, the respondents' feelings about a topic. Who among us has not taken a survey at one time or another that asked whether we "strongly agree" or "strongly disagree" with a statement such as "I would be likely to ...?"

Planning Your Survey Research

To support the rigor of your survey research, you will need a research plan that includes the following:

- One or more research questions that reflect your purpose in conducting the research
- Population and sample characteristics
- List of instructions needed to introduce and conclude the survey
- Survey questions that operationalize your research questions, ordered to appeal to the respondent
- Plan for collecting data from your sample including a pilot survey

Choosing a Population

A population is a large group of people, objects, or other entities with a set of common characteristics about which a researcher wishes to generalize the study's findings. A sample is a smaller group, selected from the population, that participates in the research. Researchers must carefully select and define both their population and their sample to best respond to a research purpose or research questions. They must also carefully describe their population and sample so that users of the research can easily confirm the reliability of the research.

As we have noted before, in rigorous research design, technical communication researchers should not generalize about a population unless the sample size and methods make it reliable to do so. Thus, choosing both your population and your sample are important tasks upon which you base the claims that result from your research. Choose a population for which you can actually design a study with confidence in your findings. For example, if you choose to study preferred pedagogies of educators, with educators as your population, you have little chance of reliably generalizing about that population. Educators where? At what level? In what field? How will you gain access to these educators? However, if you choose to study preferred pedagogies of technical communication educators in the US and you design your survey well, you have good chance of being able to reliably generalize to that population.

Constructing a Survey

Constructing the survey consists of determining what questions to ask, asking them in the appropriate format, and then arranging them in an effective order. Additionally, you need to write instructions for completing and returning the survey.

What Questions to Ask

The questions that you ask in your survey need to be driven by the research purpose and research questions. Much of the art of surveying consists of constructing questions that *operationalize* the research purpose and research questions appropriately. In other words, to operationalize your research purpose, your survey questions must break down concepts so that they are small enough to be measured and understood. For example, a researcher might have the research purpose of determining whether there is a correlation between writers' general use of the internet and their receptiveness to online collaboration tools. That researcher must construct questions that determine how much time the respondent spends on the internet, how often the respondent uses the internet, or both. The researcher must also include questions that reveal attitudes about collaboration in general and collaborating online specifically.

There is no absolute formula for writing good questions, but the following guidelines can be useful.

- **Avoid absolute terms** such as *always* or *never*.
- **Avoid statements in the negative.** For example, consider this question: "Is the following sentence true for you? 'Subject matter experts are not a source of technical information for me.'" To say that subject matter experts *are* a source, the respondent must say "No" or "False." Using a double-negative to state a positive can lead to misunderstanding and thus to unreliable data. Therefore, word your questions so that "Yes" means yes and "No" means no.
- **Keep questions unbiased.** Consider the following question: "Do you think that productivity-enhancing processes such as online editing improve your effectiveness?" This question certainly signals that the surveyor assumes that online editing enhances productivity. If the respondent does not share that perspective, this could be a difficult question for the respondent to answer honestly.
- **Focus on one concept per question.** "Double-barreled" questions combine multiple concepts and can lead to vague results. For example, the question "To what extent are indexes and tables of contents useful to you in accessing data?" combines indexes and tables of contents in the same question. Some respondents might want to provide different answers for each. It would be better if the researcher broke that question into two questions: one to probe about indexes and another to probe about tables of contents.
- **Provide a neutral or opt-out choice in any closed question.** If respondents cannot find an option that accurately represents their opinion, they may skip a question or answer it with one of the provided choices that does not accurately represent their opinion. For example, you may have asked your respondents to respond to "How

effective do you find the support of the Grants & Contracts office?" with the following choices: ineffective, somewhat ineffective, neither ineffective nor effective, somewhat effective, and effective. If the respondent has never used the Grants & Contract office, none of the answers apply. You could also have provided a "does not apply" option. In doing so, you don't leave out potential responses.

Question Formats

Survey questions can be categorized within the following formats:

- Open-ended
- Closed-ended
- Multiple-choice
- Rating
- Ranking

A well-designed survey uses the appropriate type of question to capture the data it is seeking.

Open-ended

Open-ended questions allow free-form answers. The respondents can type or write sentences or even short essays to express their answers. Open-ended questions can give rich insight into respondent attitudes and can reveal unanticipated responses or themes. For these reasons, open-ended questions are particularly well suited to qualitative research. An example of an open-ended question would be "What typically frustrates you the most when conducting research?"

The following are some advantages of open-ended questions.

- They give "voice" to the respondent; that is, they allow respondents to express their opinions in their own words. This deeper insight into the respondents' perspectives can enhance credibility in a qualitative study.
- They reduce the degree to which the researcher's framing of the problem is imposed on the respondent. In other words, they are less likely to bias the answer toward an expected response or outcome. This openness can increase the credibility and dependability of the results.

The following are some disadvantages of open-ended questions.

- They are more demanding on respondents' time and energy to answer. This extra effort could reduce the overall response rate and detract from the reliability or dependability of the study.
- They are more demanding on the researcher to analyze and report. If you expect to get 800 surveys back and are not prepared to analyze the qualitative responses, do not waste the respondents' time by asking open-ended questions.

Closed-ended

Closed-ended questions look for a single piece of data—a word or a number. They can be very useful for gathering accurate data, such as respondent age, salary, years of education, etc. An example of a closed-ended question would be "What percentage of your time do you spend on research?"

The following are some advantages of closed-ended questions.

- They gather numeric data more accurately than multiple-choice questions (which typically use numeric ranges rather than specific numbers). This precision is useful in quantitative studies where statistical tests may need to calculate a standard deviation among the responses to certain questions.
- Verbal closed-ended answers are easier to categorize and summarize than are open-ended answers.

The following are some disadvantages of closed-ended questions.

- Verbal responses could be more varied and more difficult to analyze than questions presenting a set of answers, such as multiple choice.
- Questions seeking a numeric response might ask for a greater degree of accuracy than the respondent is able to give—for example, "How many employees are in your company?" (See Multiple-choice for a better way to present that question.)

Multiple-choice

Multiple-choice questions provide an array of choices from which the respondent selects one or more answers (make sure that it is clear which is allowed). If the survey is being administered electronically, use radio buttons for the choices where only one response is allowed and use check boxes if more than one choice is allowed.

Figure 6.1 shows an example of a multiple-choice question where only one choice is allowed. Note that all choices are mutually exclusive. For example, the choices are not 100–500 and 500–1000. (Which would be the correct selection if the respondent's company had 500 employees?) Also notice that the list is exhaustive. What if it had stopped at 1000–5000 and the respondent's company had 6500 employees?

What is the size of your company? (number of employees)

- ○ Less than 100
- ○ 100 – 499
- ○ 500 – 999
- ○ 1000 – 5000
- ○ Greater than 5000

Figure 6.1 Sample of a multiple-choice question where only one selection is allowed

---

How do you search for information in a document? (check all that apply)

    ☐ Scan table of contents
    ☐ Scan index
    ☐ Use an electronic word search
    ☐ Scan the headers and footers
    ☐ Browse looking at headings
    ☐ Browse looking at illustrations

Figure 6.2 Sample of a multiple-choice question where multiple selections are allowed

Figure 6.2 shows an example of a multiple-choice question where more than one answer is allowed. Not only is that fact implied by the use of check boxes, it is explicitly stated in the question.

The following are some advantages of multiple-choice questions.

- They are easy for respondents to answer.
- They are easy for researchers to summarize.
- They can be analyzed automatically by survey software applications or optical-scanning devices.

The following are some disadvantages of multiple-choice questions.

- They can omit meaningful choices that the researcher did not anticipate. For example, in Figure 6.2, are there other ways that are not listed that a reader could use to look for information? This problem can be mitigated by allowing a choice called "Other" and letting the user provide additional choices. But in that case, the researcher gives up some of the convenience of a multiple-choice and takes on some of the inconvenience of an open-ended or closed-ended question.
- They impose the researcher's frame of reference on the respondent. The respondent's choices are constrained by the researcher's view of the question, and in fact, none of the answers may be acceptable to the respondent.

Rating

A rating is a specialized form of question that asks respondents to express their answers as a degree along an axis—such as agreement with a stated opinion or a frequency to a question that asks "How often do you ...?" Different types of scales can be used for ratings.

A Likert scale presents a statement to which respondents indicate their degree of agreement or disagreement, typically along a five-choice scale. Figure 6.3 is an example of a rating using a Likert scale.

How strongly do you agree or disagree with this statement? "Virtual teams are important to my work as an engineering professional."

☐ Strongly disagree
☐ Somewhat disagree
☐ Neutral
☐ Somewhat agree
☐ Strongly agree

Figure 6.3 Rating question using a Likert scale

The same question could have been presented in a closed-ended format: On a scale of 1 to 5, how strongly do you agree or disagree with this statement: "Virtual teams are important to my work as an engineering professional (1 = strongly disagree; 5 = strongly agree)."

A *frequency* scale asks respondents to indicate a frequency of a behavior. Figure 6.4 shows an example of a frequency scale.

A semantic differential scale usually measures a series of attitudes toward a single concept, typically using a 7-point response scale with bipolar descriptions at each end. Figure 6.5 shows a semantic differential rating.

The following are some advantages of ratings.

* They can help clarify respondents' attitudes and preferences.
* They are easy for respondents to answer.
* They are easy for researchers to summarize.

The following are some disadvantages of ratings.

* Repetitive use can habituate the respondent to keep giving the same rating.
* Numeric scales have the risk of being reversed by the respondent; that is, a respondent might use 5 to mean a top rating when the researcher intended 1 as the top rating.

How often do you include indexes in the publications you write?

☐ Almost always
☐ Often
☐ Sometimes
☐ Seldom
☐ Almost never

Figure 6.4 Frequency rating

Using the following criteria, please rate the Acme Content Management System (circle your answer):

Hard to learn	−3	−2	−1	0	+1	+2	+3	Easy to learn
Hard to use	−3	−2	−1	0	+1	+2	+3	Easy to use
Poor search capabilities	−3	−2	−1	0	+1	+2	+3	Good search capabilities

Figure 6.5 Semantic differential rating

Rank the following from 1 to 5 in the order of their helpfulness to you as a resource (1 is the most helpful).

—— Subject matter expert
—— Current documentation on previous version of product
—— Marketing collaterals
—— Manager
—— Fellow writers

Figure 6.6 Ranking question

Ranking

Ranking questions ask respondents to put items in order along a defined dimension, such as importance or desirability. Figure 6.6 shows an example of a ranking.

An advantage of ranking questions is that they can assess the relative priorities of different attributes.

The following are some disadvantages of ranking questions.

- Respondents can easily confuse ranking questions with rating questions.
- Answers are not to scale. For example, one respondent may think that the first and second items are critically important, and the other items on the list are not at all important. For another respondent, the importance might be distributed more proportionally. So a "3" from one respondent does not correlate to a "3" from another—at least as a measure of importance.

Exercise 6.1 Writing Survey Questions

You wish to design a survey that addresses the following research questions.

- Are teachers of technical writing using multidisciplinary teams to teach?
- If teachers of technical writing are using multidisciplinary teams to teach, how are these teams constructed?

Create one question for each of the following question types that will help you gather data to answer the research questions:

- Open-ended
- Closed-ended
- Multiple-choice
- Rating
- Ranking

Survey Structure

The survey needs to provide all required instructions or clarifications. At its beginning (or in an accompanying cover letter/email), a survey should explain the following in no particular order:

- What is the purpose of the study?
- Who is qualified to participate in the survey?
- How is the privacy of the respondent being protected?
- Is participation voluntary?
- Approximately how long will it take to complete the survey?
- Can respondents stop and save their responses, and complete the survey later?
- What are the consequences of participating (informed consent)?
- For academic studies, has the survey has been approved by an Institutional Review Board (IRB)?
- Are incentives being offered for participation?

Don't overdo the explanations—you don't want the respondent to get bored and abandon the survey. But make sure that you supply all needed information.

Arrange the questions into logical groupings. If the survey is long, make these groupings into sections to give respondents a sense that they are making progress through the survey. At the end of each section of a long survey, use a percentage-complete indicator to reinforce that sense of progress.

See the section in this chapter on response rate for an explanation of how the order of questions can influence the response rate. End the survey with instructions about how to return or submit the survey, and thank the respondents for their participation.

Piloting a Survey

Once you have constructed a survey that will accomplish your research purpose and produce reliable data, pilot your survey (both the instructions and the questions) to make sure that any flaws in the survey design are identified and corrected. You don't want useful data to begin to flow in and then discover flaws that might weaken your data. For example, international respondents might have trouble with the wording of a question that inadvertently contains an idiom. Perhaps you didn't include a question that respondents think is very important. Perhaps you left out a neutral or opt-out option for a question. Any number of changes might be needed to fine-tune your survey so that it can be successful.

To pilot the survey, identify a small number of people who are representative of your sample. Ask them to take the survey, and then interview them regarding their experience. You may want to ask them about clarity, time to complete, ambiguity, and any other specific questions you or they may have. If you need to revise the survey substantially based on pilot data, consider piloting it again with another small number of people until you are satisfied that the survey instrument is working well.

It is particularly easy to introduce flaws into a survey that affect the responses of international respondents and non-native speakers of a language. If you truly want international data, be sure to pilot with international participants who represent the various areas of the world from which you hope to gather data.

Implementing the Survey

As you implement your survey, consider choice of media as well as the process of delivery. You can use many of the free tools such as Survey Monkey only up to a specified number of participants. After that, you must pay for the service. Organizations often buy subscriptions for one of these services, so it is a good idea to check with your organization's research office or information technology department before choosing a tool.

Surveys today are most often launched online and linked from an email message. In this case, you should make sure that the body of your email is a concise "letter of transmittal" that addresses any of the items you would like to emphasize from the survey's instructions detailed previously as well as the following:

- Credentials of the investigators
- Importance of the study to the field
- Average time to complete the survey
- Date by which responses are needed

Remember that this email must be short so as not to lose the interest of potential respondents. As a general rule, it is best to send the original request and then follow up twice to get the best rate of response. We suggest that you send out a reminder approximately two weeks after the survey is initially sent and then again three days before the survey closes.

Surveying, like all empirical methods, should be administered systematically, so keep notes on each of the steps in your process with justification for your choices. You will need them when reporting your study results.

Reporting Survey Results

Depending on the questions asked (quantitative or qualitative), the data analysis methods and standards of rigor described in Chapters 4 and 5 also apply to surveys. For example, if a survey collects numeric data about respondent salary, gender, and years of education, the researcher could use a *t*-test of two means (as explained in Chapter 4) to see whether there is a statistically significant difference between the salaries of men and women. Alternatively, the researcher could determine whether there is a correlation between salary and years of education. If the survey contains open-ended questions that ask respondents to describe how they feel that they are perceived within their organizations, qualitative coding and categorization methods described in Chapter 5 could be applied to look for themes or models that emerge from their answers.

A method of reporting results that we have not discussed yet, and one that is common in surveys, is *frequency distribution*.

Frequency Distribution

Frequency distribution is a useful way to report how respondents answer multiple-choice questions and is generally expressed in percentages. For example, respondent answers to the question in Figure 6.1 (size of company) could be summarized as shown in Figure 6.7 showing the percentage of people who responded in each category of organization size. Frequency distributions can also be reported in tables. Each question of import could have its own frequency distribution, or related groups of questions could be combined into one frequency distribution (assuming they all had the same choices). Such a simple distribution as shown in the example here can easily be calculated manually; however, with large data sets, you can use a tool such as Microsoft Excel to calculate and chart a frequency distribution. We suggest using YouTube if needed to find instructional videos on various options for using Excel to calculate and illustrate a frequency distribution.

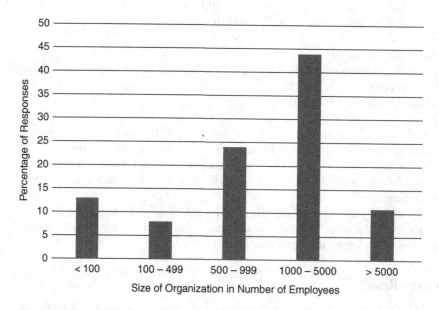

Figure 6.7 Sample frequency distribution

Measures of Rigor

Surveys are subject to the same standards of rigor discussed in Chapters 4 and 5, and they can also require analyses not yet discussed, specifically response rate, margin of error, and confidence intervals.

Response Rate

Survey reports often include the response rate. To calculate the response rate for a survey, divide the number of surveys completed by the total number you sent out. For example, if you put a survey out on a list server that had 1000 members and you got 375 back, your response rate would be 0.375 or 37.5%.

The response rate gives an indication about how representative the returned surveys are likely to be of the population surveyed. For example, if a professional association sends out a salary survey and gets a response rate of 5%, the reader of a report based on that survey's data should be cautious about how representative the results are.

In addition, a factor that can compromise the reliability of any survey is that the respondents self select whether or not to respond to the survey. This self-selection aspect undermines the principle of random or systematic sampling—an important consideration if statistical tests or analyses are being applied to the results. Methods such as snowball sampling (also called network sampling), for example, are based on self-selection. In this method, the researcher develops the respondent profile, but then respondents invite others in their own networks who fit the profile to participate. Using this sampling method, the researcher cannot calculate population size or response rate accurately. However, it is an effective way to increase the size of your sample.

The higher the response rate, the more likely that all segments of the population being surveyed are being proportionately represented. There is no hard rule for an acceptable response rate; in fact, the point is widely debated.

There are several ways that a survey designer and administrator can improve response rates.

- **Design a usable survey.** If respondents struggle with understanding how to answer or submit the survey, they are likely to abandon it.
- **Keep the length manageable.** Respondents get tired or can grow impatient if the survey is too long. Ask only what you need to.
- **Provide incentives.** Some surveys offer cash rewards for participation; some put the names of the respondents into a drawing and award a prize. Sometimes, just offering to share the results with the respondents can be a sufficient incentive for them to complete the survey.

Note: Teachers who are surveying students or managers who are surveying employees need to be sensitive about the ethical concerns of coerced participation in research.

- **Arrange the questions to pique the respondents' interest.** Don't start with the dull, demographic questions, such as age, gender, etc. Start with questions that are relevant to the goals of the survey and are likely to be of interest to the respondents. However, if some questions are likely to be sensitive, wait until you have earned the respondents' trust before asking them. For example, if a researcher is conducting a survey to assess technical communicators' ethical practices, the question "Have you ever lied to a client or employer?" might be a useful question, but it is not one that should be asked at the beginning of the survey.
- **Contact respondents three times to encourage participation.** Send out the survey, remind respondents approximately two weeks later to take the survey if they have

not done so, and then remind respondents about the closing date of the survey approximately three days before the survey closes.

- **Use mixed methods of contact if possible.** When possible, use more than one method to contact potential respondents. For example, you might email the original survey but then contact respondents via text or post with a reminder. This is often not possible but is recommended to improve response rates.

Margin of Error and Confidence Intervals

Researchers (and consumers of research) must be vigilant about generalizing to large populations based on samples. When the researcher wants to infer that the results from the survey sample are representative of the population of interest (which is often the case in research), then the report should include the *margin of error* (MOE). MOE is an estimate of how confident you can be that researched values fall within a given numeric range. Strictly speaking, margin of error can be calculated with statistical rigor only on random samples of a population. However, as we explain here, *margin of error can also be calculated usefully using the standard deviation of a sample rather than a population.* In such cases, the margin of error is useful when the sample is most representative of the population. However, realistically speaking, survey research in technical communication is seldom random.

The margin of error indicates the reliability of the inference and can be reliably estimated for the mean of a sample using the standard deviation of the sample, the size of the sample, and the level of confidence the researcher wants to obtain. The margin of error is a value, in percentage points, that essentially lets the researcher say, "I think the real value in the population falls within the reported value plus or minus this margin of error."

The confidence level is an indication of how reliable that statement is. Confidence levels of 90%, 95%, and 99% are common. So, if a researcher states that a survey has an MOE of ±3 at a 95% confidence level, that statement means "The true value for 95% of the samples that I could have taken is within ±3 of this number." In other words, the sample that the researcher actually took could be part of the 6% that "misses" the true value. Notice in Table 6.1 that the higher the confidence level for a given sample size, the wider the margin of error. The analogy is: The bigger the net you cast, the better the odds of catching a fish; the wider the margin of error is, the better odds you have that it includes the true value in the population. The trade-off is that the more confident you want to be

Table 6.1 Sample margins of error (in percent)

Sample Size	Confidence Level		
	90%	95%	99%
25	16	20	26
50	12	14	18
100	8	10	13
500	4	4	6
800	3	3	5
1000	3	3	4
2000	2	2	3

in your estimate, the less precise the estimate will be; conversely, the more precise your estimate is, the less confidence you can have in it.

The good news is that a useful approximation of the margin of error can be easily calculated. You can use Table 6.1 to get an idea of what the margins are at various sample sizes and levels of confidence.

For example, the formula for the *MOE of a sample mean* is
Margin of error is approximately equal to $z * s/\sqrt{n}$
where

- z = the appropriate z-value for your level of confidence
- s = the standard deviation (*SD*) of the sample
- $\sqrt{n}$ = the square root of the sample size

Common z values are $z = 1.645$ for 90% confidence, 1.96 for 95% confidence, and 2.58 for 99% confidence.

The bad news is that most researchers (especially student researchers) are disappointed when they see how large the margin of error is for their sample size.

For example, let's say a researcher surveys 100 technical communicators who work with usability studies. She asks them how much, on average, their organizations invest in a single remote usability study. Thirty respondents fill out the survey, and she uses Excel to calculate that the average cost, including estimated overhead and employee time, is $1,719 with a standard deviation of $972.33. (Chapter 4 includes instructions on how to use Excel to calculate the mean, sample size, and standard deviation.)

One reason that her standard deviation is high is that the data she received varied a great deal, with estimated average costs for a study ranging between $250 and $3400. Using a confidence level of 95%, the researcher sees that based on this sample of 30 technical communicators, the average remote usability test cost of $1,719 is accurate with a margin of error of ±$347.77. She was able to calculate this MOE in Excel using the alpha value of .05 for the 95% confidence level, the standard deviation, and the sample size.

The *confidence interval* is the range within which the researcher is confident that the actual value falls (once again, typically at 90%, 95%, or 99% levels). For this sample of usability study costs, the confidence interval is the mean of $1,719 ± $347.77, in other words $1,371.24 to $2,066.76. This is a wide confidence interval, and the researcher might choose to do any of the following:

- Increase her sample size to reduce the margin of error
- Lower the confidence level to 90% to decrease the margin of error
- Report the confidence as is with this wide margin of error

Unfortunately, many researchers merely report the findings and talk about them as if they were representative of the population as a whole. As a critical reader of research, you can protect yourself by calculating the margin of error yourself, as long as the researcher has reported the data. In the example above, we show you the results of the calculation of an MOE for a mean; you would select a different equation to calculate the MOE of a proportion such as that in a frequency distribution.

Exercise 6.2 Survey Exercise

You survey 32 students in a technical communication master's program and ask how much time elapsed between completing their bachelor's degree and enrolling in a master's program. You would like to use this data to make an inference about how long technical communicators typically wait before pursuing a master's degree. The average of the responses was 7.5 years and the standard deviation was 4.5. If you wish to have a 90% confidence level, what are you willing to say about the average time technical communicators wait between getting a bachelor's degree and enrolling in a master's degree program?

Is It RAD?

Remember that as you design your survey and analyze the data, you should always seek to produce research that is RAD—replicable, aggregable, and data-supported (Haswell, 2005). Your methodology must

- Be sufficiently defined so that others can repeat the study
- Be reported in sufficient detail that the data can be aggregated or combined with the results of other studies to build a body of data
- Be supported by the data, not simply be impressions or gut feelings of the researchers

Summary

Surveys can provide data on characteristics, behaviors, and attitudes.

As with any rigorous research, you need a plan to ensure that you operationalize your research questions and collect valid and reliable data.

Choose your population and sample carefully, making sure that your sample can be generalized to your population.

Construct your questions (from among five types) so that they operationalize your study's guiding research question(s).

Pilot your survey as many times as needed to ensure its reliability.

Calculate your response rate, margin of error, and confidence intervals to assess the claims that you can reliably make with your data.

Answer Key

Exercise 6.1

The answer to this exercise will be unique for each person who prepares it so there is no key to this exercise.

Exercise 6.2

"We can be 90% confident that the average time between getting a bachelor's degree and enrolling in a master's program is between 6.2 and 8.8 years."

Note: You could have calculated this manually or have used the CONFIDENCE function in Excel with an alpha of 0.1 (since the confidence level is 90%), a STDEV.S of 4.5, and a size of 32. This results in a confidence interval of 7.5 years ±1.3 years.

Reference

Haswell, R. H. (2005). NCTE/CCCC's recent war on scholarship. *Written Communication*, 22(2), 198–223.

7   Conducting Usability Studies

Introduction

Usability research is often used by general practitioners of technical communication as well as by practitioners who focus in UX (user experience) design. Usability touches every aspect of technical communication because our profession lives or dies based on effectiveness with the audience. Usability studies provide a rigorous methodology for understanding an audience's response to a product or process.

Because this book is a basic research text for technical communication, this chapter includes information on usability research methods for students and practitioners who are beginning this type of research. These beginning principles can also help consumers of usability research know what to look for in a reliable usability study. As with surveys, usability studies often collect both quantitative and qualitative data, and for this reason, we address it in its own chapter.

As a field, usability emerged during the 20th century. Human factors research appeared in the early to mid-1900s (e.g., Chapanis, Garner, & Morgan, 1949; Gilbreth & Gilbreth, 1919) and began to develop steadily. Early work was more experimental in design. Signs of modern usability methods such as small studies and user participation began to appear later with work from researchers such as Gould and Lewis (1985), Norman (1988), and Nielsen and Molich (1990), as well as with the formation of associated professional groups. Usability has developed a robust methodology with the help of many scholars and practitioners. As a result, we now have an excellent understanding of how to use a variety of methods to conduct usability research to affordably assess the usability of products and processes.

A "usable" product is one with which users can accomplish their goals with minimal frustration and error. Jakob Nielsen, a guru of usability, defines five characteristics of usability:

- Learnability: How easy is it for users to accomplish basic tasks the first time they encounter the design?

- Efficiency: Once users have learned the design, how quickly can they perform tasks?
- Memorability: When users return to the design after a period of not using it, how easily can they re-establish proficiency?
- Errors: How many errors do users make, how severe are these errors, and how easily can they [users] recover from the errors?
- Satisfaction: How pleasant is it to use the design? (Nielsen, 2012)

Certainly, other usability experts have defined "usability," but most definitions, such as Quesenbery's five Es—effective, efficient, engaging, error tolerant, and easy to learn—(n.d.) are strongly similar to Nielsen's definition.

A usability study might be constructed of multiple methods of inquiry such as usability testing, heuristic evaluation, interviews, questionnaires, and so on. However, at the core, all usability studies seek to evaluate how effectively a product or process incorporates the characteristics defined by Nielsen. Usability is assessed at the interface of user and product, whether it be a user interface with a document, a web site, or a wheelchair. Such assessments can be formative and take place throughout the development of a product, or summative and take place once development has been completed. Of course, formative assessment is more desirable so that user input into design (that is, participatory design) can contribute the most cost benefit. Increasingly, formative usability assessment is being adapted into the Agile development processes that are popular in today's software industry. In his book *Lean UX*, Gothelf (2013) shares how you can apply lean principles to a design to improve the user experience.

In addition, note that "useful" and "usable" are two entirely different concepts. A product's usefulness is determined by whether or not it serves its intended purpose. For example, a door handle might be useful if it opens the door; however, it may not be usable if it is designed in such a way that people are continually applying it in the wrong direction. For a fun video illustrating this concept of useful versus usable, look up the Norman Door video clip online.

Consider another example. First-responders in different states needed to communicate in emergency situations across state borders; however, they were unable to do so because they operated different types of radio equipment. As a result, a software network was developed to help them communicate. This network cost millions of dollars to develop and deploy. Unfortunately, because the system was somewhat complex and consistent training was not provided, the network was never widely used. In fact, in one emergency center, the computer that was dedicated to running the network ended up as a plant stand! Had the developers of this system assessed its usability throughout its development and applied what they learned about users to the system and training, it is likely that this very expensive system would be extraordinarily useful for interstate disaster communication. Such stories are plentiful; you can probably easily think of an example when a product could have benefited from usability assessment.

In this chapter, we will explain how you can design usability studies that are reliable and valid—in other words, good research!

Learning Objectives

After you have read this chapter, you should be able to:

- Construct a usability study comprised of multiple methods with appropriate rigor
- Implement the usability study
- Analyze your data to draw reliable conclusions

What Can Usability Studies Assess?

As we have noted in the introduction to this chapter, usability studies are meant to assess the usability of a product or process. Usability studies are typically constructed using multiple methods. Those methods are selected based on what you, as the researcher, wish to discover in a given study. For example, using a method called *usability testing*, you can choose to assess the performance of users as they interact with a product. In this case, you might measure such things as time on task, level of success with a task, and number of failures during which the user could not proceed without intervention. When you observe users' behavior as they interact with the interface, you can collect *manifest data*—that is, data that reflect *what people are actually doing rather than what they perceive that they are doing*. This is a strength of the usability testing method.

However, as part of usability assessment, you can and should also collect self-reported perceptions of users through methods such as questionnaires, interviews, or focus groups. For example, you might ask users to rate their responses to the following statement: "I found the system cumbersome to use" ("System Usability Scale," 2019). The data that you collect in this instance are *ideal data*—that is, data that reflect what people think or perceive that they do rather than what they actually do. Both manifest and ideal data are important to completing an effective usability assessment.

Consider the following: if I ask a student whether or not she thinks that the new updates in the medical school labs are usable, she might say "Yes! It's nice to have new equipment and furniture." I have just collected ideal data—her self-reported impression. On the other hand, I could observe this student using the new facility, and then judge whether or not she finds it usable based on a set of criteria. In this case, I would have collected and analyzed manifest data. Both types of data (ideal and manifest) are useful because together they help to capture data concerning perception and behavior.

The data collected from one method may or may not confirm the data collected from another method. In a rigorous assessment, you should seek to collect data using several methods to see whether the findings correlate. When the findings don't correlate, you should look closely to see whether the data may indicate why. If the findings do correlate, they are more reliable due to *triangulation*. That is, when a finding from one method of data collection is verified by a similar finding from one or more other methods, the finding is more reliable. The need for triangulation to ensure reliability is why rigorous usability assessment should draw on multiple methods including usability testing.

Constructing a Usability Study

As empirical research, usability studies should be conducted systematically; therefore, you need a plan. In this section, we discuss how to construct a usability study using some of the most common methods for such studies.

Research Purpose

As with any technical communication product, purpose and audience should determine content and form. So, in planning a usability study, begin with a clear identification of purpose. We favor an approach for identifying the purpose where the researcher identifies a gap in knowledge by stating the ideal state as compared to the current state. For example, in setting up a usability study, you might articulate the purpose as follows.

> Ideally, the web site will be usable for those who have common vision disabilities; however, currently the web site does not accommodate those with diabetic retinopathy. In this usability study, we seek to identify issues of usability for those with diabetic retinopathy.

Using this approach to articulate purpose can help you identify a clear gap in which to situate your study.

User and Participant Profile

Your user profile identifies the significant characteristics of the users of the interface you are assessing (that is, the population of interest). Your participant profile identifies those participants who will serve as representatives of the user group and participate in your study (that is, the sample). When you are defining your participant profile, you should address two main areas: technical expertise (with whatever technical devices, such as computers, are involved in the tasks) and subject domain knowledge. For example, if you were studying the effectiveness of computer-based training for a nursing course, the participants would need to be nurses or nursing students, and they would need the same level of computer skills that the proposed student population is expected to have.

The most reliable means to determine a potential participant's computer expertise is to ask them behavioral questions rather than ask them to rate their own skills as novice, intermediate, or expert. For example, in the case of a software product, a useful technique is to identify commercially available software products that require comparable skills to the one you will be testing and ask potential participants whether they have used those products and what they have done with them. On the other hand, if you are conducting a study that involves doing an internet search, you could ask potential participants to identify which search engines they have used and what they searched for.

Many usability researchers begin their recruiting efforts with a questionnaire that screens potential participants based on research needs. See Figure 7.1 for an example of a screening questionnaire.

Please answer the following questions in order to determine your eligibility to participate in this study.

1. Have you ever visited https://www.xxx/ before?

    ☐   Yes
    ☐   No

2. Have you ever owned a boat?

    ☐   Yes
    ☐   No

3. What is your age range?

    ☐   Less than 18 years
    ☐   18–29 years
    ☐   30–39 years
    ☐   40–49 years
    ☐   50–59 years
    ☐   60 years or above
    ☐   Prefer not to answer

4. What is your gender?

    ☐   Male
    ☐   Female
    ☐   Other
    ☐   Prefer not to answer

5. What is your education level?

    ☐   GED
    ☐   High school graduate
    ☐   Some college
    ☐   Bachelor's degree   Major _____
    ☐   Graduate degree   Major _____
    ☐   Prefer not to answer

Figure 7.1   Sample screening questionnaire

A few practical notes: The more detailed and restrictive the participant profile, the more challenging the recruiting will be. You must strike a balance between being detailed enough to ensure credibility and transferability without overly constraining yourself. If you are conducting research that you wish to transfer (generalize) to a larger population, you must also consider the impact of sample size, margin of error, and confidence intervals on your selection of participants. See Chapters 4, 5, and 6 for further information on these terms.

Finally, once you have identified your purpose and sample characteristics, you need to consider whether or not your study requires Institutional Review Board (IRB) approval so that you can begin preparing the application. As a general rule, if you plan to publish your research for academic purposes, you need to get IRB approval for your study and methodology.

Methods

After you have identified your research purpose and your user and participant profiles, you may begin to design your methods for data collection. Usability studies can be constructed using several methods, as we have already noted. In this section, we will identify some of the most commonly used methods in usability research and describe how to use them. Specifically, we focus on usability testing, heuristic evaluation, interviews, and questionnaires as they might be used in usability studies. The authors of the exemplar article presented in Chapter 12 use a more experimental methodology, so, between Chapters 7 and 12, you can consider the breadth of good usability research methods.

Usability Testing

"Usability testing" has a very specific meaning. It refers to collecting data from users as they interact in an authentic way with the product or process being studied. With that said, you must plan carefully for a systematic and consistent collection of data just as you would for any rigorous study. You will observe your participants interacting with the product or process as well as capture as much of the data regarding the interaction as possible.

As a usability researcher today, you have many choices in designing your study. Technology has enabled more options in usability testing than ever before, including moderated remote testing, unmoderated remote testing, and face-to-face testing in a lab or in the field. *Moderated remote testing* refers to a test during which the researcher and participant are not colocated but are connected by technology at the same time. They work synchronously during the test with the researcher moderating. *Unmoderated remote testing* refers to a test during which the participant works asynchronously. The participant and researcher are neither colocated nor connected by technology at the same time. In this case, the researcher receives data, including potential video, after the participant has completed the test. Finally, in *face-to-face testing in a dedicated usability lab or in the field,* the researcher moderates the test in person in a controlled lab setting or in the user's natural environment.

Each of these methods has its own advantages and disadvantages, but moderated remote testing is perhaps the most popular today because of readily available technologies. It is much less expensive to test online without the costs of travel.

To support the rigor of your test, you will need a test plan that includes the following:

- Purpose and objectives
- Participant characteristics
- Scenarios, task lists, and script
- Data to be collected and evaluation scales
- Other methods in addition to usability testing
- Test environment, equipment, and logistics
- Team roles
- Forms and checklists
- Test protocol

We discussed identifying your purpose as well as your user and participant characteristics earlier in the Methods section because those tasks must be performed for all methods, not just usability testing. In this section, we go into specific considerations for preparing a usability test plan.

Participant Characteristics

The participants in your usability study usually represent some segment of the total user population. If you seek to reliably generalize your findings to a population, you should use the material on statistical measures presented in this book to help ensure that your claims are reliable, such as margin of error calculations. However, in usability studies, you have another option to guide your claims of transferability. Nielsen and Landauer presented research that indicates that a well-designed usability study can identify 85% of problems with an interface with no more than five participants (Nielsen & Landauer, 1993). A summary of this so-called "discount" usability testing on the Nielsen Norman Group web site reads, "Elaborate usability tests are a waste of resources. The best results come from testing no more than 5 users and running as many small tests as you can afford" (Nielsen, 2000). Thus, as a usability researcher, you can generalize to a population for a specific interface with a well-designed small study. One caveat is that such generalizations can be made only for relatively homogenous groups. To return to our previous example of users who have diabetic retinopathy, one must consider that the results of a usability test with participants who have no significant disability would not apply to users who have a significant disability.

Scenarios, Task Lists, and Script

Your test plan should also include a script that supports consistency from participant to participant as well as scenarios and tasks that guide the interaction of participants with the product or process. A script discourages you from introducing an inconsistency that might compromise the validity or reliability of your study. For example, you would not want to accidentally provide one participant with some significant detail not provided to other participants. Such details might then affect their interaction with the product and introduce a variable for which you have not accounted. Your script should include all interaction with participants, including screening questionnaires, test instructions, scenarios, tasks, and any required release forms. Reading from a script can appear very stilted to participants, but you can mitigate this problem by explaining at the beginning that you will be using a script to maintain consistency within the tests that you are conducting.

The tasks that you develop for your participants to perform should relate clearly to the research goal and research questions. In addition, for your study to be credible, ensure that your tasks are authentic. For example, if you are studying how users scan a results list from a search query, it would be better to state a search goal and ask the users to identify the results that they consider most relevant. You might instruct them to "Find an article that discusses the differences between quantitative usability tests and qualitative usability tests" instead of asking them to "Search on 'Usability' and find the article written by Hughes."

In short, *do not tell the participants how to do a task*; instead, tell them what to accomplish and then observe how they go about it.

To further ensure that your participants engage authentically with the product, develop scenarios that provide context for the tasks that you ask them to do. For example, suppose that you are testing the web site of the US Coast Guard. You want to determine how usable this web site is for spouses of Coast Guard personnel as part of an effort to ensure that the organization is supporting both military members and their families. In an effort to provide authentic tasks for your participants, you might develop multiple scenarios and tasks including the following one.

Scenario

You just found out that your spouse is being transferred to the base in Honolulu, Hawaii. You have never been there and want to find out about housing options.

Task 1
Go to https://dcms.uscg.mil and find family housing information for the base in Honolulu.

Task 2
Starting from your current page, determine what costs are associated with the housing options.

Notice that a single scenario can be associated with multiple tasks, but all tasks should be associated with a scenario. Providing such context increases the authenticity of your test.

Data to Be Collected and Evaluation Scales

During a usability test, you will probably collect both quantitative and qualitative data. You have many choices of what data to collect and how to collect them. Some of the data you are most likely to collect include both performance data and preference data such as

- Think-aloud audio and notes
- Video that captures participant expressions and actions
- Real-time coded data that marks outcomes such as successes, successes with moderator intervention, and failures

One of the instructions that you should give participants before testing is how to engage in a think-aloud protocol. In thinking aloud, participants speak aloud everything that they are thinking while working with the product or process. These thoughts are lost data unless they are uttered aloud so that they become data that the researcher can analyze. It is best to explain this process to participants prior to testing and then ask them to practice it by doing a simple task in front of the moderator. It may also become necessary to prompt some participants to keep thinking aloud during testing if they lapse into silence while they are working with the product or process.

In addition, video that captures not only participants' words but also their expressions and tone of voice can be very helpful during analysis and reporting. These data are rich, adding nuance to what has been said, and because they come straight from the mouths of users, they often have a strong impact on those to whom you are reporting.

As you construct your test, consider incorporating as much real-time coding of responses as possible while you and your team are observing. You will find that doing so helps during the analysis phase both in terms of efficiency and accuracy. For example, you can observe hours of participant interactions with a product and capture all the interaction on video as well as in notes. If you wait until you have completed testing with one or all participants to do some coding of their responses, your memory will likely be challenged, and the coding itself will be very time-consuming. Suppose you are testing participants as they use an online research repository. If you create a very simple notes format that includes some simple coding as seen in Figure 7.2, you will be far advanced when you start your analysis.

Many different coding schemes can be used to aid you in collecting data as well as in analyzing them. Usability testing software such as Morae allows you to embed coding scales directly into the test, including the System Usability Scale (SUS), color-coded symbols, and other rating scales. The SUS was developed by John Brooke at Digital Equipment Corporation in 1986. It has been widely and successfully employed in usability studies and uses 10 Likert-style statements for which users provide a rating.

Web Usability Assessment
Notes: Participant 8

Front- or Back-End Assessment?
Browser?

Task 1.
Time on Task?

Start Time: End Time:

Subtask 2 (circle1, 2, or 3)
Unassisted and successful (1)   Assisted but successful (2)   Failure, did not complete (3)

Subtask 3 (circle1, 2, or 3)
Unassisted and successful (1)   Assisted but successful (2)   Failure, did not complete (3)

Notes

Task 2.
Time on Task?

Start Time: End Time:

Subtask 2 (circle1, 2, or 3)
Unassisted and successful (1)   Assisted but successful (2)   Failure, did not complete (3)

Subtask 3 (circle1, 2, or 3)
Unassisted and successful (1)   Assisted but successful (2)   Failure, did not complete (3)

Notes

Figure 7.2   Sample notes page for moderator use during testing

Other Methods

If you are incorporating other methods into your usability test session such as interviews, questionnaires, reaction cards, and so on, include them in your test plan. For example, after each task that users perform during a usability test, you might ask them to rate their experience on a Likert scale from one to five according to how well they believe the interface performed. To have them rate the consistency of design, you might say, "For the task you just performed, rate the interface on a scale of one to five, with five being most desirable, for how consistently words were used."

Test Environment, Equipment, and Logistics

Describe your test environment and all the tools/equipment that you will use, including any special logistics needed for the test. Such detail not only helps you to ensure that you have a sound plan, but it will be needed when you report on the results. When reporting, you need to provide enough detail so that your client or any consumer of your research can understand your methods and replicate them. For example, if you conduct a moderated, remote test, you must identify the software tools you use as well as where the research team and the participants are located. If you are using a web conferencing tool such as Zoom or WebEx, identify which affordances of the tool you are using: audio, video, text transcript, and so on.

Team Roles

Usability testing can be conducted by a single person or by a team. Whether you are one or many, account for the following roles in your test plan:

- Moderator
- Technician
- Note-takers

If you are doing a moderated study, the moderator's role is obviously important. This person potentially instructs participants before testing begins, guides participants through the test, and debriefs them after the test. To moderate effectively requires some thought and practice. You must guide the participants through the test without biasing their responses. That is, you must make them feel welcome and comfortable without influencing how they interact with the product. Quite a challenge!

You or another dedicated person serving as technician must be sure that the tools are ready for the test and then troubleshoot any problems that occur. Finally, you or another person serving as note-taker should gather as much data as possible using codes and descriptions. We strongly suggest that all notes be taken digitally to make extraction easier during the analysis phase.

Forms and Checklists

You will need several forms to support your study, including an Informed Consent Form whereby participants give you signed permission to use the data, including video. Your Informed Consent Form helps protect you and your participant. It should identify what data will be collected, how they will be collected, and how they will be used. It should also confirm that participation is voluntary.

Your team will also benefit from checklists that will ensure that no important details are overlooked. For example, when a team is testing, you will probably want to use a Technician Checklist that ensures that your technician has consistently set up the audio levels, webcam angles, and so on before each participant. Likewise, a Moderator Checklist can ensure that all forms, surveys, and so on are ready to go when the participant arrives (in person or online).

Test Protocol

Once you have determined all specifics of your test, it is helpful to create a test protocol (that is, an overview) of all parts of the test in order. In Figure 7.3, you can see a sample test protocol.

Test Protocol

The usability test will start with an overview and briefing on what to expect. The briefing will be followed by the tasks/scenarios outlined below. Participants will not be allowed to use any outside resources and must stay within the XXX web site during the test. This is to ensure that they are not influenced in a way that would impact the results of the usability test.
We will adhere to the following protocol for the usability test:

1. Greet participants.
2. Turn on recording of the Zoom session.
3. Test camera, microphone, and screen share for participant and moderator.
4. Conduct pre-test interview.
5. Read moderator script and practice think-aloud protocol.
6. Give participant the link to the XXX web site.
7. Give participant Scenario 1 and Tasks 1, 2, and 3.
8. Give participant Scenario 2 and Tasks 4 and 5.
9. Give participant Scenario 3 and Task 6.
10. Conduct post-test interview.
11. Ask participants if there are any final questions or comments.
12. Turn off recording and end the session.

Figure 7.3 Sample usability test protocol

Heuristic Evaluation

A heuristic is a set of guidelines that aid in analyzing a complex problem. As part of a usability assessment, you might include a heuristic evaluation. In the case of usability research, a heuristic evaluation refers to a specific process: An expert in usability, user experience, or human factors performs an assessment of a product using a set of guidelines such as Nielsen and Molich's 10 heuristics shown in Table 7.1 (or another set of heuristics appropriate to the problem).

Table 7.1 10 usability heuristics for user interface design

Visibility of system status	The system should always keep users informed about what is going on, through appropriate feedback within reasonable time.
Match between system and the real world	The system should speak the users' language, with words, phrases and concepts familiar to the user, rather than system-oriented terms. Follow real-world conventions, making information appear in natural and logical order.
User control and freedom	Users often choose system functions by mistake and will need a clearly marked "emergency exit" to leave the unwanted state without having to go through an extended dialogue. Support undo and redo.
Consistency and standards	Users should not have to wonder whether different words, situations, or actions mean the same thing.
Error prevention	Even better than good error messages is a careful design which prevents a problem from occurring in the first place. Either eliminate error-prone conditions or check for them and present users with a confirmation option before they commit to the action.
Recognition rather than recall	Minimize the user's memory load by making objects, actions, and options visible. The user should not have to remember information from one part of the dialogue to another. Instructions for use of the system should be visible or easily retrievable whenever appropriate.
Flexibility and efficiency of use	Accelerators—unseen by the novice user—may often speed up the interaction for the expert user such that the system can cater to both inexperienced and experienced users. Allow users to tailor frequent actions.
Aesthetic and minimalist design	Dialogues should not contain information which is irrelevant or rarely needed. Every extra unit of information in a dialogue competes with the relevant units of information and diminishes their relative visibility.
Help users recognize, diagnose, and recover from errors	Error messages should be expressed in plain language (no codes), precisely indicate the problem, and constructively suggest a solution.
Help and documentation	Even though it is better if the system can be used without documentation, it may be necessary to provide help and documentation. Any such information should be easy to search, focused on the user's task, list concrete steps to be carried out, and not be too large.

By Jakob Nielsen on April 24, (1994) (https://www.nngroup.com/articles/ten-usability-heuristics/).

Such heuristics can be used to contribute to a rigorous usability assessment by introducing an additional method of data collection.

Prior to testing, ask several experts in usability, user experience, or human factors, each with the same level of experience with the interface, to evaluate the product based on a set of heuristics that you have provided. For example, you might ask them to fill out a form such as the one shown in Figure 7.4 during their evaluation. Notice that in using heuristic evaluation in addition to usability testing, you are collecting data from two different types of users using two different methods.

Heuristic Evaluation Form

As a person with some knowledge of usability theory, please evaluate the interface based on the two heuristics provided here from Nielsen and Molich's 10 Usability Heuristics for User Interface Design (Nielsen, 1994).

Heuristics	Rating (1 to 5 with 1 being not effective at all and 5 being very effective)	Notes

---

Visibility of system status

The system should always keep users informed about what is going on, through appropriate feedback within reasonable time.

Match between system and the real world

The system should speak the users' language, with words, phrases and concepts familiar to the user, rather than system-oriented terms. Follow real-world conventions, making information appear in a natural and logical order.

Figure 7.4 Sample form for use in a heuristic analysis of a web site. Note that only the first two of Nielsen's guidelines are provided in this sample

In conducting a heuristic evaluation of any product, whether a web site or something else, you can often find useful heuristics on the web or in books or articles that have already been validated as part of a rigorous research methodology. You can build on that research by incorporating such a validated tool into your own research.

Exercise 7.1 Conducting a Heuristic Assessment

1. Select a web site.
2. Based on what you have learned about usability assessment in this chapter, conduct a heuristic assessment of the web site using a form such as the one presented in Figure 7.4, but using all of Nielsen and Molich's 10 heuristics.
3. Summarize your experience and perceptions in a short report. Include the completed form with your ratings.

Interviews

In Chapter 5 we covered the basics of conducting interviews for research. Because interviewing is an important method for usability research, we discuss it here through that lens.

There are many points during usability research when you might choose to use interviews to collect data. One of the most common points is during a usability test. You might decide to interview participants before they begin testing to better understand the participants' attitudes toward the product being tested, or to find out more about their background with related technology. Interviews allow you to fill in gaps in the data you are gathering. You might also choose to interview participants after they have completed a usability test so that you get reflections on what they experienced. It is often a good idea to continue collecting audio and video during post-test interviews. These data can be very rich because the interview follows directly after the testing. We have often found that the data from these interviews as well as the video clips themselves are very useful in our research and in our presentation of the research.

Questionnaires

Questionnaires can also be very helpful in collecting data as part of usability research. Like interviews, questionnaires can be used to collect usability data on their own or as part of a usability test protocol. As part of your test protocol, you can use questionnaires to screen participants, collect demographic data, and collect participant information prior to, during, or after testing. You can construct questionnaires yourself or use existing tools that best fit the questions that you seek to answer. For example, the System Usability Scale (SUS) shown in Figure 7.5 and mentioned earlier in this chapter is commonly used to guide questions related to usability testing. Because it is easy to use, reliable, and valid, it has become a standard for testing many types of products. Although gathering the data with this instrument is easy, the scoring is somewhat complex, but instructions may be found on the web at usability.gov.

You may add other methods of data collection to your usability study such as cognitive walkthroughs, focus groups, card sorting, paper prototyping, and so on. All of these tools can be used individually or in combination with or without formal usability testing. Barnum's *Usability Testing Essentials* is an excellent introductory text for test design (Barnum, 2011). Remember that your goal is to collect data systematically with standards of rigor that support your findings.

Piloting a Usability Study

Piloting a research study involves launching it in full with a small number of participants who best represent your population. Remember that part of collecting reliable and valid data is collecting it systematically so that you do not introduce unintended variables and so the research can be replicated. You do not want to find yourself midway through data collection (or, worse, actually analyzing the data from a study that you thought was complete) and discover that your study is flawed. As we described in Chapter 6, piloting your research protocol allows you to identify potential problems and correct them before you begin collecting data. If you need to revise the study substantially based on pilot data, you should consider piloting it again with another small number of people until you are satisfied that all methods used in the study are stable and working well. Discount usability testing with three to five participants usually requires piloting with one person at a time.

Question	Response				
1. I think that I would like to use this system frequently.	1 Strongly Agree	2	3	4	5 Strongly Disagree
2. I found the system unnecessarily complex.	1 Strongly Agree	2	3	4	5 Strongly Disagree
3. I thought the system was easy to use.	1 Strongly Agree	2	3	4	5 Strongly Disagree
4. I think that I would need the support of a technical person to be able to use this system.	1 Strongly Agree	2	3	4	5 Strongly Disagree
5. I found the various functions in this system were well integrated.	1 Strongly Agree	2	3	4	5 Strongly Disagree
6. I thought there was too much inconsistency in this system.	1 Strongly Agree	2	3	4	5 Strongly Disagree
7. I would imagine that most people would learn to use this system very quickly.	1 Strongly Agree	2	3	4	5 Strongly Disagree
8. I found the system very cumbersome to use.	1 Strongly Agree	2	3	4	5 Strongly Disagree
9. I felt very confident using the system.	1 Strongly Agree	2	3	4	5 Strongly Disagree
10. I needed to learn a lot of things before I could get going with this system.	1 Strongly Agree	2	3	4	5 Strongly Disagree

Figure 7.5 System Usability Scale ("System," 2019)

A final note about pilot data: The data you collect during your pilot(s) can be merged in with the other data *only if you make no changes to your methods after the pilot.* If you change your methods, you should not merge the pilot data, but you can discuss those results in the context of discussing your findings. For example, you might say, "In the pilot study, we first saw evidence of this navigation issue."

Analyzing the Data

Once you have collected your data, you can begin your analysis. You will likely have a great deal of data, even for a small study. *Analysis requires parsing the data into a form where you can begin to identify patterns.* This is the fun part! If you kept your data organized during collection and used some coding, you have already started preparing the data for analysis. As you begin, be aware of several important concepts, including differences between descriptive and inferential statistics, between small and large studies, and between top-down and bottom-up analysis. Once you understand these concepts, you will be ready to begin extracting, categorizing, and analyzing the data.

Descriptive vs. Inferential Statistics

When performing usability analysis, you will most often use descriptive statistics that help you to describe your sample based on the data. You might use summary statistics and graphics that include successes, failures, time on task, errors, preferences for layout, and so on. In this case, you are not trying to generalize to a larger population; however, you are likely to make recommendations about the product or process.

You are less likely to use inferential statistics in small usability studies. While inferential statistics allow you to infer findings to a larger population based on data collected from a sample of that population, doing so validly requires that your sample must represent an averaged view of the population and meet requirements of confidence. You are more likely to be able to use inferential statistics with larger studies.

Small vs. Large Studies

There is no specific number that is the threshold between small and large studies; however, discount usability studies that rely on three to five participants per participant profile would certainly be described as small. Many usability studies are directed toward single products and are designed using discount usability methods. However, with technology have come more choices for usability studies including unmoderated large studies.

For example, the US Department of Homeland Security attaches an automatic usability survey to many pages on their web site so that anyone who accesses the site has the option to participate. Suppose they attached a survey of three questions to the site, and after six months, they decide that it is time to analyze the responses collected. Suppose also that during that time, they had collected 1000 responses. With a study of that size, they would quite likely plan to use inferential statistics and generalize to the entire population using the web site. As another example, the usability research exemplar analyzed in Chapter 12 presents a sample of 32 students and calculates p value but does not generalize results.

Top-Down vs. Bottom-Up Analysis

You basically have two choices for how to approach your data analysis: top-down or bottom-up. In a top-down analysis, you begin with categories and place the data into them. To return to our example of the US Coast Guard web site, you might start with categories of usable web design such as the 10 usability heuristics detailed in Table 7.1. You could then analyze the data by reviewing them and placing each finding into one or more of the 10 categories represented by the heuristics. You could also research other heuristics, selecting them based on your goals for the research. The advantage of the top-down approach is that it is often faster to implement and less messy. The disadvantage is that by using pre-determined categories, you might miss something important in the data.

A bottom-up approach to data analysis is messier and somewhat slower, but it allows you to pay closer attention to new evidence that the data may be revealing and offers the potential for greater discovery. Affinity diagramming (as described below) is a very useful approach to bottom-up analysis.

Extracting the Data

Your first step in the analysis phase is to *extract the data from all sources into a common form*. When possible, you should have at least two people extract data from each source to ensure that you don't miss any items of importance. One effective method of data extraction is to review each data source (e.g., notes, video, text transcripts, etc.) and write brief statements of findings on sticky notes. This technique works well for those who think visually and benefit from physically manipulating the data. Spreadsheets also work well, as do software tools for mapping and affinity diagramming. Use any approaches that help you transcribe the data into a common format.

For example, you might be reviewing video of a participant interacting with the US Coast Guard site and notice that he stumbles on how to move from housing options to the cost of housing options. You note, "Can't find cost from housing options page." At this point, you have no notion about how this finding will fit into your analysis, and you want to let the data speak without imposing your conclusions. Later in this process, you begin to analyze what caused the failure to find the cost information, but first you want to be sure that you have extracted all of the data. You complete this extraction process for all of your data sources, and if, for example, the failure to find cost information emerges from the notes of two team members and three data sources, you might have six sticky notes that mention this one problem. Excellent. This means that multiple methods helped identify the problem and that triangulation indicates the strong reliability of this finding.

Diagramming, Categorizing, and Summarizing the Data

Once you have extracted all of your data into a common form, you should begin to categorize the findings. If you are using top-down analysis, you will begin to place the findings into pre-existing categories. If you are using bottom-up analysis, a technique called affinity diagramming will enable you to work from the data to your own categories. This method of analysis is particularly useful when you wish to let the data shape the results and when no clear categories are initially present. In other words, affinity diagramming is especially useful when you want your finding categories to emerge organically.

Let's return to the example of the US Coast Guard web site. After conducting a study that includes usability testing, heuristic analysis, and interviews, you find yourself with a large body of data in the form of audio and video from the usability tests, codes from usability tests, ratings from the heuristic analysis, and more audio/video from the interviews. Even with three to five participants, this is a large body of data. You must begin to make sense of it in some way because it seems like a jumble with no clear pattern.

Alone, or even better, as a team, begin to place the pieces of data (sticky notes in our case) into categories. With sticky notes, we suggest using a large table or a wall where the data can remain for a few days. Create category labels as you group sticky notes. If a piece of data appears to apply to two categories, make a copy of it for each category. Continue to categorize the data until clear categories emerge. If you have some data that don't fit into a category or that depart drastically from the other data, label them as such and set them aside. You will decide later what to do with them. Keep in mind that *you should be seeing both performance data (measures of how the participants performed, such as time on task and error rates) and preference data.*

When you are creating categories for the data, you are actually coding it. In Chapter 5, we discussed the importance of using inter-rater checking to be sure that categories are reliable. That is, one or more other persons use your codes to categorize a sample of data to check the reliability of the categories. If the categories that you develop for your study won't be used outside of the particular product or process that you are assessing, you probably don't need to take this extra step to confirm your coding, though doing so can help to improve reliability for any study. However, if you plan to generalize your findings to a larger population, you must check your coding.

As you identify your categories of data, you should also summarize your data with descriptive statistics and associated graphics. Consider such patterns as

- Percentage of successes on each task (based on your definition of success)
- Percentage of failures on each task (based on your definition of failure)
- Central tendencies, such as mean, mode, and median of time on task
- Ranges, such as range of time on task or range of degree of satisfaction
- Frequencies, such as frequency of hesitations or back-tracking
- Comparison of group statistics (if you have assessed more than one group)
- Comparison of product statistics (if you have assessed more than one product).

Once you have identified categories and summarized your data, you are ready to start looking at what the findings mean.

Finding Meaning and Reporting the Data

Having completed your study, you have established a rich range of categories and summarized your data from several statistical perspectives. Begin to consider what they tell you about the product or process and what your client most needs to know. Present your findings, explain them, and justify your methods.

How you report the results of a usability study can vary depending on your audience and purpose. Not all usability reporting is done formally, and priorities depend on audience and purpose. In fact, when usability studies are done as part of Agile development, you may work in short sprints, reporting your results quickly and informally and then moving on to do more testing. However, you might also report your results more formally in presentations, written reports, and articles. Regardless of how you report the results, you must do so in a way that helps your audience to understand the most important results quickly and establishes the credibility of your study.

Exercise 7.2 Conducting a Practice Usability Test (requires multiple people)

1. Using the same web site you used in Exercise 7.1, identify tasks you want participants to complete that will help you measure Nielsen's five characteristics of a usable interface.

2. Write a script for a test that takes no more than five minutes per participant.
3. Recruit three participants.
4. Run your three tests and record your data.
5. Analyze your data and prepare one PowerPoint slide with the results.
6. Present your results.

Is It RAD?

Remember that as you design your methods for a usability study, you should seek to produce research that is RAD—replicable, aggregable, and data-supported (Haswell 2005). In small studies, you will not be generalizing; however, keeping RAD principles in mind will make your study stronger. Your methodology must

- Be sufficiently defined so that others can repeat the study
- Be reported in sufficient detail that the data can be aggregated or combined with the results of other studies to build a body of data
- Be supported by the data, not simply be impressions or gut feelings of the researchers

Summary

A "usable" product is one with which users can accomplish their goals with minimal frustration and error. Nielsen defines five characteristics of usability:

- Learnability: How easy is it for users to accomplish basic tasks the first time they encounter the design?
- Efficiency: Once users have learned the design, how quickly can they perform tasks?
- Memorability: When users return to the design after a period of not using it, how easily can they re-establish proficiency?
- Errors: How many errors do users make, how severe are these errors, and how easily can they [users] recover from the errors?
- Satisfaction: How pleasant is it to use the design? (Nielsen, 2012)

A usability study might be constructed of multiple methods such as usability testing, heuristic evaluation, interviews, and questionnaires. "Usability testing" refers to collecting data from users as they interact in an authentic way with the product or process being studied. In a rigorous assessment, you should seek to collect data using several methods to see whether the findings correlate. If the findings do correlate, they are more reliable due to *triangulation*.

To support the rigor of your test, you will need a test plan that includes the following:

- Purpose and objectives
- Participant characteristics
- Scenarios, task lists, and script
- Data to be collected and evaluation scales
- Other methods in addition to usability testing
- Test environment, equipment, and logistics

- Team roles
- Forms and checklists
- Test protocol

In a top-down analysis of your data, you begin with categories and place the data into them. In a bottom-up analysis, you develop categories from the data. Affinity diagramming is a useful approach to bottom-up analysis.

Answer Key

Exercise 7.1

The answer to this exercise will be unique for each person who prepares it so there is no key to this exercise.

Exercise 7.2

The answer to this exercise will be unique for each person who prepares it so there is no key to this exercise.

References

Barnum, C. M. (2011). *Usability testing essentials: Ready, set … test!* Morgan Kaufmann.

Chapanis, A., Garner, W. R., & Morgan, C. T. (1949). *Applied experimental psychology: Human factors in engineering design.* John Wiley & Sons, Inc.

Gilbreth, F. B., & Gilbreth, L. M. (1919). *Applied motion study: A collection of papers on the efficient method to industrial preparedness.* McMillan.

Gothelf, J. (2013). *Lean UX.* O'Reilly.

Gould, J. D., & Lewis, C. (1985). Designing for usability: Key principles and what designers think. *Communications of the ACM, 28*(3), 300–311.

Haswell, R. H. (2005). NCTE/CCCC's recent war on scholarship. *Written Communication 22*(2), 198–223.

Nielsen, J. (1994). 10 Usability heuristics for user interface design. www.nngroup.com/articles/ten-usability-heuristics/

Nielsen, J. (2000). Why you only need to test with 5 users. www.nngroup.com/articles/why-you-only-need-to-test-with-5-users/

Nielsen, J. (2012). Usability 101: Introduction to usability. www.nngroup.com/articles/usability-101-introduction-to-usability/

Nielsen, J., & Landauer, T. K. (1993). A mathematical model of the finding of usability problems. *Proceedings of the INTERACT '93 and CHI '93 Conference on Human Factors in Computing Systems,* 206–213. ACM.

Nielsen, J., & Molich, R. (1990). Heuristic evaluation of user interfaces. *Proceedings of the SIGCHI Conference on Human Factor in Computing Systems,* 249–256. ACM.

Norman, D. A. (1988). *The psychology of everyday things.* Basic Books.

Quesenbery, W. (n.d.). Using the 5Es to understand users. www.wqusability.com/articles/getting-started.html

System Usability Scale. (2019). www.usability.gov/how-to-and-tools/methods/system-usability-scale.html

Part II
Exemplars and Analyses

8   Analyzing a Literature Review

Introduction

In Chapter 3, we explored literature reviews in terms of the roles they play in primary and secondary research, the purposes for preparing them, the skills that they require of the researcher, and a methodology for writing them. In this chapter, we will analyze a specific literature review to see how the authors have approached the task of exploring previous work on their topic. The chapter contains the full text of the literature review from Menno D. T. de Jong, Bingying Yang, and Joyce Karreman's article "The image of user instructions: Comparing users' expectations of and experiences with an official and a commercial software manual," which appeared in *Technical Communication* in 2017. It also includes a detailed commentary analyzing the literature review in terms of its purpose and audience, organization, use of various levels of detail in discussing sources, and handling of citations and references within the text.

Learning Objectives

After you have read this chapter, you should be able to:

- Analyze a literature review in terms of purpose and audience, organization, citations and references, and level of detail
- Apply your knowledge of these analytical techniques to preparing a review of the literature on your topic of interest

The Literature Review's Context

The literature review that we explore in this chapter examines the contributions of one book, 26 journal and magazine articles, a book chapter, and a conference paper. All were published between 1979 and 2015, and all have contributed to the body of knowledge about the use of software manuals and the image and credibility of those documents.

The first thing that you are likely to notice about this literature review is that it does not bear the section title *Literature Review*. In fact, the authors have chosen to split the review between the first two sections of their article. This approach is not unusual, and it is quite obvious from the article's second sentence that you are, in fact, reading a literature review. As we will see, the authors' approach here is straight-forward and wastes no time getting to the heart of the matter.

The Image of User Instructions: Comparing Users' Expectations of and Experiences with an Official and a Commercial Software Manual [Literature Review]

Menno D. T. de Jong, Bingying Yang, and Joyce Karreman

[This article was originally published in 2017 in *Technical Communication* 64(1), 38–49. Only the first two sections of the article that contain the literature review are reproduced here. The text of the entire article, including the list of references, appears in Chapter 9. Reprinted with the permission of the authors and of the Society for Technical Communication.]

Introduction

Although the field of technical communication has broadened considerably, providing user support remains one of the core tasks of technical communicators. Recent literature shows that the nature of user support is changing, with more attention to instructional videos (e.g., Swarts, 2012; Van der Meij & Van der Meij, 2013; Ten Hove & Van der Meij, 2015) and user forums (e.g., Frith, 2014; Swarts, 2015), but manuals and user guides are still most prevalent. For several decades, technical communication scholars and practitioners have worked on the optimization of written user instructions (Van der Meij, Karreman, & Steehouder, 2009). Despite these efforts, there is not always much optimism among practitioners about the extent to which manuals, user guides, and other types of user documentation are actually used (Rettig, 1991; Svenvold, 2015).

It is important to realize that the option to use a manual is one that has to compete with many alternatives people have, most notably exploring by themselves, asking other people for advice, and searching the Internet. Schriver (1997, p. 166), for instance, argued that "most people choose to read and to keep reading only when they believe there will be some benefit in doing so and only when they cannot get the same information in an easier way (for example, by asking someone else)." However, she also presented survey results showing that a large majority of consumers use manuals to some extent when they try out new functionality of products, although they seldom read them cover-to-cover (pp. 213–214), and that consumers see a clear instruction manual as an important asset of products (p. 223). She also found that participants quite often assign the blame to the manual when experiencing difficulties with a product (pp. 217–222), which, of course, is not a positive finding but neverthe-less suggests an important role of manuals from the users' perspectives.

These findings resonate in various earlier and later studies. Some evidence was found for the potential added value of manuals. Aubert, Trendel, and Ray (2009) showed in an experiment that pre-purchase exposure to a high-quality user manual positively affects product evaluation and purchase intention. In another experiment, Pedraz-Delhaes, Aljukhadar, and Sénécal (2010) found that users' evaluation of the documentation affects their evaluations of the product and the company behind the product. Wogalter, Vigilante, and Baneth (1998) focused on the context of reselling used consumer products and found that the availability of a manual would be an asset for selling used products, and that people are even willing to pay extra to have one. Van Loggem (2013) took an analytic approach and argued against the "persistent myth" that well-designed artifacts do not need documentation: In the case of intrinsic complexity, it is impossible that the user interface, no matter how well designed, will suffice to support all functionality.

Regarding the actual use of manuals, the available research led to varying results. Szlichcinski (1979), using a telephone system in a laboratory setting, found that the majority of the participants (83%) did not use the user instructions. Wright, Creighton, and Threlfall (1982), on the other hand, found that a majority of the consumers (66%) read at least some of the user instructions when using electric or non-electric products. Jansen and Balijon (2003) came up with even higher percentages: More than 70% of their participants indicated to always or often use the manual for products; only 8% reported never using manuals. Van Loggem (2014) provided an overview of earlier studies, with use percentages ranging between 57% and 96%, and presented new data for professionals and students that fell within this range (90% and 70%, respectively). Based on these and Schriver's results, it seems safe to assume that users at least occasionally refer to user instructions when working with products or tasks that are unfamiliar to them.

Other studies focused on determinants of using user instructions. In a study among senior users (age range 65–75), Tsai, Rogers, and Lee (2012) showed that user manuals play an important role for this age group, particularly for the purposes of better understanding the product, recalling forgotten functions, and preventing mistakes. In a comparative study, Lust, Showers, and Celuch (1992) showed that seniors use manuals significantly more than younger users. The aforementioned difference Van Loggem (2014) found between professionals and students seems to point in the same direction, as the difference between the two groups has an age dimension. Given the drastically changed media landscape, the difference between old and young users may have increased in recent years and may further increase in the future.

More intrinsically, Wright, Creighton, and Threlfal (1982) found that users' estimation of the complexity of operations strongly affects their reading intentions. In the same vein, Wiese, Sauer, and Rüttinger (2004) found that product complexity is the best predictor of manual use. Celuch, Lust, and Showers (1992) showed that prior experience and time considerations are variables distinguishing readers from nonreaders. Finally, Lust, Showers, and Celuch (1992) found a broader range of predictors, including people's general perceptions of manuals.

In the research reported in this article, we will explore people's perceptions of manuals more extensively by focusing on the source credibility of manuals. Our research was inspired by the observation that users may, to some extent, be reluctant to use the official manuals of software packages but at the same time be willing to pay for a commercial manual for their software (cf. Van Loggem, 2013). One can think of the "For Dummies" series (e.g., *Excel for Dummies*), the "Bible" series (e.g., *Excel 2016 Bible*) published by John Wiley, or the "Missing Manuals" series (e.g., *Excel 2013. The Missing Manual*) published by O'Reilly. This seems to indicate that the source of the manual (official versus commercial) plays a role in users' views of manuals. So far, the technical communication literature has not addressed how users perceive such source differences. Coney and Chatfield (1996) conducted analytical research into the differences between both types of manuals and proposed that "the determining factor in the appeal of third-party manuals

[...] is the rhetorical relationship between the authors and their audience" (p. 24). Investigating users' expectations and experiences with official and commercial manuals may shed light on user perceptions.

In two separate experiments, we investigated the effects of source (official versus commercial manual) on users' expectations and experiences. In the first experiment, we focused on the expectations users have when they are confronted with an official or a commercial manual. The second study was a 2 x 2 experiment, in which we manipulated the source of the information and the actual content to investigate the effects of perceived source on the experiences of users. The two studies were approved by the IRB of the University of Twente.

Image, Source Credibility, and User Instructions

It is very common that we form mental images of phenomena we encounter. We use such mental images to simplify and make sense of the world we live in, for instance, when making behavioral decisions. Images can be based on prior experiences, hearsay, or associations, and can have varying degrees of elaboration—from overall impressions (low), to a number of specific beliefs that lead to an overall attitude (medium), to a complex network of meanings (high) (Poiesz, 1989). In practice, we can form images at various levels. Hsieh, Pan, and Setiono (2004), for instance, distinguish between product, corporate, and country image, which may simultaneously affect purchase behavior. The assumption in our study is that software manuals will have a certain image among users, but that it may be fruitful to differentiate between official and commercial manuals.

This is related to the literature about the effects of source credibility. Research shows that source credibility plays an important role in the way people handle information. In the context of persuasive communication, research suggests that people may be more effectively convinced or persuaded when the source of information is perceived to be credible (Pornpitakpan, 2004). Johnston and Warkentin (2010) showed positive effects of source credibility on people's intentions to follow recommended IT activities—a context that is in fact closely related to that of user instructions. Other studies found that source credibility may affect people's willingness to expose themselves to information. For instance, Knobloch-Westerwick, Mothes, Johnson, Westerwick, and Donsbach (2015) showed that source credibility, operationalized as the difference between official institutions without self interest in the issue at hand versus personal bloggers, affects people's time spent on Internet messages about political issues. Winter and Kramer (2014) showed that source credibility positively affects people's decisions about which Internet sources they will read and to what extent they will read them. In their study, source credibility was manipulated by source reputation as well as by the ratings of others.

The effects of source credibility may be further explored by connecting them to the broader concept of trust (cf. Mayer, Davis, & Schoorman, 1995). Two important factors that are distinguished are competence—in the case of user manuals, this boils down to software expertise and technical communication competencies—and benevolence—perceived willingness to serve the users and their needs.

Commentary

Put briefly, the literature review in de Jong, Yang, and Karreman's article is a model literature review within a larger research study. Although the scope is modest, their treatment offers appropriately detailed and comprehensive coverage of the literature on its subject, perhaps more so than would be possible for some topics.

Their research topic, the mental images that users form of official and commercial software manuals and the credibility of the sources that produce them, is an important and extremely useful one. Practitioners know that many software users never read the instructional material provided in a manual or user guide. In fact, within the technical communication community, the assumption that many—perhaps most—users will not read the manual is commonplace. In response, de Jong and colleagues have set about to identify the reasons for this reluctance.

Purpose and Audience

Because the literature review is divided into two parts, the statement of purpose for the review is also bifurcated. The first paragraph's final sentence states the purpose of this review of literature on the use of user documentation: "there is not always much optimism among practitioners about the extent to which manuals, user guides, and other types of user documentation are actually used ...." The second purpose of the literature review is stated in the *Image, Source Credibility, and User Instructions* section. The first sentence of that section focuses on mental images: "It is very common that we form mental images of phenomena we encounter." The first sentence of the second paragraph focuses on source credibility: "This is related to the literature about the effects of source credibility."

As with other articles, we can fairly assume that de Jong and colleagues' audience is the audience of the journal in which their article appeared. *Technical Communication* is published specifically for the members of the Society for Technical Communication (STC) and more generally for others with an interest in the field and in practice-based research. Because most STC members are technical communicators working in industry, we can assume that the authors here have directed their literature review to that practitioner audience. Indeed, the title of the article ("The image of user instructions: Comparing users' expectations of and experiences with an official and a commercial software manual") establishes its very practical nature, and the works that the authors address in this literature review are mostly practitioner-oriented rather than theoretical. Furthermore, the opening paragraph of the *Introduction* situates the literature review and the entire article within what is probably the central conundrum of user documentation: How do we know that anyone uses the instructional materials that we spend long hours preparing?

Organization

The organization of the literature review is quite straightforward.

Introduction

- User support options are changing (for example, videos) but manuals are still prevalent.
- Manuals must compete with many alternatives.
- Manuals can provide added value.
- Research has yielded varying results in terms of manual use.

- Seniors tend to use manuals more than younger people.
- Perception of product complexity is a factor.
- This article will focus on source credibility.

Image, Source Credibility, and User Instructions

- Users form mental images of official and commercial manuals.
- Source credibility and trust affect the ways that readers handle information.

The *Introduction* section's eight paragraphs establish the research problem that the authors are investigating.

- Paragraph 1 quickly reviews the changing nature of user support documentation by citing eight studies, but then shifts attention to print manuals, and notes the existential problem for technical communicators: Are our products actually used?
- Paragraph 2 focuses entirely on information reported in Karen Schriver's foundational book, *Dynamics in Document Design* (1997), addressing why and when users tend to read manuals.
- Paragraphs 3 through 6 examine the added value of documentation, results of research on actual use of documentation, use by particular groups of users, and the role of product complexity in users' decisions to read the manual. These four paragraphs consider a total of 16 sources.
- The final two paragraphs of the *Introduction* tell us the specific topic of the article. Paragraph 7 begins by announcing that "we will explore people's perceptions of manuals more extensively by focusing on the source credibility of manuals," and then explains the difference between official and commercial manuals. This paragraph cites two articles and concludes with an observation that is the underlying hypothesis of the article: "Investigating users' expectations and experiences with official and commercial manuals may shed light on user perceptions."
- Paragraph 8 concludes the *Introduction* by briefly describing the two experiments that are the central focus of the entire article. It cites no sources.

The article's second section, *Image, Source Credibility, and User Instructions*, consists of three paragraphs. Paragraph 9 establishes the importance of mental images in decision-making, mentions two studies, and concludes by stating that the current article assumes that users form images of software manuals but "that it may be fruitful to differentiate between official and commercial manuals." Paragraphs 10 and 11 deal with the effects of source credibility (four citations) and the connection between that credibility and trust (one citation).

Level of Detail

As we mentioned in the introduction to this commentary, authors seldom present a lot of detail about most of the works that they include in their literature reviews. The amount of previous research to be discussed will probably require that the author devote no more than a sentence or two to most of the works included.

The authors of this literature review follow this pattern to a significant extent. Of the 29 works cited here, only two receive much more than a mention or very brief summary. Paragraph 2 focuses exclusively on findings reported by Schriver (1997), including a quotation about why users read. Similarly, Paragraph 7 includes a summary of a study by Coney and Chatfield (1996), and also includes a quotation from that article. Otherwise, this literature review tends to describe the significance of most of the works cited in a phrase or a sentence, often citing multiple articles together, most notably in the following sentence from Paragraph 1.

> Recent literature shows that the nature of user support is changing, with more attention to instructional videos (e.g., Swarts, 2012; Van der Meij & Van der Meij, 2013; Ten Hove & Van der Meij, 2015) and user forums (e.g., Frith, 2014; Swarts, 2015), but manuals and user guides are still most prevalent.

And as the number of citations in most paragraphs of the literature review suggests, few sources get more than a sentence or two of comment. If we exclude the final paragraph, which cites no sources, and Paragraph 2, which cites only Schriver, the remaining nine paragraphs cite an average of 3.67 works each.

Let's consider how the highly detailed description of Schriver might be rewritten more concisely for inclusion in a more typical standalone bibliographic essay or a literature review that is part of a report on a qualitative research project on software manual use. Here is the original paragraph that describes Schriver's findings.

> It is important to realize that the option to use a manual is one that has to compete with many alternatives people have, most notably exploring by themselves, asking other people for advice, and searching the Internet. Schriver (1997, p. 166), for instance, argued that "most people choose to read and to keep reading only when they believe there will be some benefit in doing so and only when they cannot get the same information in an easier way (for example, by asking someone else)." However, she also presented survey results showing that a large majority of consumers use manuals to some extent when they try out new functionality of products, although they seldom read them cover-to-cover (pp. 213–214), and that consumers see a clear instruction manual as an important asset of products (p. 223). She also found that participants quite often assign the blame to the manual when experiencing difficulties with a product (pp. 217–222), which, of course, is not a positive finding but nevertheless suggests an important role of manuals from the users' perspectives.

Note that de Jong, Yang, and Karreman provide three more page citations after the citation for the direct quotation even though there are no more quotations from Schriver. They do so as a favor to readers so that they will not need to search hundreds of pages of Schriver's book to find details about the information that is being summarized in a single paragraph in this literature review.

If we eliminate most of the details from this paragraph, the essential information from de Jong and colleagues' 177-word description can be reduced to 91 words.

> Users have many options other than software manuals. They can explore on their own, ask others, or search the Internet. Schriver (1997) found that people read manuals "only when they cannot get the same information in an easier way" (p. 166). Nevertheless, she reports survey results showing that most users consult

manuals when trying out new product functions but seldom read them cover-to-cover (pp. 213–214). They also tend to think that a clear manual is an important product asset (p. 223) and blame the manual when having problems (pp. 217–222)

An even briefer version—only 48 words long—can be achieved by cutting the description to the bare facts.

Schriver (1997) reports that people read software manuals only when they lack an easier alternative. They often do consult manuals, however, when learning new product functions, and they think that clear manuals are a valuable product asset. They also tend to blame the manual when they have problems.

Finally, depending on the significance of the information in Schriver's book to the primary research being described in an article, the treatment might be even terser (25 words).

Schriver (1997), Van Loggem (2014), and Wright, Creighton, and Threlfall (1982) have explored the reasons why users choose to read and not read software documentation.

In this last example, all of the specifics from Schriver's book have been removed, and it has been consolidated with two other sources on the same general topic.

Citations and References

Many academic writers produce patchworks of quotations and details from the work of other authors, especially in literature reviews. Sometimes they don't seem to have understood what they have read, or if they have understood it, they don't convey that understanding to their readers. The audience addressed by de Jong and colleagues certainly cannot make that charge about this literature review.

This two-page review of the literature (as it was originally printed in the journal) contains only two significant quotations from the sources that the authors address, and both of those quotations are less than one sentence long. (A third direct quotation is only two words.) Instead of relying on quotation, de Jong and colleagues use paraphrase and (even more often) bare-bones summary to convey to their readers the essentials of what the literature has to say about users and software manuals.

Paragraph 3 illustrates the approach that the authors take in describing and analyzing their sources:

These findings resonate in various earlier and later studies. Some evidence was found for the potential added value of manuals. Aubert, Trendel, and Ray (2009) showed in an experiment that pre-purchase exposure to a high-quality user manual positively affects product evaluation and purchase intention. In another experiment, Pedraz-Delhaes, Aljukhadar, and Sénécal (2010) found that users' evaluation of the documentation affects their evaluations of the product and the company behind the product. Wogalter, Vigilante, and Baneth (1998) focused on the context of reselling used consumer products and found that the availability of a manual would be an asset for selling used products, and that people are even willing to pay extra to have one. Van Loggem (2013) took an analytic approach and argued against the "persistent myth" that well-designed artifacts do not need documentation: In the case of intrinsic complexity, it is impossible that the user interface, no matter how well designed, will suffice to support all functionality.

In the first sentence of Paragraph 3, the authors provide a link between Schriver's findings summarized in Paragraph 2 and the four sources that they will summarize in this paragraph. The remaining five sentences provide details from those four sources, relying mostly on summary, with a two-word quotation from Van Loggem's 2013 book chapter. (Note that the authors do not include a page citation for the two-word quotation from Van Loggem, the only flaw that we can find in this literature review.) Here's a brief version of that summary.

1. Schriver's findings are echoed in both earlier and later works.
2. There is some evidence that manuals add value.
3. Aubert, Trendel, and Ray showed that manuals can affect purchase decisions, and Pedraz-Delhaes, Aljukhadar, and Sénécal demonstrated that user perceptions of manuals affect perceptions of the product and the company.
4. Wogalter, Vigilante, and Baneth explored the role of manuals in reselling products.
5. Van Loggem analyzed the "persistent myth" that well-designed products do not need documentation and argued that product complexity disproves it.

Because they use quotations and close paraphrases so sparingly, most of de Jong, Yang, and Karreman's citations in the literature review are simply the article's or paper's author(s), and year of publication. When they do quote from or closely paraphrase the literature, they include a parenthetical page reference using the APA style that is standard in *Technical Communication*.

> Schriver (1997, p. 166), for instance, argued that "most people choose to read and to keep reading only when they believe there will be some benefit in doing so and only when they cannot get the same information in an easier way (for example, by asking someone else)."

Because de Jong and colleagues have so thoroughly mastered the literature, their readers will find it very easy to understand those works in the context of this discussion. Obviously, this is not a substitute for the detailed knowledge that an expert on the subject needs to attain, but it provides those who are already experts with a reminder of what they have previously read and gives novices an overview that will prepare them for their own exploration of the sources that they cover.

The list of references at the end of the article provides full bibliographical citations for each of the works that they cover in the literature review, as well as seven other sources that they reference in their Discussion section. (See Chapter 9 for the reference list.)

Exercise 8.1 Planning a Literature Review

Using the annotated bibliography on your research topic that you prepared in Exercise 3.1, construct an outline for the literature review for the paper or article that you are planning. Use the bulleted outline in the *Organization* section of this chapter as a starting point. Indicate the section and subsection headings of your literature review. Then for each section, list the books and articles that you would address in that section, followed by brief descriptions of the major points that you would want to include about each source.

Exercise 8.2 Planning Inclusion of a New Work in Updating a Literature Review

Suppose that you are Menno de Jong, Bingying Yang, and Joyce Karreman, and you have been asked to update your 2017 article for inclusion in an anthology on software instructional materials. As part of your updating, you have surveyed the work related to your topic that has been published since your article was completed. Here's the citation and the author's abstract of one of those sources that you've located.

Citation

Homburg, R. 2017. *The influence of company-produced and user-generated instructional videos on perceived credibility and usability.* [Unpublished master's thesis]. University of Twente.

Abstract

The traditional paper manual is losing our attention, while the amount of online instructional videos grows rapidly. Many companies notice this change and produce their own online instructional videos. At the same time, users themselves are making and sharing their own instructional videos on online social platforms such as YouTube. This user-generated content (UGC) is a popular source of information for other users. A combination of the two types of instructional videos also exists: companies co-operating with users through sponsoring. The idea behind this is that users helping users could be more effective than companies helping users. People's perception of source credibility could play a role in this, as no companies with ulterior motives are involved. This study investigates the differences in people's perception of source credibility of instructional videos by different sources, and tries to determine the role of trustworthiness, competence, and goodwill as determinants. It also looks into the effect of the sources on perceived usability of the product and the instructional video. Data was gathered in an experiment with three types of software tutorial videos. Results of the study indicate that there are generally no differences in the outcomes of users' credibility or usability assessments when the source differs, but the content of the instructions is equal. This shows that letting users provide the instructions could still be as effective as letting professionals do so. The only exception is the component goodwill: Users do perceive sources of the two user-generated videos as more caring than the company as a source, with the independently produced user-generated video scoring higher on goodwill than the sponsored user-generated video. The outcomes can help organizations in the design process of instructions: they can benefit from co-operation with users, as long as they make sure that the instructional design is sufficient.

Although it deals with official and user-generated instructional videos rather than with official and commercial software manuals, Homburg cites your article and uses some of your constructs (the quality of information, language and instructions, visual elements, perceived ease of use, and source preference) in his thesis.

Based on the content of the abstract, if you were de Jong, Yang, and Karreman, would you include this 29-page thesis in updating your literature review? If so, where would you include it, and what level of detail would it require?

Summary

This chapter has expanded on the concepts and methods presented in Chapter 3 by examining the literature review from an article about official and commercial software manuals. After providing the full text of the literature review, we have explored how its authors went about writing it, examining its purpose and audience, organization, level of detail, and citations and references.

For Further Study

Select one of the articles presented in Chapters 10–12, identify its literature review, and perform the same kind of analysis as the *Commentary* section of this chapter, examining its purpose and audience, organization, level of detail, and citations and references. What similarities do you see between the literature review presented in de Jong, Yang, and Karreman's article and that in the article that you have selected? How are the two literature reviews different? What might be reasons for those differences?

Answer Key

Exercise 8.1

Because the literature review outline produced for this exercise will be unique for each person who prepares it, there is no key for this exercise.

Exercise 8.2

Responses to this question will vary, but here are some possibilities.

- Include a reference to this thesis in Paragraph 1, along with the citations of Swarts, 2012; Van der Meij & Van der Meij, 2013; and Ten Hove & Van der Meij, 2015.
- Do not include this thesis because it does not add anything that the references to Swarts, 2012; Van der Meij & Van der Meij, 2013; and Ten Hove & Van der Meij, 2015 do not already cover, and this literature review is not comprehensive of all work on this tangential subject.

References

de Jong, M. D. T., Yang, B., & Karreman, J. (2017). The image of user instructions: Comparing users' expectations of and experiences with an official and a commercial software manual. *Technical Communication*, 64(1), 38–49.

Homburg, R. (2017). *The influence of company-produced and user-generated instructional videos on perceived credibility and usability* (Unpublished master's thesis). University of Twente. https://essay.utwente.nl/72074/1/Homburg_MA_BMS.pdf

9 Analyzing a Quantitative Research Report

Introduction

In Chapter 4, we explored quantitative studies in terms of their internal and external validity, as well as their reliability. We discussed hypothesis testing as a means of determining whether an intervention (also called an independent variable) makes a difference in terms of the results (or dependent variable) measured with two or more groups. We also considered descriptive statistics that provide information about a specific set of data, as well as inferential statistics that determine whether we can draw conclusions about a larger population based on the study sample. In this chapter, we will analyze a specific report about a quantitative research project to see how its authors have approached the task. This chapter contains the full text and a detailed commentary of Menno D. T. de Jong, Bingying Yang, and Joyce Karreman's "The image of user instructions: Comparing users' expectations of and experiences with an official and a commercial software manual," which appeared in the February 2017 issue of *Technical Communication*.

Learning Objectives

After you have read this chapter, you should be able to:

- Analyze a quantitative research report
- Apply the results of your analysis to reporting a quantitative study on your topic of interest

The Article's Context

As we noted in Chapter 4, hypothesis testing involving numeric averages is a common method in research because most people can easily relate to concepts and calculations involving averages. In the simplest type of hypothesis testing, the researcher formulates a hypothesis to test—for example, that an instructional video showing how to install a network router will enable an audience to complete that task more quickly than using the same instructions delivered through a printed manual. The researcher then tests the hypothesis by studying the performance of two groups of users. One group receives the instructional video version (the independent variable or intervention), and the other receives the manual. The task completion time for each person (the dependent variable) is recorded, and the group to which the person belongs is noted.

The researcher analyzes the results of the study using descriptive and inferential statistics. The descriptive statistics used include the mean time that each group took to assemble the product, as well as the standard deviation from the mean for each group. The researcher then knows whether there is a difference between the two groups in terms of task completion time. But the researcher also wants to know whether that difference is statistically significant. To make this determination, the researcher uses an inferential statistical tool that measures the probability that the difference is due to chance or participant selection rather than to the intervention—the instructional video.

The article that we examine in this chapter is a very good example of hypothesis testing. In this case, Menno de Jong and his colleagues conducted two studies. In the first, they wanted to determine whether users have different expectations of a paid-for, commercial user manual compared to the free manual supplied with the product. In the second, they wanted to discover whether the perceived source and the actual content of the instructions affected the users' performance of tasks and their judgments. The results of these two studies and the conclusions that the authors drew from those results are reported in their article.

The article is important because the first study demonstrated a statistically significant difference in users' images and expectations of commercial and official manuals. The second study showed that the users' perceptions of the source of a manual affected their performance of tasks. Because these are preliminary studies with relatively small numbers of student users, the findings cannot be generalized, but the results are intriguing and could be confirmed by further research.

The Image of User Instructions: Comparing Users' Expectations of and Experiences with an Official and a Commercial Software Manual

Menno D. T. de Jong, Bingying Yang, and Joyce Karreman

[This article was originally published in 2017 in *Technical Communication* 64(1), 38–49. Reprinted with the permission of the authors and of the Society for Technical Communication.]

Introduction

Although the field of technical communication has broadened considerably, providing user support remains one of the core tasks of technical communicators. Recent literature shows that the nature of user support is changing, with more attention to instructional videos (e.g., Swarts, 2012; Ten Hove & Van der Meij, 2015; Van der Meij & Van der Meij, 2013) and user forums (e.g., Frith, 2014; Swarts, 2015), but manuals and user guides are still most prevalent. For several decades, technical communication scholars and practitioners have worked on the optimization of written user instructions (Van der Meij, Karreman, & Steehouder, 2009). Despite these efforts, there is not always much optimism among practitioners about the extent to which manuals, user guides, and other types of user documentation are actually used (Rettig, 1991; Svenvold, 2015).

It is important to realize that the option to use a manual is one that has to compete with many alternatives people have, most notably exploring by themselves, asking other people for advice, and searching the Internet. Schriver (1997, p. 166), for instance, argued that "most people choose to read and to keep reading only when they believe there will be some benefit in doing so and only when they cannot get the same information in an easier way (for example, by asking someone else)." However, she also presented survey results showing that a large majority of consumers use manuals to some extent when they try out new functionality of products, although they seldom read them cover-to-cover (pp. 213–214), and that consumers see a clear instruction manual as an important asset of products (p. 223). She also found that participants quite often assign the blame to the manual when experiencing difficulties with a product (pp. 217–222), which, of course, is not a positive finding but nevertheless suggests an important role of manuals from the users' perspectives.

These findings resonate in various earlier and later studies. Some evidence was found for the potential added value of manuals. Aubert, Trendel, and Ray (2009), showed in an experiment that pre-purchase exposure to a high-quality user manual positively affects product evaluation and purchase intention. In another experiment, Pedraz-Delhaes, Aljukhadar, and Sénécal (2010) found that users' evaluation of the documentation affects their evaluations of the product and the company behind the product. Wogalter, Vigilante, and Baneth (1998) focused on the context of reselling used consumer products and found that the availability of a manual would be an asset for selling used products, and that people are even willing to pay extra to have one. Van Loggem (2013) took an analytic approach and argued against the "persistent myth" that well-designed artifacts do not need documentation: In the case of intrinsic complexity, it is impossible that the user interface, no matter how well designed, will suffice to support all functionality.

Regarding the actual use of manuals, the available research led to varying results. Szlichcinski (1979), using a telephone system in a laboratory setting, found that the majority of the participants (83%) did not use the user instructions. Wright, Creighton, and Threlfall (1982), on the other hand, found that a majority of the consumers (66%) read at least some of the user instructions when using electric or non-electric products. Jansen and Balijon (2002) came up with even higher percentages: More than 70% of their participants indicated to always or often use the manual for products; only 8% reported never using manuals. Van Loggem (2014) provided an overview of earlier studies, with use percentages ranging between 57% and 96%, and presented new data for professionals and students that fell within this range (90% and 70%, respectively). Based on these and Schriver's results, it seems safe to assume that users at least occasionally refer to user instructions when working with products or tasks that are unfamiliar to them.

Other studies focused on determinants of using user instructions. In a study among senior users (age range 65–75), Tsai, Rogers, and Lee (2012) showed that user manuals play an important role for this age group, particularly for the purposes of better understanding the product, recalling forgotten functions, and preventing mistakes. In a comparative study, Lust, Showers, and Celuch (1992) showed that seniors use manuals significantly more than younger users. The aforementioned difference Van Loggem (2014) found between professionals and students seems to point in the same direction, as the difference between the two groups has an age dimension. Given the drastically changed media landscape, the difference between old and young users may have increased in recent years and may further increase in the future.

More intrinsically, Wright et al. (1982) found that users' estimation of the complexity of operations strongly affects their reading intentions. In the same vein, Wiese, Sauer, and Rüttinger (2004) found that product complexity is the best predictor of manual use. Celuch, Lust, and Showers (1992) showed that prior experience and time considerations are variables distinguishing readers from nonreaders. Finally, Lust et al. (1992) found a broader range of predictors, including people's general perceptions of manuals.

In the research reported in this article, we will explore people's perceptions of manuals more extensively by focusing on the source credibility of manuals. Our research was inspired by the observation that users may, to some extent, be reluctant to use the official manuals of software packages but at the same time be willing to pay for a commercial manual for their software (cf. Van Loggem, 2013). One can think of the "For Dummies" series (e.g., *Excel for Dummies*), the "Bible" series (e.g., *Excel 2016 Bible*) published by John Wiley, or the "Missing Manuals" series (e.g., *Excel 2013: The Missing Manual*) published by O'Reilly. This seems to indicate that the source of the manual (official versus commercial) plays a role in users' views of manuals. So far, the technical communication literature has not addressed how users perceive such source differences. Coney and Chatfield (1996) conducted analytical research into the differences between both types of manuals and proposed that "the determining factor in the appeal of third-party manuals [...] is the rhetorical relationship between the authors and their audience" (p. 24). Investigating users' expectations and experiences with official and commercial manuals may shed light on user perceptions.

In two separate experiments, we investigated the effects of source (official versus commercial manual) on users' expectations and experiences. In the first experiment, we focused on the expectations users have when they are confronted with an official or a commercial manual. The second study was a 2 x 2 experiment, in which we manipulated the source of the information and the actual content to investigate the effects of perceived source on the experiences of users. The two studies were approved by the IRB of the University of Twente.

Image, Source Credibility, and User Instructions

It is very common that we form mental images of phenomena we encounter. We use such mental images to simplify and make sense of the world we live in, for instance, when making behavioral decisions. Images can be based on prior experiences, hearsay, or associations, and can have varying degrees of elaboration—from overall impressions (low), to a number of specific beliefs that lead to an overall attitude (medium), to a complex network of meanings (high) (Poiesz, 1989). In practice, we can form images at various levels. Hsieh, Pan, and Setiono (2004), for instance, distinguish between product, corporate, and country image, which may simultaneously affect purchase behavior. The assumption in our study is that software manuals will have a certain image among users, but that it may be fruitful to differentiate between official and commercial manuals.

This is related to the literature about the effects of source credibility. Research shows that source credibility plays an important role in the way people handle information. In the context of persuasive communication, research suggests that people may be more effectively convinced or persuaded when the source of information is perceived to be credible (Pornpitakpan, 2004). Johnston and Warkentin (2010) showed positive effects of source credibility on people's intentions to follow recommended IT activities—a context that is in fact closely related to that of user instructions. Other studies found that source credibility may affect people's willingness to expose themselves to information. For instance, Knobloch-Westerwick, Mothes, Johnson, Westerwick, and Donsbach (2015) showed that source credibility, operationalized as the difference between official institutions without self interest in the issue at hand versus personal bloggers, affects people's time spent on Internet messages about political issues. Winter and Kramer (2014) showed that source credibility positively affects people's decisions about which Internet sources they will read and to what extent they will read them. In their study, source credibility was manipulated by source reputation as well as by the ratings of others.

The effects of source credibility may be further explored by connecting them to the broader concept of trust (cf. Mayer, Davis, & Schoorman, 1995). Two important factors that are distinguished are competence—in the case of user manuals, this boils down to software expertise and technical communication competencies—and benevolence—perceived willingness to serve the users and their needs.

Study 1: Users' Expectations

Our first study aimed at investigating users' expectations of official and commercial manuals. To do so, an online experiment was designed, with two conditions (official versus commercial manual), in which participants had to answer questions about their expectations of the manual. The research focused on Microsoft Excel.

Method

PARTICIPANTS

The sample consisted of 69 students from the University of Twente. The students received participant credits for participating (students in the first years of our program are required to act as participants in a number of studies). They were randomly assigned to one of the two conditions. Table 9.1 provides the background characteristics of the two groups of participants. Differences between the two groups were tested, and there was no significant difference found for any of the background variables.

MANIPULATION

One group of participants was exposed to the official manual (*Microsoft Excel 2010 Official Offline Help Manual*), and the other group to a commercial instruction book (*Excel 2010 for Dummies*). The manipulation consisted of three images, presented side by side: the cover, the table of contents, and a random page of the instructions. Both versions were equally long, contained one color page and two black-and-white pages, and were representative for the look and feel of both types of manuals.

Table 9.1 Background characteristics of the two groups of participants (Study 1)

		Official manual	Commercial manual
Gender	Male	10	9
	Female	22	26
Age	M (SD)	19.3 (1.8)	20.2 (2.0)
Education level	Bachelor's student	25	30
	Master's student	7	5
Experience with Excel	Yes	31	33
	No	1	2
Estimated Excel skills	Beginning user	27	26
	Advanced user	4	7
	Expert user	0	0
Percentage of Excel functionalities used	M (SD)	18.5 (15.5)	20.2 (19.4)
Experience with manuals in general	Yes	5	2
	No	26	31

PROCEDURE

Data were collected using Qualtrics, a tool for online surveys and experiments. The first screen provided an introduction to the research. Participants were told that they would be exposed to a software program plus manual, and that their thoughts about the manual were the focus of the research. They were informed that they would not have to read the manual for the research. The second screen provided basic information about Excel, including a screenshot. On the third screen, the manual version was presented. After that, questions about the manual were asked, chunked on different screens.

INSTRUMENT

All constructs in the questionnaire were measured using five-point Likert scales. Above all sets of questions, the following overall instruction was given: "Imagining using this manual, what do you think the manual will be like?" Seven constructs were included in the research: connection to real-life problems, ease of locating information, expertise, language and instructions, layout, redundancy, and source preference. These constructs were meant to reflect different aspects of competence and benevolence, and proved to be statistically distinguishable in a factor analysis (with varimax rotation). Three constructs that were originally included in the questionnaire (assumptions about users, empathy, and quality of the information) were removed because they did not appear to be statistically distinguishable constructs.

The construct *connection to real-life tasks* involved the extent to which the manual was expected to support realistic tasks that people want to perform with the software. It was measured with two items (two other items were deleted based on the factor analysis) (Cronbach's alpha = .70). The two items were "This manual connects the functionality of Excel to real-world tasks of users," and "This manual focuses strongly on what users want to do with Excel."

The construct *ease of locating information* focused on participants' expectations of the findability of information in the manual. It was measured with four items (two items were deleted based on the factor analysis) (Cronbach's alpha = .81). Examples of items are "I will find the answers to my questions without much effort in this manual," and "This manual is clearly structured."

The construct *expertise* focused on the manual's writers' knowledge of the software, and was measured with four items (one item was deleted based on the factor analysis) (Cronbach's alpha = .77). Examples of items are "The authors of this manual are experts in using Excel," and "The authors of this manual know different solutions to achieve the same goal in Excel."

The construct *language and instructions* focused on the expected quality of the (textual and visual) instructions in the manual. It was measured with five items (three items were deleted based on the factor analysis) (Cronbach's alpha = .85). Examples of items are "The text of this manual is easy to understand," and "This manual contains figures and illustrations where necessary."

The construct *layout* involved participants' expectations of the visual appearance of the manual, and was measured with three items (two items were deleted based on the factor analysis) (Cronbach's alpha = .71). Examples of items are "The layout of this manual is user friendly," and "The layout of this manual is inviting."

The construct *redundancy* focused on participants' expectations of irrelevant information and wordiness of the manual. It was measured with four items (one item was deleted based on the factor analysis) (Cronbach's alpha = .78). Examples of items are "This manual contains a lot of information that is not relevant to users," and "This manual contains too much information."

The construct *source preference* involved the extent to which participants would prefer the manual over other possible sources, and was measured with three items (two items were deleted based on the factor analysis) (Cronbach's alpha = .79). Examples of (negatively worded) items are "I would prefer to use Google instead of using this manual," and "I would prefer to use the online help instead of using this manual."

Results

The data were analyzed using a multivariate analysis of variance, with the manual version as independent variable, and the seven expectation constructs as dependent variables. The first step in the analysis involves the multivariate test results, which focuses on the effects of the independent variable on the conglomerate of dependent variables. A significant multivariate test result is required before the univariate effects on separate dependent variables can be examined. This appeared to be the case (Wilks' lambda = .61, $F(7,59) = 5.48$, $p < .001$, partial $\eta^2 = .39$). Overall, the manual version had a strong effect on participants' expectations.

The second step focuses on the effects of the manual version on the seven dependent variables. Table 9.2 presents the mean scores in the two conditions and the univariate test results. As can be seen, participants had more positive expectations of the commercial manual for four of the seven dependent variables: connection to real-life tasks, language and instructions, layout, and source preference. The commercial manual was

Table 9.2 Participants' expectations regarding the manual

Constructs	Official manual M (SD)	Commercial manual M (SD)	Univariate test result
Connection to real-life tasks	3.3 (.7)	3.7 (.6)	$F(1,65) = 7.59, p < .01$, partial $\eta^2 = .11$
Ease of locating information	3.5 (.7)	3.7 (.7)	$F(1,65) = 1.06, p = .31$
Expertise	4.1 (.6)	3.8 (.7)	$F(1,65) = 3.57, p = .06$, partial $\eta^2 = .05$
Language and instructions	3.6 (.6)	4.1 (.5)	$F(1,65) = 10.04, p < .005$, partial $\eta^2 = .13$
Layout	3.3 (.7)	3.8 (.6)	$F(1,65) = 9.99, p < .005$, partial $\eta^2 = .13$
Redundancy	2.9 (.4)	3.2 (.8)	$F(1,65) = 2.73, p = .10$
Source preference	2.4 (.7)	3.1 (1.0)	$F(1,65) = 11.83, p < .005$, partial $\eta^2 = .15$

Note: Variables measured on five-point scales (1= negative; 5= positive).

expected to have a stronger connection to real-life tasks, to be more effective in language use and instructions, to have a more appealing layout, and to be a stronger competitor to other possible sources of information than the official manual. The effect sizes (as indicated by the partial η^2) refer to practically meaningful effects.

On the other hand, participants tended to have relatively high expectations of the writers' expertise in the case of the official manual, as compared to the commercial manual. They expected the writers of the official manual to know more about the Excel software. No differences were found regarding the expected ease of locating information and the amount of redundant information in the manual.

Among the seven dependent variables, source preference can be seen as an indicator of behavioral intentions, as it does not focus on specific aspects of a manual but involves participants' preferences for the manual compared to other information sources. To test the relationship between the six expected manual properties and source preference, we conducted a linear regression analysis. Surprisingly, the six constructs did not have any predictive value for the participants' scores on source preference ($R^2 = .01$, $F(6,60) = 1.05$, $p = .40$).

Conclusions

The results of the first study suggest that the commercial manual has a better image than the official manual. The findings acknowledge that the writers of the official manual are closer to the software, and thus may have more software expertise, but think that the writers of the commercial manual will do a better job providing them with the information they want. The added value of the commercial manual involves content (connection to real-life tasks) and formulation and visuals (language and instructions, and layout) but not the structuring (ease of locating information) and redundancy. In general, participants would be more willing to use the commercial manual compared to other possible sources than to use the official one compared to other sources (source preference).

Study 2: Users' Experiences

Our second study aimed at investigating users' experiences using an official or commercial manual. These experiences may be triggered by the actual content of the manual and by the perceived source. We therefore conducted a 2 x 2 experiment, with content (official versus commercial) and perceived source (official versus commercial) as independent variables (see Table 9.3). We included both official and commercial content in our experiment, because it is conceivable that the congruence between perceived source and actual content may affect users. Limiting our research to either official or commercial content would then lead to biased results. It must be stressed that a comparison of the effects of official and commercial content is beyond the scope of our research, as we cannot be sure of the representativeness of the specific combination of tasks and instructions for the complete manuals. The dependent variables involved both task performance (effectiveness and efficiency) and participants' judgments.

Method

PARTICIPANTS

The sample consisted of 83 students of the University of Twente. The participants either received participant credits or a small gift for their participation. They were randomly assigned to one of the four conditions. Table 9.4 shows the participants' background characteristics. Only one significant difference in background characteristics was found: Participants in the two conditions with the commercial manual content estimated the percentage of their usage of the Excel functionality significantly higher than participants in the two conditions with the official manual content did ($F(1,73)$ = 5.14, $p < .05$). However, this background variable appeared to have no significant correlation with any of the dependent variables.

MANIPULATION

The experimental materials were based on the same two manuals as used in the first experiment: the *Microsoft Excel 2010 Official Offline Help Manual* (official manual) and *Excel 2010 for Dummies* (commercial manual). We selected the content of both manuals that was relevant for the two tasks. We also added a section on conditional

Table 9.3 Experimental design

	Perceived Source	
Content	Official	Commercial
Official	1	2
Commercial	3	4

Table 9.4 Background characteristics of the four groups of participants (Study 2)

		Perceived official source		Perceived commercial source	
		Official content	Commercial content	Official content	Commercial content
Gender	Male	10	14	14	10
	Female	10	8	7	10
Age	M (SD)	23.3 (3.4)	22.1 (2.7)	22.6 (2.9)	22.5 (1.7)
Education level	Bachelor's student	14	15	14	17
	Master's student	6	7	6	4
Experience with Excel	Yes	18	21	19	19
	No	2	1	2	1
Estimated Excel skills	Beginning user	12	12	11	11
	Advanced user	5	9	8	7
	Expert user	1	0	0	1
Percentage of Excel functionalities used	M (SD)	16.4 (14.2)	29.4 (24.1)	18.4 (14.6)	24.0 (17.3)
Experience with manuals in general	Yes	1	2	7	4
	No	17	19	21	15

formatting to somewhat complicate the sub task of locating the right information. This amounted to 12–15 pages of text and images. Four versions of the manual were made by combining the communicated source (official versus commercial) and the content (official versus commercial). Text, images, layout, and structure were exactly the same as in the original manuals. Only one small layout change was made: In the version combining the content of the commercial manual and the source of the official manual (condition 3 in Table 9.3), the font of the headings was replaced with a more official font, to give it a more official look and feel.

PROCEDURE

This experiment was conducted in individual sessions in separate, quiet rooms at the university campus, in the presence of a facilitator. The participants were given two tasks with a fictitious "student data" file in Excel. The first task involved restructuring the entire file using the students' ages as the primary sorting criterion and their last name as the secondary. The second task involved making sure that the name columns (first and last name) and the headings row would always be visible, no matter how far you would scroll down or to the right (this is called freezing panes in Excel).

Participants were encouraged to use the manual when working on the tasks. They were not allowed to use other sources of information, such as online help or the Internet. The maximum time for completing the tasks was 30 minutes. The facilitator kept track of the time during the session. If participants were not able to finish the tasks within 30 minutes, they were asked to stop. After the task execution, participants filled out an online questionnaire regarding their experiences.

INSTRUMENT

Two task performance indicators were collected during the task execution: the number of correct tasks—participants could get 0, 1, or 2 points—and the time taken. Participants' effectiveness was measured using the number of correct tasks; their efficiency was based on their task-time ratio (the number of correctly executed tasks divided by the time taken in minutes).

For the participants' judgments, the same seven construct questionnaire was used as in Study 1, this time with Cronbach's alphas in the range of .61 and .83. One new construct was included, focusing on participants' overall experience using the manual (four items, Cronbach's alpha = .86). For this construct, participants were asked to rate their overall experience with the experimental manual using semantic differentials, such as "very positive" versus "very negative," or "very efficient" versus "very inefficient."

Results

TASK PERFORMANCE

Table 9.5 presents the results regarding the task performance of the participants with the four manual versions. As the two dependent variables are related, we used two separate univariate analyses of variance to test the results for significance. For effectiveness (number of correct tasks), two significant main effects were found. There was a significant effect of perceived source ($F(1,79) = 4.23$, $p < .05$, partial $\eta^2 = .05$): Participants who believed that the manual they used was a commercial one outperformed participants who thought they worked with the official manual. There was an opposite main effect of manual content ($F(1,79) = 4.23$, $p < .05$, partial $\eta^2 = .05$): Participants working with the official manual content outperformed the participants working with the commercial manual. As said earlier, this result must be treated with caution, as we cannot be sure of the representativeness of the official and commercial manual excerpts for the complete manuals. No interaction effect was found ($F(1,79) = .22$, $p = .43$).

Table 9.5 Task performance indicators (Study 2)

	Perceived official source		Perceived commercial source	
	Official content	Commercial content	Official content	Commercial content
Number of correct tasks	1.4 (.6)	1.3 (.7)	1.8 (.4)	1.4 (.6)
Task-time ratio	.09 (.09)	.07 (.06)	.14 (.06)	.13 (.11)

Note: Task-time ratio was calculated by dividing the number of correctly executed tasks by the time taken in minutes.

For efficiency (the task-time ratio), one significant result was found, regarding the perceived source of the manual (F(1,79) = 8.15, p < .01, partial η^2 = .09): Participants who thought to work with the commercial manual outperformed the participants who thought they had an official manual. No main effect for manual content (F(1,79) = .63, p = .43) and no interaction effect (F(1,79) = .06, p = .81) were found.

PARTICIPANTS' JUDGMENTS

The results of participants' judgments regarding the manual can be found in Table 9.6. In general terms, there were some remarkable differences and similarities compared to the expectation scores in Study 1. As shown by a series of t-tests comparing the scores of Study 1 and Study 2, the actual judgments were significantly lower than the expectations in Study 1 for the variables expertise (p < .001), language and instructions (p < .005), layout (p < .001), and redundancy (p < .01). The variables connection to real-life tasks, ease of locating information, and source preference had similar scores as those in Study 1.

The scores on overall satisfaction, being an overarching construct, were tested separately using a univariate analysis of variance. No significant effects were found of perceived source (F(1,79) = 2.03, p = .16) and manual content (F(1,79) = 1.84, p = .18), and no interaction effect was found (F(1,79) = .97, p = .33).

The scores on the seven remaining evaluation constructs were tested using a multivariate analysis of variance. The multivariate test results indicate a significant effect of manual content (Wilks' lambda = .83, F(7,73) = 2.15, p < .05, partial η^2 = .17). However, the univariate tests did not result in any significant difference between the two manual versions regarding the seven evaluation constructs. In the multivariate test, no significant effect for perceived source (Wilks' lambda = .94, F(7,59) = .72, p = .66) was found and no interaction effect was found (Wilks' lambda = .91, F(7,59) = 1.00, p = .44).

Table 9.6 Participants' judgments regarding the manual (Study 2)

	Perceived official source		Perceived commercial source	
	Official content	Commercial content	Official content	Commercial content
Overall experience	3.1 (.7)	3.1 (.8)	2.6 (.9)	3.1 (.8)
Connection to real-life tasks	3.6 (.7)	3.3 (.8)	3.6 (.7)	3.4 (.8)
Ease of locating information	3.5 (.9)	3.5 (.6)	3.7 (.8)	3.4 (.8)
Expertise	3.5 (.5)	3.6 (.4)	3.5 (.6)	3.5 (.5)
Language and instructions	3.5 (.7)	3.3 (.6)	3.8 (.6)	3.5 (.5)
Layout	2.9 (.8)	3.2 (.7)	3.1 (1.0)	2.8 (.7)
Redundancy	2.9 (.8)	2.7 (.7)	2.6 (1.0)	2.5 (.8)
Source preference	2.5 (.9)	2.5 (.9)	3.0 (.9)	2.5 (.8)

Note: Variables measured on five-point scales (1 = negative; 5 = positive).

Conclusions

The results of the second study partially confirm that the image of commercial manuals is better than that of official manuals. We did not find significant differences regarding participants' judgments of the manual versions. Both manual content and perceived source did not appear to matter. However, we found significant differences in the participants' performance: Participants performed significantly better, in terms of effectiveness and efficiency, when they believed the manual they used was a commercial manual.

Discussion

Main Findings

The two studies described in this article show that there may be image differences between official and commercial manuals. The first study showed that these image differences manifest themselves in users' expectations. Participants had significantly more positive expectations about several aspects of the commercial manuals. For one, they expected a better connection between the manual and real-life tasks. On the spectrum between a system orientation and a user orientation, they expected the commercial manual to be closer to the user. This is plausible, as the writers of commercial manuals can be seen as outsiders and expert users, instead of representatives of the company responsible for the software. This connects to the "sense of otherness," which Coney and Chatfield (1996) distinguish in commercial manuals, and which may be cultivated by a strong authorial voice in commercial instructions.

Participants also had more positive expectations about the language and instructions and the layout of the commercial manual. They expected the commercial manual to be more effective in providing instructions and to have a more inviting and user-friendly layout. This may be related to the fact that commercial writers are credited with authorship and may be expected to have a good reputation, whereas the writing team of official manuals remain anonymous. It may also have to do with assumptions about the importance attached to the quality of manuals in the two contexts, as distinguished by Carliner's (2012) business models. Commercial manuals are perceived as core products of the publishers, whereas official manuals may primarily be seen as supporting products for the software company.

Finally, participants would consider the commercial manual to be a more serious competitor to other sources of information than the official manual. This difference does not focus on specific aspects of the manuals but is related to behavioral decisions about using or not using the manual. It may have to do with quality expectations, but our regression analysis showed that participants' expectations regarding the other six constructs did not have any predictive value for their source preference score. It may therefore relate to a less rational inclination to prefer paid advice. A possible explanation for this finding may be found in a more general psychological mechanism, according to which people may be more inclined to use paid-for support than free support. Gino (2008) described this mechanism, which she called the "paid-advice effect," in a series of experiments. Participants could earn money by answering questions but were sometimes offered (either voluntary or obligatory) free or paid advice. Even though they were informed that the advice was a random selection of possible recommendations, they consistently used paid advice more often than free advice. Such behavior may be fueled by a desire to optimally benefit from invested money (this explanation would not be valid in our research, as the participants did not pay for either manual version) but may also be a general notion that paid advice must be better than free advice.

Study 2 showed that the image of the manual remains relevant when people are actually working with the instructions. Participants performed significantly better when they thought their instructions were based on the commercial manual. This is remarkable, as the content of the official manual in fact proved to be more effective for this particular set of assignments. Of course, the positive score of the content of the official manual should be treated with caution. The research was primarily designed to focus on image, not on content quality. The manual versions were only small excerpts from both manuals; we cannot say anything about their representativeness for the entire manuals, and the tasks the participants had to do were only two of a multitude of possible tasks. However, the discrepancy between the positive effects of a commercial source and the lower effectiveness of the commercial content, within the boundaries of this experiment, provides a clear indication of the image problem of official manuals and the favorable image of commercial manuals. This is the type of image that may become a self-fulfilling prophecy: The mere source effect of a commercial manual leads to more effectiveness in user performance, which in turn may contribute to its positive image.

The fact that we did not find any significant differences in the second study regarding participants' judgments is remarkable. The clear image differences that emerged from the first study disappear when participants are actively working with the manual. The significant effects that manual version had on actual performance was not reflected in the participants' judgments. The phenomenon that experimental research with manuals does result in significant differences in performance but fails to show significant differences in self-report measures, however, is not new. In their research into the effects of motivational elements in user instructions, Loorbach, Karreman, and Steehouder (2007, 2013) also found results combining significant effects on task performance and no effects on the self-reported variables.

The same phenomenon can also be found in De Jong's (1998) research into the effects of pretesting and revising brochures: The revised versions appeared to be better in terms of comprehensibility or persuasiveness but did not show significant improvements in overall appreciation. An explanation for this lack of significant differences may lie in participants' cognitive workload: When concentrating on performing tasks and/or processing information, it may be too much for participants to also focus on a detailed evaluation of the document they use.

Limitations and Future Research

Of course, it should be noted that our findings regarding users' expectations and experiences are based on single experiments, with one particular software package, and one representative of the available gamut of commercial user guides. Future research could verify whether the effects found in our studies are generalizable to official and commercial software manuals in general.

In the first study, we could not convincingly connect the image findings to participants' intentions to use the manual. Future research that further explores the connection between image differences and use intentions would be interesting. Such research could also take the reverse route by offering participants the choice between an official and a commercial manual, and asking for a preference and motivations.

In the second study, only two specific tasks were selected, corresponding to 12–15 pages in the manuals. Our findings regarding the effects of perceived source may be expected to be robust, as these effects do not depend on the characteristics of the specific content used. Our findings regarding the effects of manual content, however, cannot be seen as generalizable, because the set of tasks and the selection of manual texts may not be representative. These findings merely served as reference point, indicating that the official manual content, at the very least, was not worse than the commercial content.

Future research may also aim to shed light on the underlying mechanism of the differences found in our study. Is the paying an important factor? Or the authorship? Or the publisher or book series? Or a general sense of quality assurance? Experiments that systematically manipulate different versions of official and commercial manuals may help to further explore such factors.

Practical Implications

A limitation to the practical implications of our study is that official manuals will always be official manuals, and, for that matter, commercial manuals will always be commercial manuals. However, our overall finding that image matters, not only for the intention to actually use a manual but also for its effectiveness in use, may inspire manufacturers of products to start paying attention to the image aspects of manuals. It seems to be interesting to explore ways of improving the image of particular manuals, for instance by using quality marks or making usable documentation one of the explicit assets of a product. This connects to earlier discussions in the technical communication literature regarding the value added by technical documentation (Mead, 1998; Redish, 1995). Two prominent ways of reaching added value are enhancing the user experience and reducing costs. Image may affect both and is to date an underused phenomenon.

Conclusion

In sum, our research shows that the source of a manual (official versus commercial) matters for users. In our first study, we showed that the source affects users' expectations of a manual. Their expectations of commercial manuals are significantly more positive in several respects. In our second study, we showed that the source also affects users' task performance. Users are more effective when they think they work with a commercial manual. More research is needed to further explore these intriguing findings.

References

[This article appeared before the 7th edition of the *APA Publication Manual* was published, and the citations follow the style of the 6th edition.]

Aubert, B., Trendel, O., & Ray, D. (2009). The unexpected impact of user manual at the pre-purchase stage on product evaluation and purchase intention: An exploratory study. *Advances in Consumer Research, 36,* 944–946.

Carliner, S. (2012). Using business models to describe technical communication groups. *Technical Communication*, 59, 124–147.

Celuch, K. G., Lust, J. A., & Showers, L. S. (1992). Product owner manuals: An exploratory study of nonreaders versus readers. *Journal of Applied Social Psychology*, 22, 492–507.

Coney, M. B., & Chatfield, C. S. (1996). Rethinking the author-reader relationship in computer documentation. *ACM SIGDOC Asterisk Journal of Computer Documentation*, 20(2), 23–29.

De Jong, M. D. T. (1998), *Reader feedback in text design. Validity of the plus-minus method for the pretesting of public information brochures*. Dissertation University of Twente. Amsterdam, The Netherlands: Rodopi.

Frith, J. (2014). Forum moderation as technical communication: The social Web and employment opportunities for technical communicators. *Technical Communication*, 61, 173–184.

Gino, F. (2008). Do we listen to advice just because we paid for it? The impact of advice cost on its use. *Organizational Behavior and Human Decision Processes*, 107, 234–245.

Hsieh, M.-H., Pan, S.-L., & Setiono, R. (2004). Product-, corporate-, and country-image dimensions and purchase behavior: A multicountry analysis. *Journal of the Academy of Marketing Science*, 32, 251–270.

Jansen, C., & Balijon, S. (2002). How do people use instruction guides? *Document Design*, 3, 195–204.

Johnston, A., & Warkentin, M. (2010). The influence of perceived source credibility on end user attitudes and intentions to comply with recommended IT actions. *Journal of Organizational and End User Computing*, 22(3), 1–21.

Knobloch-Westerwick, S., Mothes, C., Johnson, B. K., Westerwick, A., & Donsbach, W. (2015). Political online information searching in Germany and the United States: Confirmation bias, source credibility, and attitude impacts. *Journal of Communication*, 65, 489–511.

Loorbach, N., Karreman, J., & Steehouder, M. (2007). Adding motivational elements to an instruction manual for seniors: Effects on usability and motivation. *Technical Communication*, 54, 343–358.

Loorbach, N., Karreman, J., & Steehouder, M. (2013). Verification steps and personal stories in an instruction manual for seniors: Effects on confidence, motivation, and usability. *IEEE Transactions on Professional Communication*, 56, 294–312.

Lust, J. A., Showers, L. S., & Celuch, K. G. (1992). The use of product owner manuals: A comparison of older versus younger consumers. *Journal of Business and Psychology*, 6, 443–463.

Mayer, R. C., Davis, J. H., & Schoorman, F. D. (1995). An integrative model of organization trust. *Academy of Management Review*, 20, 709–734.

Mead, J. (1998). Measuring the value added by technical documentation: A review of research and practice. *Technical Communication*, 45, 353–379.

Pedraz-Delhaes, A., Aljukhadar, M., & Sénécal, S. (2010). The effects of document language quality on consumer perceptions and intentions. *Canadian Journal of Administrative Sciences*, 27, 363–375.

Poiesz, T. B. C. (1989). The image concept: Its place in consumer psychology. *Journal of Economic Psychology*, 10, 457–472.

Pornpitakpan, C. (2004). The persuasiveness of source credibility: A critical review of five decades' evidence. *Journal of Applied Social Psychology*, 34, 243–281.

Redish, J. (1995). Adding value as a professional technical communicator. *Technical Communication*, 42, 26–39.

Rettig, M. (1991). Nobody reads documentation. *Communications of the ACM*, 34(7), 19–24.

Schriver, K. A. (1997). *Dynamics in document design: Creating text for readers*. New York, NY: John Wiley.

Svenvold, M. (2015). Instructions not included: What the vanishing manual says about us. The disappearance of the instruction manual. *Popular Science*. Retrieved from: www.popsci.com/instructions-not-included

Swarts, J. (2012). New modes of help: Best practices for instructional video. *Technical Communication, 59*, 195–206.

Swarts, J. (2015). What user forums teach us about documentation and the value added by technical communicators. *Technical Communication, 62*, 19–28.

Szlichcinski, K. P. (1979). Telling people how things work. *Applied Ergonomics, 10*, 2–8.

Ten Hove, P., & Van der Meij, H. (2015). Like it or not. What characterizes YouTube's more popular instructional videos? *Technical Communication, 62*, 48–62.

Tsai, W.-C., Rogers, W. A., & Lee, C.-F. (2012). Older adults' motivations, patterns, and improvised strategies of using product manuals. *International Journal of Design, 6*(2), 55–65.

Van der Meij, H., Karreman, J., & Steehouder, M. (2009). Three decades of research and professional practice on printed software tutorials for novices. *Technical Communication, 56*, 256–292.

Van der Meij, H., & Van der Meij, J. (2013). Eight guidelines for the design of instructional videos for software training. *Technical Communication, 60*, 205–228.

Van Loggem, B. (2013). User documentation: The Cinderella of information systems. In A. Rocha, A. M. Correia, T. Wilson, & K. A. Stroetmann (Eds.), *Advances in information systems and technologies* (Vol. 206, pp. 167–177). Heidelberg, Germany: Springer.

Van Loggem, B. (2014). 'Nobody reads the documentation': True or not? In *Proceedings of ISIC: The Information Behaviour Conference, Leeds, 2–5 September, 2014, Part 1*. Retrieved from www.informationr.net/ir/19-4/isic/isic03.html#.VmAs9k2hdaQ

Wiese, B. S., Sauer, J., & Rüttinger, B. (2004). Consumers' use of written product information. *Ergonomics, 47*, 1180–1194.

Winter, S., & Kramer, N. (2014). A question of credibility—Effects of source cues and recommendations on information selection on news sites and blogs. *Communications, 39*, 435–465.

Wogalter, M. S., Vigilante, W. J., & Baneth, R. C. (1998). Availability of operator manuals for used consumer products. *Applied Ergonomics, 29*, 193–200.

Wright, P., Creighton, P., & Threlfall, S. M. (1982). Some factors determining when instructions will be read. *Ergonomics, 25*, 225–237.

Commentary

We selected this article by de Jong and colleagues for three primary reasons. First, it is a brief, straightforward example of an article reporting the results of hypothesis testing. There are no significant problems with the design or execution of the two studies, nor with the analysis or reporting of the results.

Second, although the article reports the results of hypothesis testing, the complex designs of the two experiments rule out the use of the *t*-test for most of the analysis. Instead, it uses univariate and multivariate analysis of variance (ANOVA and MANOVA) rather than the *t*-test. It also uses another statistical measure, Cronbach's alpha, the most widely used measure of reliability. We hope that by seeing an article using different statistical tests, you will see that the principles of critically evaluating a quantitative study remain the same.

Finally, we chose this article because it deals with a topic that is relevant to so many technical communicators—user perceptions of official and commercial manuals and user expectations and experiences.

Purpose and Audience

The purpose of this article is to explore the authors' research question, stated in the next-to-last paragraph of the *Introduction* section.

In the research reported in this article, we will explore people's perceptions of manuals more extensively by focusing on the source credibility of manuals. Our research was inspired by the observation that users may, to some extent, be reluctant to use the official manuals of software packages but at the same time be willing to pay for a commercial manual for their software . . . .

Because the article was published in *Technical Communication*, we can safely assume that its primary audience is technical communication practitioners, with a secondary audience of technical communication teachers and researchers. The studies are exploratory and cannot be generalized beyond the bounds of their design. They were conducted with a population of university students, the experiments addressed only a single product, and they utilized limited content from the official manual for that product and from one commercial manual. Nevertheless, technical communication practitioners will be interested in the findings because they address a common concern about use and user perceptions of value of the software documentation that so many technical communicators produce on the job.

Organization

Because it reports on two related studies, the organization of this article is a variation on the standard IMRAD format for research reports: Introduction, Methods, Results, and Discussion.

* **Introduction:** As we saw in Chapter 8, the article's first two sections review the literature, providing the context for the two experiments that they designed and conducted. We won't consider these two sections further in this chapter.
* **Methods and Results 1:** In the *Study 1: Users' Expectations* section, the authors describe their experimental method (the participants, manipulation or intervention, procedure, and instrument), results, and conclusions for the first experiment.
* **Methods and Results 2:** In the *Study 2: Users' Experiences* section, the authors describe their experimental method (the participants, manipulation or intervention, procedure, and instrument), results, and conclusions for the second experiment.
* **Discussion:** The *Discussion* section reviews the main findings, the limitations of the two studies, the potential for further research, and the practical implications of the two studies. A brief conclusion summarizes the implications of the two experiments for technical communication practitioners.

We should note that although the authors and we ourselves refer to the studies described here in experimental terms, they are actually quasi-experimental. Unlike a true experiment, the authors do not state a hypothesis for either study that they wish to validate or refute.

The IMRAD format is commonly used by authors of quantitative research papers and articles in medicine, the physical sciences, and engineering, as well as in the social sciences. That de Jong and his co-authors use this format is not surprising. In fact, using an adapted IMRAD structure gives the article ethos or credibility by observing the conventions for presenting quantitative research results in a form that readers will understand and even expect.

Design of the Two Studies

You'll recall that one of our examples in Chapter 4 proposed to examine how incorporating the results of usability testing of a web site reduced the mean registration time between users of the original design and users of the revised design. We formulated a null hypothesis ("The mean registration time will not be reduced between the original web site design and a second design that resulted from applying information learned during usability testing"), collected data with users of both versions of the site, calculated the mean for both sets of users, and tested the difference between the means to determine whether it was statistically significant. In this article, de Jong and colleagues took a similar approach in the research that they report, but the differences between our simple example in Chapter 4 and the "real world" example reported in the article that we are examining here are instructive.

Hypotheses

In Chapter 4, we explained that you should begin a research project by formulating a hypothesis based on your research question that states the independent variable (the intervention that you will test), the dependent variable (the result that you will measure), and the predicted direction of the change you expect to see ("mean registration time ... will be reduced," for example). We also noted that the hypothesis is usually stated in the null form: "The mean registration time will not be reduced between the original web site design and a second design that resulted from applying information learned during usability testing." This hypothesis is often stated explicitly in the research report.

Although it is important for researchers to understand that statistical reliability comes from testing the null hypothesis, many published reports do not explicitly state the test hypotheses in either original or null form. One reason is that published reports must speak somewhat plainly, especially when writing to an audience not familiar with the conventions of research design. Such is the case with the article by de Jong and his colleagues. Indeed, although this article describes two experiments and the testing of multiple hypotheses, the hypotheses are never stated explicitly. Instead, the last paragraph of the *Introduction* section includes the following implicit description of the experiments' hypotheses:

> In two separate experiments, we investigated the effects of source (official versus commercial manual) on users' expectations and experiences. In the first experiment, we focused on the expectations users have when they are confronted with an official or a commercial manual. The second study was a 2 x 2 experiment, in which we manipulated the source of the information and the actual content to investigate the effects of perceived source on the experiences of users.

Explicit versions of these implicit hypotheses might be the following.

H_1: Users have greater expectations of commercial manuals than of official manuals.

H_2: Users' perceptions that they are using a commercial manual positively affect their experience.

Of course, these implicit original hypotheses are not in null form. The null versions would be the following.

H_{0-1}: There is no difference in users' expectations of official and commercial manuals.

H_{0-2}: The manual's source makes no difference in users' experience.

The fact that the null hypotheses are not explicitly stated is not a flaw in the article. Many research reports omit the formal hypothesis statement, and including it here might have negatively affected the tone and readability of the article for its intended audience of practitioners. Moreover, it's most unlikely that a reasonably intelligent reader would fail to infer them from the context. Still, there would be no question that readers would understand exactly what the authors were testing if they had stated their hypotheses explicitly.

We should note that H_1 could be divided into seven hypotheses to reflect the seven expectation constructs—the dependent variables—in the questionnaire that participants completed regarding the version of the manual that they were presented with. For example, the original hypothesis and its null version for the first expectation construct would read as follows,

H_{1a}: Users have greater expectations of commercial manuals than of official manuals regarding the manuals' support for realistic tasks that they want to perform.

H_{0-1a}: There is no difference in users' expectations of official and commercial manuals regarding the manuals' support for realistic tasks that they want to perform.

Similarly, H_2 could be divided into two hypotheses to reflect users' ability to perform tasks correctly and their opinion of the manual that they used.

H_{2a}: Users' perceptions that they are using a commercial manual positively affect the number of tasks that they perform correctly.

H_{0-2a}: Users' perceptions of a manual's source do not affect the number of tasks that they perform correctly.

H_{2b}: Users' perceptions that they are using a commercial manual positively affect their opinion of the manual.

H_{0-2b}: Users' perceptions of a manual's source do not affect their opinion of the manual.

Tests

STUDY 1

The hypotheses for the seven expectation constructs were tested by presenting the control group with three side-by-side images from the official manual and presenting the experimental group with three corresponding images from the commercial manual. Participants in each group were asked to respond to a series of questions that were scored using a 5-point Likert scale and that asked their impressions of the manual in terms of the expectation constructs based on the sample that they were given. Responses were recorded using a software tool, and the means and standard deviations for each of the two groups were calculated (see Table 9.2).

This test was straightforward, although it was more complex than our example in Chapter 4 because it included seven dependent variables. One criticism of the article is that the authors' descriptions of the various components would have been significantly improved by including figures illustrating the two groups of sample images. Such figures would have greatly assisted the reader in understanding exactly what was tested.

STUDY 2

In Study 2, the authors wanted to investigate how users' experiences with the official or commercial manual might be affected by their perception of the source. This experiment used a 2 x 2 design in which two groups used the official or the commercial manual to perform some basic tasks with Excel. The other two groups used either the commercial manual content, believing that it was from the official manual, or the official manual content, believing that it was from the commercial manual. They were given 30 minutes to complete two tasks. The means and standard deviations of the number of correct tasks and task-time ratios for each group were calculated (see Table 9.5). Participants in each group were then asked to respond to a post-test questionnaire scored using a 5-point Likert scale asking their impressions of the manual that they used in terms of the expectation constructs and their overall experience based on the manual that they were given. Responses were recorded using a software tool, and the means and standard deviations for each of the four groups were calculated (see Table 9.6).

Again, although more complex than our Chapter 4 example because it included four groups, each with a different condition, this test was straightforward. And again, the authors' descriptions of the test would have been significantly improved by including figures with samples from the four manuals to help the reader understand exactly what was tested.

Exercise 9.1 Formulating Hypothesis Tests

Using the same null hypotheses as de Jong and colleagues, suggest a different way of testing the hypotheses.

Sample Selection

You'll remember that we talked about samples of convenience in Chapter 4—populations to which the researchers have ready access. A sample of convenience (in this case, students at the authors' university) compromises the principle that every member of the general population should have an equal chance of being randomly selected for inclusion in the study. As a result, the researchers must either demonstrate how the sample is still representative of the general population or temper their findings with appropriate limitations.

In the first paragraphs of the *Method* sections of the descriptions of the two experiments, the authors describe the samples that they used for each experiment. For Study 1, 69 University of Twente students were recruited, and each was randomly assigned to one of the two conditions tested (samples from the official or the commercial manual). For Study 2, 83 students from the University of Twente were recruited, and each was assigned to one of the four conditions tested (the content of the official versus commercial manual as the actual source, and official versus commercial manual as perceived source). For Study 1, participants received participation credit (students at the University of Twente are required to participate in a certain number of studies). For Study 2, participants received either participation credit or a small gift. Information about the participant demographics and prior use of and familiarity with Microsoft Excel is provided in Table 9.1 for Study 1 and in Table 9.4 for Study 2.

The authors make no claims that the members of the two participant groups are representative of either the entire University of Twente student body or the general population of potential manual users. To begin with, the number of participants in both experiments is relatively small, and most seem to be drawn from bachelor's or master's students enrolled in a single department at the university. However, the studies reported in this article are preliminary and exploratory in nature, and the authors do not attempt to generalize the results beyond the small populations that they studied.

Exercise 9.2　Adapting Research Populations

Assume that you want to replicate the research that de Jong and colleagues report in this article using a professional rather than a student setting. How would you change their research design to accommodate this change in population? How would you change the population of your new study to make it more representative of the professional population and thus allow you to generalize the results?

Reports of Results

The results of the two experiments are reported in tables in the article, and explained and expanded upon in the text of the *Results* subsection for each study.

In the *Results* subsection for Study 1, de Jong and colleagues compare the responses of participants to two manual types (commercial and official, the independent variable) along seven dependent variables (the expectation constructs). Because the authors note that "A significant multivariate test result is required before the univariate effects on separate dependent variables can be examined," they first use a MANOVA to determine that "manual version had a strong effect on participants' expectations." Note that the p value for this multivariate test is less than 0.001 (less than one chance in 1000), an indication that the probability that the results were the result of random chance or participant selection is extremely low.

They then use a univariate ANOVA to examine how the manual version affected the seven dependent variables (the expectation constructs). Table 9.2 (Participants' Expectations Regarding the Manual) summarizes the results for Study 1. The mean scores show that participants had more positive expectations of the commercial manual for four of the seven dependent variables, and a more positive expectation of the official manual for one of the seven. There was no significant difference in expectations for the other two dependent variables. The p values in Table 9.2 show that the effects for five of the seven dependent variables were almost certainly not the result of coincidence or participant selection.

For the second experiment, de Jong and colleagues use the 2 x 2 design shown in Table 9.3. This research design involved two independent variables: actual content (official versus commercial) and perceived source (official versus commercial). In other words, the official manual's content was presented to one group in its original form and to a second group as though it were from a commercial manual. Similarly, the commercial manual's content was presented to one group in its original form and to a second group as though it were from an official manual. Specifically, the authors manipulated the content as follows.

> Four versions of the manual were made by combining the communicated source (official versus commercial) and the content (official versus commercial). Text, images, layout, and structure were exactly the same as in the original manuals. Only one small layout change was made: In the version combining the content of the commercial manual and the source of the official manual (condition 3 in Table 9.3), the font of the headings was replaced with a more official font, to give it a more official look and feel.

Unfortunately, the complexity of this aspect of the research design is not easy to describe in words. It would be far easier to understand if the authors had provided illustrations from the four manual versions. More precisely, it is not really clear exactly how the source was communicated, given the fact that the "Text, images, layout, and structure were exactly the same as in the original manuals." Was the source simply indicated to participants in the instructions given to them or on the sample pages themselves? Or were page headers altered? A few sample images would have significantly clarified the research design.

The research design also included two dependent variables: effectiveness (number of correct tasks) and efficiency (task-time ratio). These values were recorded, and means and standard deviations were calculated. Table 9.5 provides summarized data for task performance.

The authors used two separate ANOVAs and regression analyses to test the significance of the results for task performance. They found that "Participants who believed that the manual they used was a commercial one outperformed participants who thought they worked with the official manual" and that "Participants working with the official manual content outperformed the participants working with the commercial manual." The p value was less than 0.05 and the partial η^2 (comparable to the r^2 value in regression analysis described in Chapter 4) was 0.05, both indicators of low probability that random chance or participant selection was responsible for the results.

The authors also found a significant result for efficiency. "Participants who thought to work with the commercial manual outperformed the participants who thought they had an official manual." The p value was less than 0.01, and the partial η^2 was 0.09, again indicators of low probability that random chance or participant selection was responsible for the results.

Participants completed the same Likert scale-scored questionnaire used in Study 1, indicating their judgments of the manuals that they used. Study 2 also used the same expectation construct questionnaire as Study 1. The means and standard deviations for number of correct tasks and task-time ratio were recorded for each condition. Table 9.6 provides summarized data for Study 2 for participants' judgments for each of the expectation constructs and each condition. The authors compared the results for Study 2 with those from Study 1 using a series of *t*-tests and discovered that

> the actual judgments were significantly lower than the expectations in Study 1 for the variables expertise ($p < .001$), language and instructions ($p < .005$), layout ($p < .001$), and redundancy ($p < .01$). The variables' connection to real-life tasks, ease of locating information, and source preference had similar scores as those in Study 1.

The Study 2 participants' overall satisfaction scores were tested with a univariate ANOVA, but "No significant effects were found of perceived source ..." The other expectation constructs from Study 2 were tested with a MANOVA, but only one significant difference—for manual content—was found (p value was less than 0.05, and partial η^2 was 0.17).

Analysis

Based on the results that they report from their first experiment, de Jong and his colleagues conclude that users' expectations of official and commercial manuals are different, and that they have more positive expectations of commercial manuals than of those that are supplied with the product. In the second experiment, users performed better if they thought that the manual that they were using was a commercial manual, but there were no differences in their judgments of the manuals that they used.

We need to examine their analysis to ensure that it meets the standards of rigor for inferences based on measurements: the validity of the measurement and the reliability of the inference.

Validity

INTERNAL VALIDITY

You'll remember from Chapter 4 that internal validity in a quantitative study addresses the question "Did you measure the concept that you wanted to study?" The concepts that de Jong and his co-authors wanted to study were the differences in user expectations and user experiences between official and commercial software manuals.

The first experiment examined the concept of user expectations of a manual based on users' examinations of sample pages. The independent variable or intervention was the difference between two sources, official and commercial manual pages. There were seven dependent variables, the seven expectation constructs in the questionnaire that de Jong and his co-authors used. Because the experiment was carefully designed and the concept was validly operationalized (that is, the authors defined a valid means of measuring user expectations through the questionnaire), and because the questions asked the users about their expectations, it is clear that de Jong and colleagues did measure the concept that they wanted to study in Study 1.

The second study was more complicated. In this 2 x 2 experiment, the authors manipulated two independent variables: the perceived source of the instructions (official and commercial) and actual content (official and commercial). Performance on sample tasks and users' judgments about the sources of the instructions that they used were the dependent variables. Again, because the experiment was carefully designed and the concepts (performance and judgments about the sources of instructions) were validly operationalized, and because task performance and user judgments were clearly defined, it is obvious that de Jong and colleagues did measure the concepts that they wanted to study in Study 2.

EXTERNAL VALIDITY

As you will remember, external validity answers the question "Did what you measured in the test environment reflect what would be found in the real world?" We can manage external validity by taking care when we design our test that the conditions in the test environment match the conditions in the general environment as much as possible and that the sample group itself is a fair representation of the general population of interest.

Because de Jong and his co-authors make no claims about the generalizability of their findings, the question of external validity in this case is essentially moot. In the *Discussion* section, the authors state that their experiments provide only preliminary answers to their research questions and that further testing with more examples of different software and a larger sample of commercial manuals is needed. Nevertheless, their findings make us as readers think carefully about how users perceive our documentation products and how readily they will use them.

We looked earlier at the sample population for this study, and we remember from Chapter 4 that participant selection must be random. That is, the selection and assignment of test participants should be independent and equal. Selection is independent if selecting a participant to be a member of one group does not influence the selection of another participant. Selection is equal if every member of a population has an equal chance of being selected. The test population for de Jong and his colleagues was clearly a sample of convenience, and the members of that sample were not selected independently. Students self-selected whether to volunteer to participate in the experiments, and at least some of that self-selection was based on the requirement that they participate in a certain number of studies. Assignment of the volunteers to one or another of the experimental and control groups, however, was random and therefore equal.

Reliability

Again, you'll recall from Chapter 4 that the reliability of a study involves the likelihood that the results would be the same if the experiment were repeated, either with a different sample or by different researchers. The measure of the study's reliability is the statistical significance of the results. In other words, how confident can we be that the difference in performance between the control group and the experimental group is due to the intervention (the independent variable) introduced rather than to coincidence?

In inferential statistics, we judge the reliability of our conclusions about a set of data based on two principles.

- The smaller the variance in the data, the more reliable the inference.
- The bigger the sample size, the more reliable the inference.

When we examine the variance in data, we want to know the probability that the results were caused by the intervention and not by sampling error or coincidence. You'll recall that in Chapter 4 we observed that most research in our field is considered rigorous enough when results have a p value of 0.1 or lower; in other words, there is no more than 1 chance in 10 that the results were caused by coincidence or sampling error. For more conservative situations where harm could be done by accepting a false claim, then a p value of less than 0.05 or even 0.01 might be more appropriate; in other words, there are no more than 5 chances in 100 or 1 chance in 100 that the difference is not statistically significant.

The MANOVA test used in Study 1, "which focuses on the effects of the independent variable [the manual used] on the conglomerate of dependent variables [the seven expectation constructs]," yielded a p value of less than 0.001 (see the *Results* subsection for Study 1). This is a very high level of significance—that is, less than 1 chance in 1000 that the difference was the result of sampling error or coincidence. Similarly, five of the seven univariate ANOVAs in Study 1 resulted in acceptably low p values (see Table 9.2).

In terms of task performance in Study 2, the participants who thought that they were using the commercial manual completed more tasks successfully than those who thought that they were using the official manual, with a p value less than 0.05. However, those using content from the official manual completed more tasks successfully than those using the commercial manual, also with a p value less than 0.05. Participants who thought that they were using the commercial manual also had a lower task-time ratio than participants who thought that they were using the official manual, with a p value of less than 0.01. See the *Results* subsection for Study 2 for details.

In terms of participants' judgments in Study 2, there were no significant differences. However, a series of t-tests comparing results from both studies showed that

> the actual judgments were significantly lower than the expectations in Study 1 for the variables expertise ($p < .001$), language and instructions ($p < .005$), layout ($p < .001$), and redundancy ($p < .01$). The variables connection to real-life tasks, ease of locating information, and source preference had similar scores as those in Study 1.

See the *Results* subsection for Study 2 for details.

Another measure of reliability besides the p value is Cronbach's alpha, which is a number from 0 to 1, with higher values indicating a higher degree of consistency or reliability. The Cronbach's alpha measures that de Jong and colleagues calculated for the seven constructs tested in Study 1 range from 0.70 to 0.85, indicating a very high level of reliability. The Cronbach's alphas for the seven expectation constructs in Study 2 range from 0.61 to 0.83; the Cronbach's alpha for the eighth construct, overall user experience, measured only in Study 2, was 0.86. Again, these measures indicate a relatively high to very high level of reliability. See the *Instrument* subsections for both studies for details.

Aside from the lack of illustrations of pages from the sample manuals, if we can fault de Jong and his co-authors at all, it is that they don't really address the limitations imposed on their study by the convenience sample of participants. As noted earlier, convenience samples are often drawn from student populations, and these population aren't very diverse. Indeed, the demographic data reported in Tables 9.1 and 9.4 show that the participants in de Jong and his co-authors' studies are younger, better educated, more experienced with Microsoft Excel, and (for Study 1) more likely to be female than typical users, although the self-reported skill levels with the product are probably more in line with the general population.

As for sample size, the experimental and control groups comprised a total of 67 for Study 1 and 83 for Study 2. These samples are small, but since the results are not generalized and the research is preliminary and exploratory, the small samples are not a significant problem. However, further research should use much larger and significantly more diverse populations.

Conclusions

The final subsection of the article is brief but effective. The results of the first experiment indicate that users' expectations of a manual are affected by its source, with a preference for commercial manuals over official documentation. The second experiment's results indicate that users perform tasks better if they believe that they are using a commercial manual.

Finally, the researchers note that their exploratory experiments should be followed up by further studies. Although not explicitly mentioned here (although it is elsewhere in the article), future research should test more than one software application and more than one commercial manual for each. Not mentioned at all in the article is the problem of the population studied. Again, future studies would be most beneficial if they used participants that are more representative of typical users of software—more diverse in age, background, and experience.

Is It RAD?

As we saw in Chapter 2, RAD research as defined by Haswell (2005) must be replicable, aggregable, and data-supported. In other words, the methodology must be sufficiently defined that others can repeat the study. The results from the study must be reported in sufficient detail that they are capable of being aggregated or combined with the results of those repeated studies to build a body of data that can be compared and further built upon. And the conclusions of the study must be supported by those data, and not be simply the impressions or gut feelings of the researchers.

It is clear from our analysis that de Jong, Yang, and Karreman's article meets these requirements. The methodologies of both experiments are described in enough detail to allow them to be replicated, though illustrations from the manual pages used in both experiments would have made repeating the experiments easier and more precise. The data are reported in great detail, enabling the data to be aggregated with data collected in repetitions of the experiments. And finally, the conclusions drawn in the article are based entirely on the data reported. As a result, we can conclude with confidence that this is a RAD article.

Summary

This chapter has expanded on the concepts and methods presented in Chapter 4 by examining an example of a quantitative research report. Following the full text of the sample article, we have explored its purpose and audience, its structure, the study design, the results, and the inferences drawn from those results. We have seen that the quantitative study conducted by de Jong and his co-authors meets the requirements of internal validity because it operationalizes the concepts that it wants to examine.

Their two studies also meet the requirements of external validity because the test environment, excluding the sample population studied, is sufficiently like the "real world" environment. Finally, the inferences that the authors have drawn from their results are reliable because of the very low probability of error.

For Further Study

Select one of the quantitative research articles that you have collected as part of the annotated bibliography that you prepared for Exercise 3.1 or any other quantitative research article. Perform the same kind of analysis of that article as the *Commentary* section of this chapter, examining the article's purpose and audience, organization, study design, report of results, analysis, and conclusions.

1.   What similarities do you see between the approach used in de Jong, Yang, and Karreman's article and that in the article that you have selected?
2.   How are the two quantitative analysis articles different?
3.   What might be reasons for those differences?

Answer Key

Exercise 9.1

The answer to this exercise will be unique for each person who prepares it, so there is no key to this exercise.

Exercise 9.2

The answer to this exercise will be unique for each person who prepares it, so there is no key to this exercise.

References

de Jong, M. D. T., Yang, B., & Karreman, J. (2017). The image of user instructions: Comparing users' expectations of and experiences with an official and a commercial software manual. *Technical Communication, 64*(1), 38–49.

Haswell, R. H. (2005). NCTE/CCCC's recent war on scholarship. *Written Communication, 22* (2), 198–223.

10 Analyzing a Qualitative Research Report

Introduction

In Chapter 5, we discussed qualitative research methods such as interviews, focus groups, usability tests, field observations, and document analysis—techniques that draw on research methods frequently used in the social sciences. We noted that these qualitative methods typically involve three phases: observing behavior, recording data, and analyzing the data. We defined the standards of rigor and ways of ensuring rigor in qualitative studies, and we also described coding and categorization schemes used to record and analyze qualitative data.

In this chapter, we will examine a sample article reporting the results of qualitative research. The chapter contains the full text of Jessica Smith and Tom van Ierland's "Framing controversy on social media: #NoDAPL and the debate about the Dakota Access Pipeline on Twitter," which appeared in the September 2018 issue of the *IEEE Transactions on Professional Communication*, as well as a detailed commentary about the article.

Learning Objectives

After you have read this chapter, you should be able to:

- Analyze a qualitative research report
- Apply the results of your analysis to reporting a qualitative study on your topic of interest

The Article's Context

Because qualitative research methods are so commonly used in the social sciences, it's not surprising that Smith and van Ierland's article is in many ways a sociological study of attempts to influence public opinion about an extremely controversial oil pipeline. The study investigates the content of influential tweets that used the #NoDAPL hashtag during two Twitter outbreaks associated with executive orders issued by two US presidents in December 2016 and January 2017.

The article is significant for several reasons. It is one of a relatively small number of qualitative research reports that combines a focus on social media, public affairs, and ethics. And "Framing controversy on social media" is especially noteworthy because it received the IEEE Professional Communication Society's Rudoph J. Joenk Award for Best Paper in the *Transactions* for 2018.

The article is timely because it reflects some of the principal concerns with environmental, political, and social issues that affect engineers and professional communicators today. It analyzes 140 influential tweets using a framing typology for scientific and environmental debates to determine how the frames employed in those tweets influenced public debate of those issues. And by examining the tweets in the context of those frames, it attempts to determine how the frames contributed to polarizing the debate.

In other words, "Framing controversy on social media" is not only a really interesting example of an important type of research report, it's also a really interesting article.

Framing Controversy on Social Media: #NoDAPL and the Debate about the Dakota Access Pipeline on Twitter

Jessica M. Smith and Tom van Ierland

[This article was originally published in 2018 in the *IEEE Transactions on Professional Communication* 61(3), 226–241. Reprinted with the permission of the Institute of Electrical and Electronics Engineers.]

In 2016, the Dakota Access pipeline (DAPL) sparked a vociferous activist movement and broad public debate in the US about oil and the environment, climate change, and Native American sovereignty. The debate was notable for the crucial role played by social media: Activists flocked to the Standing Rock Reservation in North Dakota to attempt to halt construction of the pipeline and used social media to share first-hand information from the protests, while social media users from around the world used these platforms to demonstrate their support for the protests. Indeed, tweets about the pipeline dominated Twitter each time a major political decision was made in the case of whether the $3.8 billion USD pipeline would be finished. On December 4, 2016, use of #NoDAPL, the preferred hashtag of the antipipeline activists, skyrocketed in response to news that the US Army Corps of Engineers had denied the easement for completion of the pipeline. Energy Transfer Partners, the company in charge of the pipeline's construction, claimed that each extra month of delay would cost the company more than $80 million USD and asked the court to allow building to continue until easement was granted. That day, 300,000 #NoDAPL tweets were posted, representing more than a third of all tweets made in the US on that date. By the end of the #NoDAPL controversy, total tweets soared into the millions.

Emerging as an increasingly more important platform for discussions about energy and other controversial projects, Twitter is changing the way that protests are organized and online debates unfold. People tweeting about these oil- and gas-related projects often use hashtags specifically generated for these projects, with #NoDAPL being the most notable of these. Even the proponents of the pipeline, including Republican groups that had made support for the oil and gas industry part of their party platform, used #NoDAPL rather than the more neutral hashtag #DAPL, we suspect, to make their views visible in the debate.

Throughout the controversy, the proponents and opponents who turned to Twitter empowered a larger debate that spilled into and shaped protests on the ground as well as coverage of the event in print and television news sources. For example, major outlets such as CNN covered events in which protesters around the country created #NoDAPL banners and signs, such as at the 2017 New Year's Day football game between the Minnesota Vikings and Chicago Bears. Examining the use of #NoDAPL over the course of the debate reveals a series of "outbreaks" that accompanied major activities or political decisions. After the first outbreak in December when the easement was denied during the Obama administration, the second major one occurred on January 24, 2017, when, in his first days in office, President Donald Trump delivered on a campaign promise and signed an executive order allowing pipeline construction to proceed. The Tweets per Minute maximum increased from 300 on January 23 to 5500 on January 24. Though each decision sparked an outbreak, the ways in which the most popular tweets framed—or packaged and presented—information changed between the two administrations.

The debate over the pipeline provides a particularly dramatic example of engineering decisions and practices sparking a broader controversy and international debate, though we could point to many others such as hydraulic fracturing or "fracking," genetically modified organisms, self-driving cars, transportation infrastructure, and medical technologies. Public perception plays an increasingly powerful role in shaping the ability of companies to do their business, especially in the mining and energy industries [1]. Social media represents a key arena in which public perceptions of engineering-related controversies are formed, transformed, and publicized. Minimally, these channels provide crucial information for engineers, either as individuals or as employees of companies in the spotlight, to better understand public perceptions of their companies or industries, even if they themselves do not tweet or create posts. Going a step further, engineers who advise a company's public relations department or create social media content could assist their employers and professions by meaningfully engaging stakeholders and critics through social media [2]. In our own research, we have found that multiple oil and gas companies, for example, now encourage their employees to use their own social media sites to share information about their industry and respond to online debates.

To understand the context of their work—and perhaps even intervene constructively in public debates—engineers must be able to understand the perspectives of various stakeholders, including people opposed to their work. This paper provides one approach for how to do so by investigating how the #NoDAPL tweets that received large numbers of likes and retweets framed their posts. For example, we will show that influential tweets during the Obama administration hailed the halting of the pipeline as a triumph of social progress, while those during the Trump administration framed his executive order as a lack of public accountability.

Because framing shapes the interpretation of information, examining the frames present in the most popular tweets sheds light on which frames were the most compelling to social media users. Energy Transfer Partners and other petroleum advocacy organizations appealed to the safety of the pipeline, the necessity of its products, and the potential for the project to generate economic development—an approach that resonates with many engineers. Yet, these arguments failed to gain a foothold in the wider debate, likely because they did not engage the broader questions of morality that dominated antipipeline activism online and on the ground. If engineers are to effectively engage the public on controversial issues—or assist the externally facing people inside of their companies to do so—our research suggests that they must not simply rely on scientific defenses of the safety of their practices, but engage the frames being used by their critics.

Background

The Dakota Access pipeline is a feat of engineering, currently carrying crude oil from the Bakken shale play in northwestern North Dakota 1172 miles southeast to a storage hub in south central Illinois. It has the capacity to transport 570,000 barrels of oil per day. The oil is then sent to refineries on the East Coast by train or to areas in the Midwest and Gulf Coast through other pipelines. The project was estimated to cost $3.8 billion USD and was led by Energy Transfer Partners from Dallas, TX, with interests held by Philips 66 from Houston, TX, and Sunoco Logistics Partners from Philadelphia, PA. The US Army Corps of Engineers was also a key factor in the pipeline construction and ensuing debate, as the federal agency tasked with approving interstate pipelines and providing permits for water crossings, such as underneath Lake Oahe near the Standing Rock Sioux Reservation.

In April 2016, members of the Standing Rock Sioux Tribe began protesting in an attempt to halt construction of the pipeline, which was set to cross land that was promised to the tribe in the 1851 Treaty of Fort Laramie but never granted to them. It was also set to cross their primary source of drinking water, even though an alternative path closer to the state's capital of Bismarck was rejected because of its proximity to water supplies for that city. After a group of approximately 200 Native Americans rode horseback on April 1 to protest the proposed routing of the pipeline, online protests on social media quickly followed. #NoDAPL became the most prominent hashtag, with celebrities such as Leonardo Di Caprio, Shailene Woodley, and Mark Ruffalo declaring themselves opponents of the pipeline, and influential politicians such as Bernie Sanders, a frontrunner for the Democratic presidential nomination, following quickly. Native Americans from other tribes, college students from around the country, and other political activists joined the protests, creating a large camp at Standing Rock that used social media to share updates on the protests with a wider public.

The Standing Rock Sioux sued the Corps in July, arguing that the agency did not properly consult them before approving the section near their reservation. The protests gained worldwide attention on December 3 when guard dogs owned by private security companies attacked the protesters in North Dakota when they refused to leave the site of the pipeline. To protect the protesters, about 2000 US military veterans decided to join the protest camps on December 3, calling their group Veterans Stand for Standing Rock. The next day, the US Army Corps of Engineers, under direction from the Obama administration, denied the easement for the construction of the pipeline underneath Lake Oahe. This decision resulted in the largest #NoDAPL outbreak, when almost 300,000 tweets containing #NoDAPL were sent out in the US.

The pipeline became even more politicized as presidential candidate Donald Trump promised that he would guarantee that it would be finished if he were elected, even as the US Army Corps of Engineers said that it was investigating alternative routes for the final construction. Trump made good on his promise in his first days in office when, on January 24, 2017, he signed an executive order allowing construction of the Dakota Access pipeline and the Keystone XL pipeline to proceed. Critics of Trump

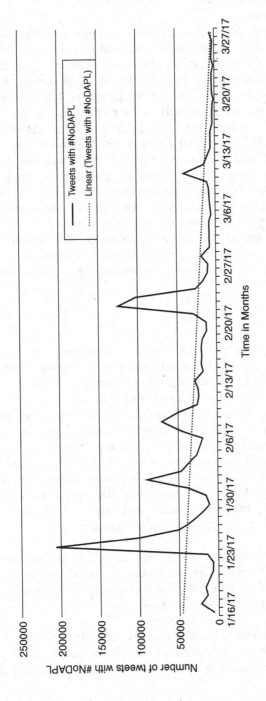

Figure 10.1 Decline of #NoDAPL tweets over time

attacked the decision by pointing out that he had investments in Energy Transfer Partners and that his Secretary of Energy, Rick Perry, sat on the board of Directors of Sunoco Logistics, which is a Dakota Access LLC co-owner. After a spike, #NoDAPL Twitter activity then waned, exhibiting a slow downward trend with slight increases every seven days (see Figure 10.1). Following Yin et al. [3], we refer to high increases in the use of the hashtag as outbreaks. Examining the collected data, we found five of these significant increases.

Engineers seeking to understand or participate in public debates about issues central to their profession should use this information to recognize and then to engage the frames being used by the public in a particular controversy relevant to their profession. In the sections that follow, we begin by situating our research in relation to broader academic literature on framing and social media. We then present the research questions that guided our work, as well as the methods we used to analyze the data. In particular, we used Nisbet's [4] typology of frames to characterize the primary frames of each #NoDAPL tweet that received 1500 or more likes or retweets. Using this categorization, we find that the most-used frames were morality/ethics and conflict/strategy, and that there was no noticeable middle path frame. We suggest that these frames contributed to political polarization and the echo-chamber effect [5], in which people access and believe information that already corresponds with their opinions. This tendency reduces the potential constructive nature of Twitter for promoting dialogue to reach consensus on controversial topics. But identifying the dominant frames does shed light on how the public interprets information, providing guidance for engineers and other technical professionals seeking to constructively engage their critics.

Literature Review

This section brings together academic literature on framing and social media to contextualize our research and consider the impact of the framing strategies animating the most influential #NoDAPL tweets. Drawing on Bateson [6], Goffman [7] provided the foundational theoretical basis for understanding frames as devices that organize thinking and make meaning. Frames encourage certain interpretations while discouraging others. For example, news articles that cover the DAPL controversy by referring to the growing demand for petroleum products frame the pipeline as necessary for consumers and their standard of living. In contrast, news articles that cover DAPL by referencing a long string of pipeline leaks frame the pipeline as susceptible to failure and therefore a source of environmental risk. Frames allow the people conveying and receiving information to "locate, perceive, identify and label" events, ideas, and issues [4, p. 44]. Framing, then, is an "exercise of power" since by

> defining the terms of debate, groups and advocates can influence the amount of attention an issue receives, the arguments or considerations that are considered legitimate or out of bounds, and the voices who have standing to express their opinion or participate in decisions. [8]

In his foundational analysis of nuclear energy, climate change, and evolution, Nisbet adapts a typology of frames originally developed by Gamson and Modigliani [9] to categorize how "various actors in society define issues in politically strategic ways …

and how diverse audiences differentially perceive, understand and participate" in those debates [4, p. 43]. Framing thus shapes but does not determine how audiences connect with and think about issues. Indeed, framing is so powerful that different frames result in different audience responses to the same content [10].

Given the importance of frames for shaping how debates unfold and a variety of stakeholders participate in them, engineers and science communicators should become more conscious of and deliberate in the use of framing techniques. As Nisbet argues, effective framing can generate interest and concern in topics, and shape policy preferences and actors' behaviors. Of particular interest to this study is his proposal that effective framing can "go beyond polarization and unite various publics around common ground" [4, p. 43] (see also [11]). Research already shows that journalists have a particular role to play, as they translate—and in the process, condense—complex events and findings into a relatable format for the public and policymakers [12]. Our research expands the scope of *who* frames an issue to include users of popular media and to highlight a role for engineers and other technical professionals.

The skyrocketing use of social media merits an expansion of how scholars analyze framing, media, and social controversies. Platforms such as Twitter empower a wide range of actors to publish short posts (140 characters during the #NoDAPL debate, 280 as of February 2018) that can be read and shared far beyond an individual user's contacts. By transforming the public from consumers of media to creators of it, social media platforms have widened the range of people who shape public discourse through framing strategies [8]. Of particular concern is the possibility that online social media amplify polarization and the echo-chamber effect, since they are "passed along and pre-selected by peers and opinion leaders who are likely to share an individual's worldviews and political preferences" [8, p. 371]. Research with Italian and German Twitter users suggests that the possibility for social media to amplify the echo chamber effect is conditioned by a user's broader patterns of political conversation, including offline networks, and their own habits in social media use [5].

Communication research on Twitter is growing in response to its increasing influence. One major area analyzes the "sentiment" of posts to predict a wide range of phenomena, from consumer practices and brand loyalty [13] to stock market performance [14], [15] and political opinion and elections [16]–[18]. Turning from the public to politicians themselves, a rhetorical analysis of Trump's tweets [19] illuminates patterns in how he used Twitter to rally his base and attack his opponents: whereas the *deliberative*, forward facing mode is preferred by many politicians, Trump used the *forensic* mode to judge others and their activities.

Researchers also assess organizations' stakeholder engagement practices by investigating their Twitter practices. Although Twitter provides a framework for two-way communication, for example, many nonprofits use the platform to share information in a one-way manner, missing opportunities to create dialogues and closer relationships [20]. An analysis of ConocoPhillips' social media response to their oil spill in China also suggests that the company failed to effectively use this platform to respond to mounting anger and criticism from the public, resulting in significant reputational damage [3]. In contrast, scholars suggest that some corporations such as BP have learned to use Twitter to respond more effectively to critics, especially during times of crisis [2].

Focused attention to crisis communication considers the role of Twitter in reputation management [21]. Research suggests that the medium—traditional versus "new" sources such as social media—matters more than the message or content for perceptions of reputation [22]. Moreover, social media is used more frequently than traditional news sources during times of crisis [23]–[25] and can be more trusted than traditional news sources during those times as well [26], [27]. Yin et al. [3] point out a substantial gap in this literature: Attention to how members of the public, rather than public relations professionals or organizations, use social media platforms such as Twitter. Our study begins to address that gap by exploring how a wide range of Twitter users tweeted about the Dakota Access pipeline controversy. Specifically, we identify the framing strategies utilized in the most influential tweets and the ways that these changed over time in relation to dramatic shifts in the political landscape.

This study thus builds on growing research on Twitter activism, especially related to antifossil fuel movements. Twitter plays an important role in these movements, providing a dispersed network of activists with a platform for coordination as well as a sense of belonging [28], [29]. Frames are central to such activism, not only providing organizing themes and motivations, but in framing the kind of activism itself. Hopke's [28] analysis of #Global Frackdown activism identifies frames that encompass collective action, personal action, and a hybrid of both. Hydraulic fracturing is a technique used to extract oil and gas from unconventional sources, such as tight shale plays. Opponents refer to this kind of oil and gas development as "fracking." Analyzing the #fracking hashtag, Hopke and Simi [30] found that its use was dominated by activist individuals and organizations. In contrast, our analysis of #NoDAPL tweeting found more significant roles for nonactivists as influential participants in the debate. The most influential #NoDAPL tweeters included individuals with no evident activist orientation, such as Sabreigha and Sinamonnroll, in addition to known figures like Bernie Sanders and established news sources such as CNN and the *New York Times*. The influence of nonactivists in the #NoDAPL suggests that #NoDAPL reached a wider national audience of people who were not otherwise involved in antifracking movements but made themselves stakeholders in the pipeline debate. This point should call the attention of engineers to the influence that members of the public can have in shaping and growing activism surrounding environmental and scientific controversies.

Research Questions

Our analysis was motivated by three key research questions:

> **RQ1.** Which framing strategies are present in the most influential (determined by the number of retweets and "likes") posts using #NoDAPL on Twitter?
>
> **RQ2.** How do the framing strategies used in the most influential #NoDAPL tweets change in relation to major political events?
>
> **RQ3.** Do the framing strategies used in the most influential #NoDAPL tweets amplify the echo-chamber effect and polarization on Twitter?

Seeking to bring together research on framing techniques and social media communication, we designed our study to identify which frames were most prevalent in the most popular #NoDAPL tweets. Using the most popular tweets would allow us to capture the framing techniques undertaken by Twitter users rather than by journalists and other formal media makers, who otherwise dominate the existing literature. To add nuance to our analysis, we analyzed the changes in the most prevalent frames over time, to explore whether those framing techniques changed in relation to major political events. Finally, we sought to gauge the possibility for frames to amplify political polarization and the echo chamber effect by investigating the prevalence of frames that would lend themselves to dialogue, particularly the middle way frame. Identifying the primary frames used in the most influential tweets provides engineers a window into understanding public perception, especially the topics and stances that were most compelling to social media users participating in the debate.

Methods

We collected tweets containing #NoDAPL on a daily basis from Trendsmap.com during the height of activity from December 1 to December 16 and from January 12 to April 3 (as confirmed by our colleague who ran the #NoDAPL script, there were no outbreaks between December 16 and January 12). Daily reports from Trendsmap about the use of #NoDAPL in the US contained the necessary information for this study because they provided the time of tweeting, number of likes and retweets per tweet, and other relevant data. Though this paper examines general trends in the US as a whole, we also collected information on #NoDAPL tweets in North Dakota and Colorado because the physical protests against the Dakota Access pipeline took place in North Dakota near the Standing Rock Sioux Reservation and because we sought to analyze how communities experiencing controversial oil and gas development, such as those in Colorado, used nationally prominent tweets such as #NoDAPL.

To assure the trustworthiness and credibility of our data collection, we triangulated our data with that collected by computer scientist Dr. Thyago Mota, who developed and ran a script tracking #NoDAPL use. He collected results similar to those of Trendsmap.com and is now investigating the role "advocates" (people with a determined position on an issue) play in the debate about the hashtag. We also worked with a developer of Trendsmap.com to assure credibility of the collected data and to learn more about how Trendsmap processes the raw data. The data collected from Trendsmap had already been processed through their algorithms based on factors such as location, mobile device, and gender. Nevertheless, these processed data were used to further analyze the key players in the online twitter about the #NoDAPL.

The daily reports from Trendsmap contained what Trendsmap determined to be the most influential tweets of the day based on the number of retweets and likes, along with accompanying information on those tweets' usernames and other user data. From these daily reports during the top two outbreaks, we selected the most influential by using a boundary of either 1500 retweets or 1500 likes. For example, if a tweet had 1200 likes and 800 retweets, it was not selected as part of the results for these data. If a tweet had 9 likes and 1600 retweets, it was selected.

The sample size for Twitter research varies widely. Studies with very large sample sizes either randomly select the tweets analyzed rather than analyzing each one or they employ automated computer coding programs [18], [19], [27], [29]. Given our desire to gain an in-depth understanding of the most influential tweets, we opted for a sample size that would enable fine-grained analysis of all of the most influential tweets, rather than a random selection. The criteria for 1500 retweets or likes resulted in a sample of 140 tweets, which falls within the range of tweets per coder in the existing literature.

Table 10.1 Frames used with #NoDAPL on Twitter

	Number of Times Used	Percentage
Social Progress	5	4
Economic Development	0	0
Morality/Ethics	33	24
Scientific Uncertainty	1	1
Pandora's Box	9	6
Public Accountability/Governance	15	11
Middle Way	0	0
Conflict/Strategy	56	40
Social Progress and Morality/Ethics	2	1
Morality/Ethics and Public Accountability	4	3
Morality/Ethics and Conflict/Strategy	8	6
Pandora's Box and Conflict/Strategy	2	1
Pandora's Box and Morality/Ethics	2	1
Public Accountability/Governance and Conflict/Strategy	3	2

To analyze the most influential tweets, we used the framing typology for scientific and environmental debates [4], [8] to categorize each of the influential tweets that we collected (see Table 10.1). The categorizing was done jointly by a team of researchers working together, with the second author categorizing all tweets and the first author and two other researchers discussing and verifying those codes.

The sentiments of the majority of the tweets could be categorized using one frame. For those in which a second frame was also evident, we included that frame as well. In our analysis, we have separated the double-framed tweets into distinct categories of their own (see Figure 10.2) to gain a better understanding of which framing strategies tended to be used together.

One of the limitations of our study is its focus on the two largest outbreaks of Twitter activity using the #NoDAPL hashtag, each of which followed in the wake of a major presidential decision on the pipeline. Future research on the frames used in the #NoDAPL debate in general, including the days with fewer total tweets, could determine whether more middle ground views were expressed throughout the entire debate, rather than just in its most feverish stages.

A second limitation is that we analyze only tweets using the #NoDAPL hashtag, though we also read the #DAPL tweets during our initial collection of data. The #DAPL hashtag appears more neutral and could be a source of less polarized opinions, as it does not explicitly state a positive or negative stance on the pipeline, but it was used almost exclusively by pipeline proponents to express their support of the project. We did not include #DAPL tweets because none of them reached comparable levels of influence (as measured by likes and retweets) as the #NoDAPL tweets. Future research could examine the frames used by the #DAPL tweets and compare them with those used in the #NoDAPL tweets to investigate whether they were using similar or different framing techniques and whether the #DAPL hashtag opened up more opportunities for either the middle way or the scientific uncertainty frames.

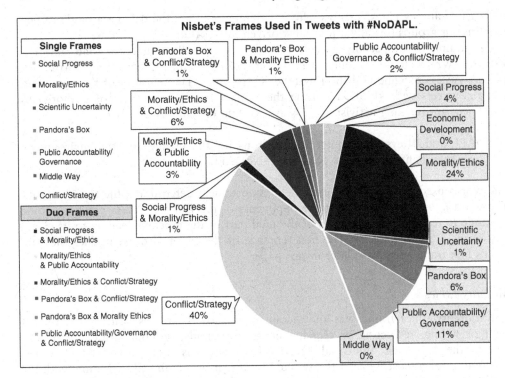

Figure 10.2 Frames used in influential #NoDAPL tweets

Results

Two frames dominated the most influential tweets: conflict and strategy, and morality and ethics. Examples of tweets employing these frames follow in the sections below. The dominant frames also changed over time. The Obama executive order had a marked influence on the frames used by the #NoDAPL tweeters. When the US Army Corps of Engineers denied the easement on December 4, opponents of the Dakota Access pipeline exuded a sense of victory, which we find resulted in an increase of the use of the conflict and strategy frame, as well as the social progress frame. This sense of victory changed a month later with the Trump executive order: The use of the public accountability and Pandora's box (invoking uncontrolled disaster) frames increased in the hours and days after that executive order, changing the landscape of the policy debate about the Dakota Access pipeline.

Conflict/Strategy

The frame most commonly used by the opponents of the Dakota Access pipeline was the conflict and strategy frame, appearing in 72 of the 140 most influential tweets, or approximately half of all analyzed #NoDAPL tweets (see Figure 10.2). This number includes 59 tweets that exhibited only the conflict and strategy frame, as well as 13 that combined this frame with another one. This frame defines a science-related issue as "a game among elites, such as who is winning or losing the debate; a battle of personalities or groups (usually a journalist-driven interpretation)" [4, p. 46].

This frame was especially prevalent in the tweets following the second outbreak after the White House announced President Trump's executive order supporting the Dakota Access pipeline and the Keystone XL pipeline. The majority of the opponents sent out tweets with #NoDAPL using the conflict frame, including one authored by prominent antifracking activist actor Mark Ruffalo on January 23, 2017:

> DAPL and #Keystone would be disastrous for the people and the environment. Let's keep fighting brothers & sisters. #NoDAPL

The words used in a tweet are extremely important in determining the tweet's frame. The word *fighting* indicates that the opponents of the Dakota Access pipeline should continue to resist the construction of the final part of the pipeline, leading to its classification as a conflict frame. The first sentence of the tweet also exemplifies the Pandora's box frame, discussed later in this paper.

Morality/Ethics

The morality/ethics frame was the second most-used. It frames issues "in terms of right or wrong; respecting or crossing limits, thresholds, or boundaries" [4, p. 46]. Out of the 140 most influential #NoDAPL tweets, this frame was used 49 times, accounting for almost 34% of the selected tweets. If we combined this frame with the public accountability and governance frame (see below), which implies ethics and morality, then the number would increase to 64. The opponents of the Dakota Access pipeline were especially likely to use the frame to convey their message that the construction of the pipeline was unethical in the light of the energy transition and that construction and pipelines in general are dangerous.

This frame was used to categorize the tweets that argued that something about the pipeline was moral or ethical on one hand, or was immoral or unethical on the other. Our data show that in the days after the US Army Corps of Engineers denied the easement for the construction of the Dakota Access pipeline underneath Lake Oahe, large numbers of pipeline opponents called the Corps' decision the moral and ethical thing to do. Many tweeted using this frame, and their tweets often contained a message of gratitude toward President Obama and the US Army Corps of Engineers, calling the decision to deny the easement ethical. An example of a tweet using #NoDAPL and the moral frame was tweeted by @MisterPreda on December 5, the day after the US Army Corps of Engineers denied the easement:

> The Dakota pipeline will no longer destroy precious land & water. Sending [heart symbol] to the beautiful Native Americans who stood & fought. #NoDAPL

The user believes that denying the easement was the right and moral thing to do, and the tweet also expressed gratitude toward the people who worked for this effort.

After President Trump signed his executive order on January 24, 2017, the number of tweets using the morality frame increased, but this time to deem the president's decision immoral. The total number of tweets using #NoDAPL also increased drastically, as can be seen in Figure 10.1. Many opponents of the pipeline were shocked by

President Trump's "unethical" decision, as portrayed in many tweets using #NoDAPL. An example tweet demonstrating this sense of indignation is one sent by @ShaunKing on January 24, 2017:

> The Dakota Access pipeline must be stopped. It is immoral and unethical. I stand with Standing Rock. #NoDAPL

Tweeted a few hours after President Trump signed the executive order, this tweet clearly shows the strong stand that @ShaunKing took by calling President Trump's actions immoral and unethical. The widespread use of the morality and ethics frame differs from Nisbet's [4] findings that it especially appeals to a religious, conservative audience.

The morality and ethics frame was frequently used in conjunction with the conflict and strategy frame. A total of 21 tweets were coded as using two different frames simultaneously. The most common combination of frames used from these tweets was the morality/ethics and conflict/strategy duo (eight occurrences). An example of this combination is the following:

> Trump's executive order on #DAPL–violates the law and tribal treaties. We will be taking legal action. #standwithstandingrock #noDAPL

This tweet was posted by @StandingRockST, the official Standing Rock Sioux Tribe's twitter account on January 24. The tweet accuses President Trump of acting immorally by violating the law and tribal treaties (morality and ethics frame), before stating that the Standing Rock Sioux Tribe will be taking legal action to fight the President's decision (conflict and strategy frame).

Public Accountability/Governance

The third most commonly used framing strategy was public accountability and governance. For Nisbet, this frame draws out questions of transparency, corruption, and public versus private interests. These questions inherently imply ethics and morality, but merit their own frame because they focus specifically on public accountability. As noted in the previous section, because of this implication, all of the tweets that used the public accountability and governance frame could have also been categorized as morality and ethics. We did not do so because we wanted to maintain the analytic usefulness of distinguishing critiques of public accountability. The few tweets that we classified as using both frames explicitly used the words *ethics* or *morality*, rather than simply insinuating an ethical or moral issue.

Twitter users drew on the public accountability and governance frame to question President Trump's objectivity after he signed the executive order. An example of this questioning of his objectivity is a tweet from @funder:

> Trump Owns Stake in Oil pipeline. This should disturb everyone. #NoDAPL.

This tweet, sent out on January 25, questions President Trump's neutrality in the process of granting the easement for the final construction of the pipeline. So many similar tweets using this frame and accusing President Trump of having special interests emerged in the wake of the executive order that they generated a marked increase in the use of the public accountability and governance frame compared to the weeks before (see Figure 10.3). Pipeline opponents called his decision immoral and speculated about corruption in the new White House. The tweet below was sent out by actor Mark Ruffalo on January 26 showing speculation about corruption.

> Congress: Require Trump to prove he has no Dakota Access pipeline conflicts of interest: #NoDAPL

This tweet is another classic example of the public accountability frame and shows the doubts about President Trump not serving any special interests.

Pandora's Box

The Pandora's box frame is used as a "call for precaution in face of unexpected consequences or catastrophe" that references fatalism or out-of-control consequences [4, p. 46]. It was the fourth most-used frame, with pipeline opponents employing it to convey their opinion that the consequences of an oil spill underneath Lake Oahe, due to a potential failure of the Dakota Access pipeline, would result in catastrophic consequences for the environment and the Sioux Tribe living nearby. The use of the Pandora's box frame drastically increased after President Trump signed his executive order on the Dakota Access pipeline, as did the public accountability/governance frame, and continued to be frequently used in the following weeks.

The use of the Pandora's box frame had previously been limited because in December, when the easement for the pipeline was denied by the US Army Corps of Engineers, opponents of the pipeline were convinced that the pipeline would be rerouted or even completely cancelled. As a result, other frames were more commonly used than Pandora's box during that time period, but that trend changed dramatically after President Trump's executive order. The use of Pandora's box can be seen as a scare tactic used by pipeline opponents by arguing that the construction and operation of the pipeline underneath Lake Oahe could result in a catastrophe.

An example #NoDAPL tweet using the Pandora's box frame was tweeted by Mark Ruffalo a day before President Trump's executive order on the construction of the pipeline:

> DAPL and #Keystone would be disastrous for the people and the environment. Let's keep fighting brothers & sisters. #NoDAPL

Ruffalo's tweet, which garnered more than 1200 retweets and more than 2800 likes, warns that finalizing the construction of DAPL could result in dire consequences for people living close to it as well as for the environment. In the context of the wider debate, he suggests that a leak in the pipeline could result in endangering the water supply of the nearby Standing Rock Sioux Tribe and that it would destroy wildlife. It is a good example of tweets using the Pandora's box frame because it contains the characteristics of a message that warns of a catastrophe if the path of the construction of the Dakota Access pipeline is chosen. It also contains elements of the conflict/strategy frame discussed above.

Another was posted by the "unofficial resistance team" of NOAA (the National Oceanic and Atmospheric Administration) on February 2.

When the last tree is cut down, the last fish eaten, and the last stream poisoned, you will realize that you cannot eat money #NoDAPL

The use of Pandora's box is evident due to the apocalyptic images that indicate that there will be a catastrophe waiting when the pipeline is in use.

Social Progress

The fifth most-used frame was social progress, or "improving quality of life or a solution to problems" or "harmony with nature instead of mastery, sustainability" [4]. Out of the 140 #NoDAPL tweets that were examined in this research, seven were categorized under the social progress frame. Though this is a relatively small number, the most retweeted and liked of the 140 tweets used the social progress frame and was tweeted by @Sabreigha on December 4:

> The oil pipeline is being rerouted! Don't ever let them tell you your voice/protest doesn't matter #NoDAPL #DAPL

This tweet was categorized as using a social progress frame because it combines a "solution to a problem" according to @Sabreigha with a message of sustainability. Overall, this tweet characterizes social progress because the user believes that rerouting the pipeline offers a solution to problems. The other six tweets categorized as using a social progress frame are similar to @Sabreigha's in that they all believe that there is a solution offered to the problems. Five out of seven tweets using the social progress frame were sent in December in the days before and after the US Army Corps of Engineers decided to not grant an easement for the Dakota Access pipeline, leaving only one social progress tweet responding to the Trump executive order.

Another example of a tweet using the social progress frame was sent out by Senator Bernie Sanders on December 3, 2016:

> Native Americans have been cheated for hundreds of years. The time to change that is now. #NoDAPL

This tweet was posted a day before the US Army Corps of Engineers decided to deny the final permit for the Dakota Access pipeline. It shows indignation about how Native Americans have been treated and calls for social progress.

Scientific Uncertainty

Nisbet defines this frame as calling on the authority of "sound science" and expertise of scientists, and can either invoke or undermine expert consensus. This frame is often used in other policy debates, including the notable example of climate change skeptics arguing that the science forming the consensus on human-caused global warming is not trustworthy. The policy debate about the Dakota Access pipeline did not exhibit an expert consensus on the social and environmental impacts of the pipeline, perhaps because of the highly politicized nature of the debate. The tweet that comes closest to the scientific uncertainty frame is Mark Ruffalo's tweet sent out on January 26, 2017:

> Don't believe "alternative facts" magical pipeline job growth numbers.
> #ClimateFacts #NoKXL #NoDAPL @realDonaldTrump. The real
> facts:https://t.co/SPcZ5eUuk1

However, this tweet more strongly questions the job growth numbers President Trump claimed that the Dakota Access and Keystone XL pipeline would create. Although it does not question any science associated with the pipeline, it does question the White House's internal research on potential job growth numbers.

This frame is often used in the debate about climate change, where research suggesting man-made climate change is often described as alarmism by critics. In the online debate about the Dakota Access pipeline on Twitter, this frame has been used only sporadically because other frames were found to be more effective.

Economic Development

The economic development frame is characterized by Nisbet as "economic investment, market benefits or risks; local, national or global competitiveness" [4]. We regarded this frame as the economic argument often used to convey the message of economic stimulus and thus market benefits on a local, national, or global scale. This argument is often used by the right side of the political spectrum to convey the message that a new measurement or development will create more jobs in the area. Indeed, the oil and gas industry frequently uses economic development to rally support for new developments, including the Dakota Access pipeline project. President Trump himself stated that the construction of the Keystone XL pipeline would create 28,000 new jobs, while signing his executive orders on the Dakota Access pipeline and Keystone XL.

Yet the economic development frame was not seen in any of the 140 most influential tweets with #NoDAPL; not a single tweet using this frame accumulated the required 1500 retweets or likes, the limit for our collected data. It is possible that this frame did not surface as influential in the #NoDAPL tweets because proponents of the Dakota Access pipeline frequently used the neutral #DAPL or no hashtags at all. Though it did not use the requisite #NoDAPL hashtag, an example of a tweet that would have been categorized using the economic development frame would be President Trump's tweet following the signing of his executive order on January 24.

> Signing orders to move forward with the construction of the Keystone XL and
> Dakota Access pipelines in the Oval Office.

President Trump later added that the pipelines would create thousands of new jobs and combining that sentence with his tweet would have been a good example of the economic development frame. Other Republican politicians sent out similar tweets about the Dakota Access pipeline but either used a different hashtag or no hashtags at all, or they did not get the required 1500 retweets or likes. This fact suggests that the economic development argument frequently used by industry is not influential in addressing their opponents.

The opponents of the Dakota Access pipeline were extremely unlikely to use the economic development frame, but there were instances in which they tweeted to comment on tweets that used the economic development frame. For example, Mark

Ruffalo sent out a tweet on January 26, two days after Trump signed his executive order and promised thousands of new jobs due to the construction of the pipeline:

> Don't believe "alternative facts" magical pipeline job growth numbers. #ClimateFacts #NoKXL #NoDAPL @realDonaldTrump. The real facts:https://t.co/SPcZ5eUuk1

This tweet received more than 10,000 retweets and 21,700 likes in the following days. However, we did not categorize it using the economic development frame because the tweet explicitly argues against that frame.

Middle Way

Nisbet's middle way frame, which centers on "finding a possible compromise position, or a third way between conflicting/polarized views or options" [4, p. 46] would appear in the policy debate about the Dakota Access pipeline as a tweet written to unite both sides of the debate or offer a neutral solution beneficial to both sides. Of the 140 tweets that received at least 1500 retweets or likes, not one could be categorized as a middle way frame. This fact indicates that the middle way frame is either not the first choice of the proponents or opponents of the pipeline, or that this framing strategy is not effective in garnering retweets or likes.

The absence of the middle way frame is striking, suggesting that the two sides of the debate are polarized and that a middle ground is missing. A more neutral middle ground could potentially offer solutions that appeal to both sides of the debate and also make the debate less aggressive.

Discussion

Our research found that the frames characterizing the most influential tweets using the #NoDAPL hashtag changed in response to dramatic political shifts and decision making (see Figure 10.3). The last two weeks of 2016 and the first two weeks of 2017 are not included in the graph because they were not included in this study, as there were no major outbreaks during that period.

When Obama's administration halted construction of the pipeline in response to public outcry, the most influential tweets celebrated the event using the social progress frame, argued that it was the ethically correct decision, and put conflict in a positive light. When Trump reversed that decision less than two months later, Twitter users condemned the executive order by continuing the emphasis on conflict but also raising questions about the corruption of the administration by invoking the public accountability frame and using the ethical frame to argue that Trump's decision was immoral. Remaining the same across both outbreaks was a lack of tweets using the scientific uncertainty frame. Indeed, scientific evidence played a relatively minor role in the debate, even though this was the go-to response of the pipeline proponents who took part in the debate and is perhaps the most professionally comfortable frame for engineers. Our research suggests that sharing technical information about risk was not sufficient to quell debate. That information might be more persuasive if it were explicitly entered into the dominant ethical framing of the debate, rather than proposed separately.

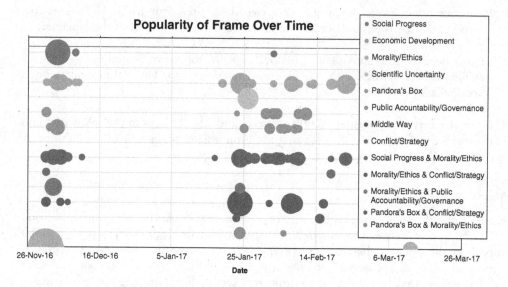

Figure 10.3 Popularity of framing techniques used over time, with the number of tweets represented by the size of the bubble. For example, the large bubble in the upper-left corner represents almost 40,000 likes and retweets that used a social progress frame during the first outbreak in December. The smallest bubbles are tweets that barely passed the threshold of 1500 retweets or likes

The other frame that was absent was the middle way frame, which plays a role in reifying political difference. Our research suggests that Twitter activism on the Dakota Access pipeline strongly contributes to the echo-chamber effect and polarization during periods of intense media scrutiny. The echo-chamber effect applies to #NoDAPL in the way that the users of the hashtag chose frames that are characteristic of their side of the debate: Opponents couched their critiques in moral/ethical arguments, speculating about disaster and the corruption of the new Republican administration, while the few proponents relied on established yet unconvincing appeals to economic development. Moreover, the use of the conflict and strategy frame dominated the most influential tweets during both outbreaks, encompassing almost half of the total influential tweets analyzed for this study, while the middle way frame was entirely absent. Future research should investigate whether middle way frames were more prevalent in the influential tweets that occurred outside of the outbreaks, as it is possible that intense media scrutiny amplifies polarization.

While the middle way frame could combat the echo-chamber effect by creating space to learn and change perspective, it was not employed by influential users of #NoDAPL. The absence of the middle way during the outbreaks indicates that proponents and opponents of the pipeline did not see the middle way frame as an effective way to convey their message via Twitter, that the tweets using the middle way frame were not effective in getting large numbers of retweets or likes, or both. The echo-chamber effect was mitigated only partially by pipeline proponents using the opponents' #NoDAPL hashtag, instead of or in addition to their own preferred #DAPL hashtag, to make their tweets more visible in the online debate (given the dominance of the #NoDAPL hashtag over the #DAPL one). However, the use of the #NoDAPL hashtag by pipeline proponents did not necessarily result in decreased polarization, as participants used the hashtag to express strongly positive views and discredit their opponents.

Implications of the Research

Our research suggests both a gap and an opportunity for engineers and other technical professionals. Social media use is growing across major demographic categories in the US [31]. Though we could not find peer-reviewed research on the prevalence of Twitter use among engineers, some studies suggest that engineers use Twitter infrequently: The highest rate found in an IEEE study was 20% among engineers aged 18–34, while only 10% of those 49 and older did so [32]. Yet Twitter is playing an increasingly prominent role in shaping public perception of the industries, companies, and projects that involve engineers, such as the Dakota Access pipeline. At a minimum, Twitter provides engineers useful information on public perception of their work. Engineers seeking to intervene in that perception—either by generating content themselves or advising the public relations personnel who do so—could use this information to more effectively share their professional knowledge with the public by using the same influential frames as their interlocutors.

In the outbreaks that we analyzed, it is striking that the frame for scientific uncertainty played a marginal role in the debate, even though it tends to be the preferred frame for engineers. This frame invokes the authority of sound science and the expertise of scientists, either to reinforce or undermine expert consensus. The absence of this frame in the influential tweets suggests that the debate participants were less interested in or concerned by the scientific and engineering questions that could have been brought to bear on issues central to those fields, such as the likelihood of pipeline leaks and strategies to prevent them, or the effect of oil contamination in water and soil. Instead, debate participants were motivated by broad questions of struggle, morality, and ethics. This study suggests that the appeals made by Energy Transfer Partners and other petroleum advocacy organizations to defend the pipeline by appealing to safety and economic development were ineffective. This case cautions that for engineers' expertise to be persuasive in debates, they should explicitly link that expertise to the questions motivating their interlocutors.

Conclusion

Twitter and other social media platforms are changing the ways in which activists and members of the public engage controversies, as they empower a wider range of actors to frame debates and their stakes. Yet rather than opening up debates about the Dakota Access pipeline and proliferating a variety of perspectives on its positive and negative effects, our study of the most influential #NoDAPL tweets pointed to amplified online polarization through the creation and reinforcement of an echo chamber effect. This finding contrasts with more hopeful views [4], [11] that framing can decrease polarization. In the case of #NoDAPL "hashtagging politics" [28, p. 1], the most influential tweets were those taking a strongly critical stand on the pipeline. This finding raises the question of whether Twitter is or could be a constructive platform where users come together to debate and dialogue in the spirit of being willing to learn from others' conflicting opinions to form new viewpoints. Use of the middle way frame would be an ideal way to make these policy debates on Twitter more constructive, though it is unclear why this frame was absent in the online outbreaks during the #NoDAPL debate. Future research could explore whether the frame was simply not used at all, whether the tweets using it did not garner substantial likes and retweets to circulate through a wider audience, or whether it was used outside of the outbreak periods. Our case provides a cautionary note that some platforms, such as Twitter, might be predisposed to propagate particular frames over others. Twitter is popularly viewed as a platform for aggressive stance-taking. Future research could explore whether the absence of the middle way frame for the #NoDAPL hashtag was specific to that debate or whether it also holds true for Twitter activity in general.

Finally, we signal an opportunity for engineers and other technical professionals to communicate more effectively with people involved in broad public debates. Dialogue about the risks posed by pipelines would benefit from a critical analysis of how they are constructed, what measures are taken to protect safety, and how they can be sited more responsibly. Engineers can contribute that expertise to debates, but we speculate that people involved in online protests would be more interested in that information if it is presented and packaged—or *framed*—in a way that resonates with their own concerns. We believe that responses using the same frames used by their opponents would have been more effective, since they would have engaged in a dialogue encompassed by the same frame rather than broadcasting information into an echo chamber. As more petroleum companies lead a wider field in turning to social media to respond to crises [18], [19], engineers from many disciplinary backgrounds and in a variety of industrial sectors have a crucial role to play in bringing their expertise to bear on controversies. But to do so effectively, they must communicate their expertise to the public in ways that resonate with the dominant concerns being used by nonengineers, and identifying and engaging the same frames as their critics is a first step in doing so.

Acknowledgment

This work was supported in part by the ConocoPhillips Center for a Sustainable WE^{2}ST (Water-Energy Education, Science & Technology), which was established at the Colorado School of Mines to promote the joint sustainability of unconventional energy development and water resources in arid regions. ConocoPhillips does not direct the research or have access to the research data. Findings, opinions, and conclusions in this work are those of the authors and are not a statement or representation of the views, policies, or opinions of ConocoPhillips or its employees or representatives. The authors would like to acknowledge colleagues from WE^{2}ST for their support and guidance throughout this study. The authors also gratefully acknowledge J. Leydens for his guidance and feedback throughout the research and writing, as well as T. Mora for his collaboration early in the research. The authors also thank the peer reviewers and editor for their generative feedback.

References

[1] R. Davis, and D. M. Franks, "Costs of company-community conflict in the extractive sector," Corporate social responsibility initiative report no. 66, John F. Kennedy School Government, Harvard Univ., Cambridge, MA, USA, 2014.

[2] S. Muralidharan, K. Dillistone, and J.-H. Shin, "The Gulf Coast oil spill: Extending the theory of image restoration discourse to the realm of social media and beyond petroleum," *Public Rel. Rev.*, vol. 37, no. 3, pp. 226–232, Sep. 2011.

[3] J. Yin, J. Feng, and Y. Wang, "Social media and multinational corporations #x2019; Corporate social responsibility in China: The case of ConocoPhillips oil spill incident," *IEEE Trans. Prof. Commun.*, vol. 58, no. 2, pp. 135–153, Jun. 2015.

[4] M. C. Nisbet, "Framing science: A new paradigm in public engagement," in *Communicating Science: New Agendas in Science Communication*, L. Kahlor & P. Stout, Eds. New York, NY, USA: Routledge, 2009, pp. 40–67.

[5] C. Vaccari, A. Valeriani, P. Barberá, J. T. Jost, J. Nagler, and J. A. Tucker, "Of echo chambers and contrarian clubs: Exposure to political disagreement among German and Italian users of twitter," *Social Media Soc.*, vol. 2, no. 3, pp. 1–24, Sep. 2016.

[6] G. Bateson, "A theory of play and fantasy," In *Steps to an Ecology of Mind: Collected Essays in Anthropology, Psychiatry, Evolution, and Epistemology*. London, UK: Chandler, 1972, pp. 177–193.

[7] E. Goffman, *Frame Analysis: An Essay on the Organization of Experience*. New York, NY, USA: Harper & Row, 1974.

[8] M. C. Nisbet, and T. P. Newman, *Framing, the Media, and Environmental Communication*. London, UK: Routledge Handbooks Online, 2015.

[9] W. A. Gamson, and A. Modigliani, "Media discourse and public opinion on nuclear power: A constructionist approach," *Amer. J. Sociol.*, vol. 95, no. 1, pp. 1–37, 1989.

[10] D. Kahneman, "Maps of bounded rationality: Psychology for behavioral economics," *Amer. Econ. Rev.*, vol. 93, no. 5, pp. 1449–1475, Dec. 2003.

[11] M. D. Barton, "The future of rational-critical debate in online public spheres," *Comput. Comp.*, vol. 22, no. 2, pp. 177–190, Jan. 2005.

[12] D. Scheufele, "Framing as a theory of media effects," *J. Commun.*, vol. 49, no. 1, pp. 103–122, Mar. 1999.

[13] B. J. Jansen, M. Zhang, K. Sobel, and A. Chowdury, "Twitter power: Tweets as electronic word of mouth," *J. Amer. Soc. Inf. Sci. Technol.*, vol. 60, no. 11, pp. 2169–2188, Nov. 2009.

[14] C. Simões, R. Neves, and N. Horta, "Using sentiment from Twitter optimized by genetic algorithms to predict the stock market," in *Proc. IEEE Congr. Evol. Comput.*, 2017, pp. 1303–1310.

[15] V. S. Pagolu, K. N. Reddy, G. Panda, and B. Majhi, "Sentiment analysis of Twitter data for predicting stock market movements," in *Proc. Int. Conf. Signal Process., Commun., Power Embedded Syst.*, 2016, pp. 1345–1350.

[16] D. Ayata, M. Saraçlar, and A. Özgür, "Political opinion/sentiment prediction via long short term memory recurrent neural networks on Twitter," in *Proc. 25th Signal Process. Commun. Appl. Conf.*, 2017, pp. 1–4.

[17] R. Castro, L. Kuffó, and C. Vaca, "Back to #6D: Predicting Venezuelan states political election results through Twitter," in *Proc. 4th Int. Conf. eDemocracy eGovernment*, 2017, pp. 148–153.

[18] A. Hernandez-Suarez et al., "Predicting political mood tendencies based on Twitter data," in *Proc. 5th Int. Workshop Biometrics Forensics*, 2017, pp. 1–6.

[19] A. Watt, C. Carvill, R. House, J. Livingston, and J. M. Williams, "Trump typhoon: A rhetorical analysis of the Donald's Twitter feed," in *Proc. IEEE Int. Prof. Commun. Conf.*, 2017, pp. 1–7.

[20] R. D. Waters, and J. Y. Jamal, "Tweet, tweet, tweet: A content analysis of nonprofit organizations' Twitter updates," *Public Rel. Rev.*, vol. 37, no. 3, pp. 321–324, Sep. 2011.

[21] E. Lozano, and C. Vaca, "Crisis management on Twitter: Detecting emerging leaders," in *Proc. 4th Int. Conf. eDemocracy eGovernment*, 2017, pp. 140–147.

[22] F. Schultz, S. Utz, and A. Göritz, "Is the medium the message? Perceptions of and reactions to crisis communication via twitter, blogs and traditional media," *Public Rel. Rev.*, vol. 37, no. 1, pp. 20–27, Mar. 2011.

[23] T. Johnson, and B. Kaye, "Choosing is believing? How web gratifications and reliance affect internet credibility among politically interested users," *Atlantic J. Commun.*, vol. 18, no. 1, pp. 1–21, Jan. 2010.

[24] C. H. Procopio, and S. T. Procopio, "Do you know what it means to miss New Orleans? Internet communication, geographic community, and social capital in crisis," *J. Appl. Commun. Res.*, vol. 35, no. 1, pp. 67–87, Feb. 2007.

[25] K. K. Stephens, and P. C. Malone, "If the organizations won't give us information: The use of multiple new media for crisis technical translation and dialogue," *J. Public Rel. Res.*, vol. 21, no. 2, pp. 229–239, Apr. 2009.

[26] T. Seltzer, and M. A. Mitrook, "The dialogic potential of weblogs in relationship building," *Public Rel. Rev.*, vol. 33, no. 2, pp. 227–229, Jun. 2007.

[27] M. Taylor, and D. C. Perry, "Diffusion of traditional and new media tactics in crisis communication," *Public Rel. Rev.*, vol. 31, no. 2, pp. 209–217, Jun. 2005.

[28] J. E. Hopke, "Hashtagging politics: Transnational anti-fracking movement Twitter practices," *Soc. Media Soc.*, vol. 1, no. 2, pp. 1–12, 2015.

[29] W. L. Bennett, and A. Segerberg, *The Logic of Connective Action: Digital Media and the Personalization of Contentious Politics*. Cambridge, UK: Cambridge Univ. Press, 2013.

[30] J. E. Hopke, and M. Simis, "Discourse over a contested technology on Twitter: A case study of hydraulic fracturing," *Public Understanding Sci.*, vol. 26, no. 1, pp. 105–120, 2017.

[31] "Social media fact sheet," Pew Res. Center, Internet, Sci. Technol., Washington, DC, USA, Feb. 5, 2018.

[32] J. Don, "How industrial engineers use social media," *ISA Interchange*, Aug. 31, 2016. [Online]. Available: https://automation.isa.org/2016/08/how-industrial-engineers-use-social-media/. Accessed on: Mar. 4, 2018.

Commentary

If you have any interest in social media, politics, social justice, or the environment—or any combination thereof—you probably found this article very compelling. The debate over the Dakota Access Pipeline, at the center of Smith and van Ierland's study, is chock full of controversies, from the differing approaches of two political parties to protecting the environment, to the very different consideration given to the safety of water supplies of Native American and primarily European-American communities, to the injustices perpetrated on the indigenous peoples from the time of European colonization of North America until today.

One of the hallmarks of good qualitative research reports is the rich grounding in the area of interest that they usually provide, regardless of their subject. The fact that Smith and van Ierland's article addresses such a central topic as the effect of communication frames in shaping opinion about the common life of one nation (and by extension, the entire world) makes it even more compelling to technical communicators. Add the increasing awareness and importance of Twitter and social media in the years since the controversy, and the article's significance looms even larger than it did when it was originally published.

Studies based on document analysis research like this one typically contain extensive quotations and paraphrases, and this article is no exception to that rule. The tweets quoted in the article are largely responsible for the sense of reality and immediacy that pervades it.

Purpose and Audience

The article's three research questions state its purpose. They are restated and elaborated upon in the final paragraph of the Research Questions section (emphasis added).

> Seeking to bring together research on framing techniques and social media communication, *we designed our study to identify which frames were most prevalent in the most popular #NoDAPL tweets*. Using the most popular tweets would allow us tö capture the framing techniques undertaken by Twitter users rather than by journalists and other formal media makers, who otherwise dominate the existing literature. To add nuance to our analysis, *we analyzed the changes in the most prevalent frames over time*, to explore whether those framing techniques changed in relation to major political events. Finally, *we sought to gauge the possibility for frames to amplify political polarization and the echo chamber effect by investigating the prevalence of frames that would lend themselves to dialogue*, particularly the middle way frame. Identifying the primary frames used in the most influential tweets provides engineers a window into understanding public perception, especially the topics and stances that were most compelling to social media users participating in the debate.

In pursuing these three goals—to identify the most prevalent frames, analyze how frames changed over time, and determine how frames affected the polarization of thought—the authors make a significant contribution to social media research. Engineers and the scientific community in general need to understand the relationship between social media and public perceptions of technical issues. This need is particularly vital at a time when "alternate science" is a distressing component of the alternate reality that those media have played such a large part in creating.

Because it was published in the *IEEE Transactions on Professional Communication* published by the Institute of Electrical and Electronics Engineers Professional Communication Society, we can assume that this study is directed toward that journal's dual audience of professional communicators and engineers. Indeed, the final sentence of the paragraph just quoted specifically calls out the engineering audience's centrality for the authors, who themselves are practicing engineers. As its title indicates, the *IEEE Transactions* is notable among the major technical and professional communication journals in that its articles often have a dual audience of professional communicators and engineers.

Organization

This article uses the IMRAD structure that we discussed in Chapters 2 and 9.

- The untitled introductory section and the *Background*, *Literature Review*, and *Research Questions* sections comprise the lengthy introduction to the article. The authors situate the research in its political and social context, examine relevant previous research, and state the three research questions that the authors explore in the rest of the article.
- The *Methods* section describes how the data were collected, how the most influential tweets were identified, and how those tweets were categorized.

- The *Results* section examines the prevalence of tweets according to each of eight frames. This section includes multiple quotations of illustrative tweets.
- The *Discussion, Implications of the Research*, and *Conclusion* sections point out the conclusions that the authors draw from their study, as well as ways that engineers can use the results in speaking out on matters of public controversy to facilitate dialogue rather than stoking the fires of public opinion.
- The *Acknowledgments* recognize the assistance of a grant and several individuals who aided the authors in their work.
- The *References* include full details on all of the sources that the authors used.

Qualitative Methodology

As noted above, the *Methods* section describes the data collection, categorization, and analysis techniques that Smith and van Ierland used in their study. Their qualitative methodology is document analysis, but because the subjects of the analysis are Twitter tweets, the texts analyzed are exceptionally brief. At the time when the subject tweets were written, Twitter imposed a 140-character limit on each tweet, though that limit has subsequently been doubled. The authors use their analysis of tweet "documents" to categorize the content using a frame typology that they discussed extensively in the literature review. As discussed in Chapter 5, top-down coding using a relevant existing typology saves time and avoids "reinventing the wheel."

The tweets analyzed in this study were selected using four criteria:

- They used the #NoDAPL hashtag.
- The tweets' authors were located in the US.
- The tweets were posted between December 1 and December 16, 2016 and between January 12 and April 3, 2017, and they responded to executive orders regarding the pipeline issued by Presidents Barack Obama and Donald Trump.
- Each tweet included in the study received a minimum of 1500 retweets or 1500 "likes" because the authors wanted to focus on the most influential tweets.

The researchers used Trendsmap.com to collect tweets each day during these two outbreaks (no #NoDAPL outbreaks appeared between December 16, 2016 and January 12, 2017).

The tweets collected were neither self-selected by their authors nor randomly selected by the authors of the study, but resulted from the application of the selection criteria. Smith and van Ierland explain those criteria as follows.

> Given our desire to gain an in-depth understanding of the most influential tweets, we opted for a sample size that would enable fine-grained analysis of all of the most influential tweets, rather than a random selection. The criteria for 1500 retweets or likes resulted in a sample of 140 tweets, which falls within the range of tweets per coder in the existing literature.

Although the authors should have been more specific and cited at least one source for their "like" and retweeting criteria, the 140 tweets in their sample provide an adequate corpus of tweets for the preliminary study in this article.

Analysis

Coding and Categorizing

As with virtually all qualitative studies, the coding and categorizing of data produced in this research project was done "off stage." The reader doesn't see the researchers doing this work, just the results of that effort (see Table 10.1 in the article text). On the basis of these results, we know that the researchers used the following frame categories to classify the data from the 140 tweets that they analyzed:

- Social Progress
- Economic Development
- Morality/Ethics
- Scientific Uncertainty
- Pandora's Box
- Public Accountability/Governance
- Middle Way
- Conflict/Strategy

These categories were derived from Matthew C. Nisbet's framing typology for scientific and environmental debates cited in the text (publication details can be found in the *References* at the end of the article).

In addition, while categorizing the tweets, the authors decided to separately code 21 double-framed tweets using the following combined categories:

- Social Progress and Morality/Ethics
- Morality/Ethics and Public Accountability
- Morality/Ethics and Conflict/Strategy
- Pandora's Box and Conflict/Strategy
- Pandora's Box and Morality/Ethics
- Public Accountability/Governance and Conflict/Strategy

Because the codes were predefined, the authors combined the coding and categorizing stages of analysis. This top-down approach makes sense because the "documents" are so brief and the categories were predefined.

Exercise 10.1 Coding Qualitative Data

Choose a scientific, technological, or environmental issue that has been trending on Twitter during the past month, and select the most commonly used hashtag that characterizes it. Collect a minimum of 20 tweets that used that hashtag during that period. Then analyze and categorize those tweets using the eight framing strategies that Smith and van Ierland used in their article.

Exercise 10.2 Coding Qualitative Data into Multiple Categories

Using your response to Exercise 10.1, can you identify any double-framed tweets among those that you selected? (You need not limit yourself to the six double frames that Smith and van Ierland identified—any combination of the eight framing strategies is acceptable.) Write a brief explanation for why you categorized any double-framed tweets that you identified.

Did you find any middle ground tweets? If so, can you explain why you did?

Detecting Patterns or Modeling

In this study, the need to discern patterns or construct models from the data that have been coded and categorized is unnecessary. The dataset of 140 tweets collected by the researchers was restricted by the 140-character limit for tweets, the selection criteria that they used, and the framing structure that they adopted from Nisbet. Moreover, the direction of the Twitter discussion was determined by the two executive orders that sparked the tweet outbreaks. In short, the patterns in the data are identical to the frame categories.

What Smith and van Ierland discovered as a result of their analysis is that two frames, conflict/strategy with 56 tweets and morality/ethics with 33 tweets, accounted for 64% of the total tweets. If we add the tweets that were dual framed by one or both of the conflict/strategy and morality/ethics frames, the total accounts for fully 78% of the tweets. This fact is not surprising given the heated nature of the debate.

On the other hand, none of the tweets studied used the middle way frame, which would reflect a search for possible compromise. As Smith and van Ierland note,

> The absence of the middle way frame is striking, suggesting that the two sides of the debate are polarized and that a middle ground is missing. A more neutral middle ground could potentially offer solutions that appeal to both sides of the debate and also make the debate less aggressive.

Standards of Rigor and the Conclusions of the Study

The counts and percentages of tweets characterized by the eight single frames and six double frames lead Smith and van Ierland to draw a number of conclusions about the two tweet outbreaks using the #NoDAPL hashtag.

Not surprisingly, President Obama's order to halt pipeline construction led to celebratory tweets using the social progress frame, while President Trump's reversal resulted in tweets emphasizing conflict and corruption. There were no tweets that used the middle ground frame and only one that used the scientific uncertainty frame. This fact indicates the strength of the opposition to the pipeline and the polarization of opinion, contributing to an echo-chamber effect in which people hear only a single viewpoint.

Unlike the findings of two studies in their literature review, the framing used in these two Twitter outbreaks did not decrease polarization. Perhaps most illuminating is Smith and van Ierland's observation in their *Conclusion* section.

> Our case provides a cautionary note that some platforms, such as Twitter, might be predisposed to propagate particular frames over others. Twitter is popularly viewed as a platform for aggressive stance-taking. Future research could explore whether the absence of the middle way frame for the #NoDAPL hashtag was specific to that debate or whether it also holds true for Twitter activity in general.

The authors acknowledge that the conclusions that they reach in this article could be affected by the selection criteria that they chose. They focused only on the two largest outbreaks of tweets, and they included only tweets that used the #NoDAPL hashtag. They suggest that their results might have been different had they looked at days with fewer tweets than those included in this study. Results might also have differed if they had included tweets with the neutral #DAPL hashtag.

Credibility

In Chapter 5 we noted that in a quantitative study we assess the internal validity by asking whether the researchers have measured the concept that they wanted to study, whereas in a qualitative study we look more to the credibility of the data. Did the participants truly represent the population or phenomenon of interest, and are their behavior and comments typical of that population? The example that we cited was a study of how help desk personnel use a product's technical documentation, based on interviews of help desk supervisors. We concluded that help desk supervisors might not be credible sources of information about how help desk employees use documentation.

The question then is whether the tweets selected for analysis in Smith and van Ierland's article are credible representatives of the "population" of interest. Because the researchers have described their criteria for selecting tweets—specifying a single hashtag used, origination in the US, appearance within strict time limits, and a defined number of likes or retweets to denote influence—we can conclude that these participant tweets give the study credibility because their authors are the most influential participants in the debate, the very population that Smith and van Ierland wanted to study.

It's important to note that when we talk about credibility of the population of a qualitative study, we don't mean that this is a representative sample of a much broader population. The researchers don't intend to generalize about all tweets about the Dakota Access Pipeline. Instead, they are examining a relatively small number of tweets to determine the answers to their research questions concerning only those tweets. Furthermore, this study used document analysis methodology credibly to discover the tweeters' opinions, motives, and reactions, not to learn about their behavior.

Transferability

The requirement of external validity in quantitative research (whether the thing or phenomenon being measured in the test environment reflects what would be found in the real world) is comparable to the need for transferability in qualitative research (whether what we observe in the test environment reflects what would be found in the real world). Are the tweets that Smith and van Ierland report on authentic? As we observed in Chapter 5, the burden here is on the reader of the study, but the fact that the tweets that the authors examined are indeed "real-world" examples makes the issue of transferability moot. The tweets studied were not generated for the purposes of the study but rather were selected from tweets written by participants in an actual social media debate.

Dependability

Just as quantitative research must be reliable, qualitative research must be dependable. How confident can we be that the conclusions reached in the research project could be replicated if the study were conducted by different researchers? As we noted in Chapter 5, we make this determination based on the researchers' depth of engagement, the diversity of their perspectives and methods, and their staying grounded in the data.

DEPTH OF ENGAGEMENT

By depth of engagement, we mean that the more opportunities researchers give themselves to be exposed to the environment and to observe the data, the more dependable their findings will be. The analysis here is limited to a study of only 140 of many thousands of very brief tweets sent during the two periods studied, so depth of engagement is limited. Nevertheless, it's obvious from the *Results* section that the authors have engaged deeply with their subject data.

DIVERSITY OF PERSPECTIVES AND METHODS

Seeing data from multiple perspectives—for example, using multiple researchers or multiple data collection techniques—increases the rigor of a qualitative study. Because tweets were originally categorized by van Ierland and those categorizations were discussed and verified by Smith and two additional people, the study is characterized by a good degree of diversity of perspective. Diversity of methods is more problematic, however.

The authors state in the *Methods* section that they triangulated the data that they collected from Trendsmap with the data collected by Dr. Thyago Mota, who "collected results similar to those of Trendsmap.com ...." Strictly speaking, however, this was not an instance of triangulation but more a "sanity check" that confirmed that the data collected in similar studies were in fact similar.

Using multiple methods such as field observations and interviews in addition to document analysis would have allowed the researchers to truly triangulate by comparing the data collected using the various techniques to determine whether it is internally consistent and whether the conclusions suggested by the data gathered from document analysis, for example, can also be drawn from the data collected from the interviews and field observations. Smith and van Ierland used only a single method, however.

We have no way of knowing who or how diverse most of the tweeters were. The tweets included in this study were selected on the basis of their influence on the debate. Each of the included tweets received a minimum of 1500 "likes" or was retweeted at least 1500 times. And though we know the Twitter handles or usernames for the tweets quoted in the article, only three of the authors—actor and activist Mark Ruffalo, US Senator Bernie Sanders, and President Donald Trump (a tweet quoted in the article but not part of the study because it didn't use the #NoDAPL hashtag)—are people whose names are likely familiar to most readers of the article. Moreover, the number of tweets quoted in the article is limited, and two of them are repeated. So we must ask whether there are other ways that this study achieved diversity of perspective.

Using another technique called member checking, researchers can ask the participants to review their analysis of the data and the conclusions based on it. But Smith and van Ierland did not employ member checking. Given the nature of the data and their authors, however, member checking would have been difficult if even possible at all.

Yet another way of ensuring diversity of perspective is peer review—asking other researchers whether the conclusions reached in the study make sense based on their own experience. This type of peer review is typically performed during the study by asking colleagues to examine interview transcripts, field notes, or usability test results, as well as tentative conclusions based on those data. In this case, Smith and van Ierland asked two other people to discuss and verify their characterizations of the subject tweets. And of course the manuscript itself was peer reviewed and approved by two reviewers prior to publication in the *IEEE Transactions on Professional Communication*.

Thus, diversity of perspectives and methods is a weakness in this study. Nevertheless, the data are unarguably authentic, and the analysis was straightforward.

STAYING GROUNDED IN THE DATA

The final determinant of dependability is staying grounded in the data. All conclusions and statements must be traceable back to directly observed data within the study. This is certainly not a problem with Smith and van Ierland's article. All of the results, discussion, and conclusions derived from the study can be mapped to the tweets and the framing strategies that were used to categorize them.

Is It RAD?

As we saw in Chapter 2, RAD research as defined by Haswell (2005) must be replicable, aggregable, and data-supported. In other words, the methodology must be sufficiently defined that others can repeat the study. The results from the study must be reported in sufficient detail that they are capable of being aggregated or combined with the results of those repeated studies to build a body of data that can be compared and further built upon. And the conclusions of the study must be supported by those data, and not simply be the impressions or gut feelings of the researchers.

It is clear from our analysis that Smith and van Ierland's article meets these requirements. The methods are described in enough detail to allow them to be replicated. The data are reported in sufficient detail to enable them to be aggregated with data collected in repetitions of their project. And finally, the conclusions drawn in the article are entirely based on the data reported. As a result, we can conclude with confidence that this is a RAD article.

Summary

In this chapter we have analyzed a sample article reporting the results of a qualitative research study using the concepts and methods presented in Chapter 5. Following the text of the exemplar article, we have examined its purpose and audience; its structure; its methodology; the analytical method used to interpret the data; and the standards of rigor—the credibility, transferability, and dependability—of the study's conclusions.

The conclusions resulting from the analysis of tweets reported by Smith and van Ierland meet these standards of rigor to an acceptable degree. The tweets in fact are the "population" of interest, though that population is significantly limited by the selection criteria that the authors used. Moreover, the discussion that they report is authentic, and the study's conclusions could be duplicated by a different researcher.

For Further Study

Read one of the following articles reporting the results of a qualitative study:

> Boiarsky, C. (2018). Communication between government agencies and local communities: Rhetorical analyses of primary documents in three environmental risk situations. *IEEE Transactions on Professional Communication, 61*(1), 2–21.

> Malone, E. A. (2019). "Don't be a Dilbert": Transmedia storytelling as technical communication during and after World War II. *Technical Communication, 60*(3), 209–229.

When you have completed a careful reading of the article, analyze it using the Commentary in this chapter as a model, as well as the information about qualitative studies in Chapter 5.

Answer Key

Exercise 10.1

The answer to this exercise will be unique for each person who prepares it, so there is no key to this exercise.

Exercise 10.2

The answer to this exercise will be unique for each person who prepares it, so there is no key to this exercise.

References

Haswell, R. H. (2005). NCTE/CCCC's recent war on scholarship. *Written Communication, 22* (2), 198–223.

Smith, J., and van Ierland, T. (2018). Framing controversy on social media: #NoDAPL and the debate about the Dakota Access Pipeline on Twitter. *Technical Communication, 66*(3), 226–241.

11 Analyzing a Report on the Results of a Survey

Introduction

Chapter 6 explored surveys in terms of what population they sample, what they can measure, what kinds of questions they can ask, how questions can be formatted, and how they should be structured. In Chapter 6 we also discussed how to report results, looking especially at frequency distribution, response rate, and margin of error and confidence intervals. In this chapter, we examine in detail a journal article that reports the results of a survey. The chapter contains the full text of Sarah Read and Michael Michaud's "Hidden in plain sight: Findings from a survey on the multi-major professional writing course," which appeared in the Summer 2018 issue of *Technical Communication Quarterly*, along with a commentary on the article.

Learning Objectives

After you have read this chapter, you should be able to:

- Analyze an article reporting the results of a survey
- Apply the results of your analysis to preparing your own report of survey results

The Article's Context

We saw in Chapter 6 that surveys ask a defined population of respondents to answer questions about specific characteristics, attitudes, or behavior. We explored how to define the population of interest and select a sample from that larger population. We also described what types of questions a survey can include—open-ended, closed-ended, multiple choice, ranking, and rating—and how to avoid problems in phrasing questions to ensure that you operationalize the research purpose. Finally, we discussed the importance of piloting the survey, and we explored ways of reporting results and measures of rigor (response rate, margin of error, and confidence intervals) when you wish to generalize from the sample to the larger population of interest.

For teachers of technical and professional communication, one of the biggest pedagogical challenges that they face is teaching the introductory course to the field. Read and Michaud have chosen this course—the instructors who teach it, the students who take it, the pedagogical approaches used, and various institutional characteristics—as their research subject. They are interested in the multi-major professional writing course because it is taught in many colleges and universities of all varieties, from community colleges to major research institutions, and thus enrolls a diverse range of students. Similarly, the course is taught by a wide range of staff, including graduate teaching assistants, part- and full-time instructors, as well as tenured and tenure-track professors. Some schools use a common syllabus, texts, pedagogical approaches, and assessment techniques, while others allow the various people teaching the course to define their own approaches. And the students enrolling in these courses are as diverse as the institutions that they attend.

Thus, the multi-major professional writing course is a fertile topic for an exploratory study conducted using a survey. We will see that Read and Michaud carefully define the population that they want to study, describe the way that they selected their sample, and report on their formulation of questions and piloting of the survey. They then report the results, using frequency distribution of responses for most questions (number and/or percentage; see Tables 11.1–11.8, 11.13–11.15), and averages for some questions (see Tables 11.9–11.12). They avoid the problems associated with generalizing from their sample to the larger population and thus do not report response rate, margin of error, or confidence intervals.

Read and Michaud's article is a good model for student researchers to emulate because their survey is carefully constructed and the results are sensibly—if limitedly— analyzed. After reading the article, we will examine its components carefully in our Commentary.

Hidden in Plain Sight: Findings from A Survey on the Multi-Major Professional Writing Course

Sarah Read and Michael Michaud

[This article was originally published in 2018 in *Technical Communication Quarterly,* 27(3), 227–248. It is reprinted by permission of the publisher, Taylor & Francis Ltd. In this reprint, the questions in the *Appendix* have been numbered as they were in the Qualtrics survey to assist the reader in referring to them in the tables within the article and in the Commentary later in this chapter. We gratefully acknowledge the assistance of the authors in providing this information. In addition, several typographic errors have been silently corrected.]

Introduction

With some variation, the following course title and description can be found in college and university catalogs across the broad institutional landscape of U.S. higher education:

> Introduction to Technical/Professional Writing: Introduces technical and professional communications. Students compose, design, revise, and edit effective letters, memos, reports, descriptions, instructions, and employment documents. Emphasizes precise use of language and graphics to communicate complex technical and procedural information safely, legally, and ethically.

Descriptions such as this are recognizable by students, faculty, and administrators as a normal part of the writing curriculum at most colleges and universities. Its broadness and muddiness of scope are neatly signified by the "slash" construction of the course title, which conflates two related but distinct subfields: technical and professional writing. Although the course this description describes promises broad recognition in terms of its value to the curriculum, the ease with which it fulfills the expectations of its stakeholders has tended to foreclose the possibility of critical conversation about it.

One of the consequences of the normalization of descriptions like the one above is a dearth of disciplinary attention paid to the *status* of courses like Introduction to Technical/Professional Writing (beyond pedagogical innovations and/or local institutional issues and concerns). By invoking the notion of status, we mean to address more than whether this course is afforded low or high prestige by whichever stakeholders are in a position to proffer such a judgment. Asking about status is really a way of asking, "What is going on with this course given the present circumstances?" For example, a project team would prepare a status update of a large and complex project.

The notion of status is useful because it is broad and capacious, allowing inquiry into contemporary circumstances related to the course that have not previously been studied in a systematic way. Such inquiry can provide a foundation for a focused disciplinary conversation that we feel is long overdue. In this article, we initiate this conversation by reporting on results of a survey of writing instructors who teach what we call the "multi-major professional writing course" (MMPW). (Our survey instrument can be found in the Appendix.) In what follows, we report on findings from our survey and marshal them to make arguments about the status of the MMPW course in U.S. higher education.

What Is the MMPW Course?

The term *multi-major professional writing*, or MMPW, is our adaptation of what Kain and Wardle (2005) have termed "multi-major (or perhaps multi-professional) communication courses" (p. 114). The MMPW course exists in a wide range of forms in many, if not most, postsecondary settings. Although it appears in many instantiations, we understand it to be a survey/introduction to professional writing as a mode of communication that is decontextualized from any specific knowledge or professional domain (e.g., engineering, medicine, or a specialized business profession). MMPW courses are usually taken as service classes by students from across the university at all levels of degree completion and often also as electives by English or Writing majors and minors.

The broadness of our definition of the MMPW course reflects the difficulty previous researchers have encountered when trying to categorize or define curricula or programs in business, professional, and technical writing. Sullivan and Porter (1993), for example, acknowledge the difficulty of defining the boundaries between business, professional, and technical writing and fashion a definition for professional writing that is nearly as inclusive as ours: "a course or courses offered, usually by the department of English, as a service to other disciplines in the university; often, loosely equated with business and/or technical writing" (p. 392). Yeats and Thompson (2010) developed a continuum for identifying the curricular focus of technical and professional communication (TPC) programs yet, not surprisingly, found that that the idiosyncratic nature of programs based on local variation overwhelmed their attempt to categorize program types more systematically. Our own review of the 152 MMPW course titles and descriptions provided to us by our survey respondents revealed that of the 114 titles and descriptions that deployed at least one of the terms *business, technical, scientific, workplace, and/or professional*, 57, exactly half, used more than one of these terms, often deploying them interchangeably.

Given these definitional problems, we assumed that the variation Sullivan and Porter (1993) and Yeats and Thompson (2010) found in their studies is reflected in the introductory MMPW course. To avoid the inevitable conclusion that variation in the course across institutional types and variation in the institutional history of the course overwhelms systematic categorization or study, we chose to acknowledge this variation as a baseline for this survey and to set the definition for our unit of analysis to be broad enough to accommodate most of this variation. We felt that despite the variation in this introductory service course, including its name (business? technical? professional?), curricular focus, and institutional situation, enough commonality exists to warrant a study that assumes a singular entity within which variation is expected and normalized.

Why Study the MMPW Course?

One of the main reasons to study the MMPW course is because it is likely the second-most ubiquitous writing course in U.S. higher education, after first-year composition (FYC). And yet, despite its ubiquity, the MMPW course lacks most of the institutional recognition and professional infrastructure that has grown up around FYC. Although real numbers are not available for the total number of sections taught annually of either the MMPW course or FYC, we have made educated estimates based on data that is available from Carnegie Classification of Institutions of Higher Education (CCIHE; 2016a), the National Census of Writing (2013), and this survey. We conservatively estimate that there are about 40,000[1] (rounded to the nearest 10,000) sections of the MMPW course taught each year in the United States across all institutional types and about 200,000[2] (rounded to the closest 10,000) sections of FYC. Even though FYC sections outnumber MMPW sections by approximately 5 to 1, we can't think of another writing course that approaches this level of ubiquity, with the possible exception of academic English courses for second-language students. MMPW courses consume a substantial amount of institutional and faculty investment—an amount that more than warrants a study of the course's status.

What Is Already Known About the MMPW Course?

Despite its ubiquity and despite the fact that the origins of technical and professional communication (TPC) in the academy are in the MMPW course (Staples, 1999), little is known about the status of the course laterally across institutional contexts. This is because, historically, the MMPW course has existed at the periphery of the curricular and intellectual projects of English departments and other institutional sponsors (i.e., engineering schools or independent departments of rhetoric and writing). Although the ubiquity of the course expanded rapidly during the post-WWII era with the growth of engineering programs in higher education, the course suffered from "status-driven polarities" (Staples, 1999, p. 161) that ensured that it was taught by low-status members of English departments such as graduate students and adjuncts who had little to no incentive to do research on the course. Data gathered in 2009 suggested that as much as 83% of TPC service courses were taught by contingent faculty (Meloncon & England, 2011, p. 405). Whether because of lack of stakeholders' motivation in the course or its low status in English studies, systematic research into the MMPW course remained a risky endeavor for scholars in the emerging field of technical and professional communication. As a result, the MMPW course has not often been studied as a unit of analysis across the diversity of institutional contexts in which it is taught.

Since the postwar period of expansion, the field of TPC has, with success, invested a great deal in the development of graduate and undergraduate degree programs and certificates to raise the field's institutional and professional profile. This investment has resulted in dramatic growth of academic programs in TPC (Meloncon, 2012; Meloncon & Henschel, 2013). This investment is also evidenced by the focus of previously published survey-based studies, which chart the development of disciplinary apparatus, such as undergraduate majors and minors, graduate programs, and certificates (Allen & Benninghoff, 2004; Reave, 2004; Yeats & Thompson, 2010), and the profiles of the members of the Association of Teachers of Technical Writing (ATTW) professional organization (Dayton & Bernhardt, 2004). In none of these studies, however, is the MMPW the unit of analysis. The previous survey studies that are most closely related to the MMPW course focus on TPC textbooks (McKenna & Thomas, 1997; Warren, 1996), which are primarily in use in introductory-level courses, and the status of contingent faculty in TPC (Meloncon & England, 2011). These studies, however, are not based on the systematic collection of survey data from stakeholders but, instead, are "surveys" in the sense of overviews based on the historical development of the course or data that is available online, such as an online schedule of classes.

We also want to acknowledge that there is an extensive literature developing the curriculum and pedagogy for the TPC service course in journals such as *Technical Communication Quarterly, Journal of Business and Technical Communication, Journal of Technical Writing and Communication, Technical Communication,* and, to a lesser extent, *College Composition and Communication* and other composition journals. The depth and variety of this literature (e.g., Blakeslee, 2001; Kain & Wardle, 2005; Read & Michaud, 2015; Spinuzzi, 1996; Wickliff, 1997) lends the impression that the MMPW course has been a focus of research interest; however, the majority of this literature focuses on pedagogical practices or curricular developments and not on the status of the course as a unit of analysis unto itself. As important as the study and development of teaching practices and curricula are, these types of studies do not document or account for aspects of the course that we report on in this article, such as institutional situation and instructor confidence in the course.

This survey study assumed that the MMPW course is a unit of analysis that can be studied laterally across institutional types and contexts. Because of this assumption, this study was a robust, but preliminary, effort to survey instructors and to investigate the question, "What is the status of the multi-major professional writing (MMPW) course?" The results of such a study provide a macrolevel view of who teaches the course, its institutional situation, common pedagogical practices, and perceptions of the effectiveness of the course. The value of this view is that it can motivate discussion at the level of the field's professional organizations about ways to develop and invest in the MMPW course so as to ensure that it remains relevant and responsive to changes in the 21st-century academy and economy.

Methodology

Given the ubiquity of the MMPW course across all contexts of higher education, we chose to make the principal design choice of our methodology a commitment to proportional representation among respondents by the type of institution where they primarily teach an MMPW course (e.g., associate's colleges, master's colleges and universities, doctorate-granting institutions, tribal colleges). This methodology generally reflects that of Meloncon and England's (2011) study of contingent faculty in TPC, which drew data on the rank of instructors teaching the TPC service course across six Carnegie classifications of 4-year institutions. However, we wanted this survey data to reflect the experiences and views of MMPW instructors across the diversity of 2- and 4-year postsecondary contexts. Toward these ends, we used data published by the CCIHE (2016b) to set quotas for respondent recruitment by institutional type from the overall population of MMPW instructors. We used CCIHE's six postsecondary institutional types for 2-year and 4-year institutions and reported proportional representation of student enrollment at each institutional type (see Table 11.1). In addition, we set our recruitment goal at 150 completed surveys. Random sampling, the ideal technique in social science research, was not possible because this study recruited from a population that is not centrally documented. Nonrandom techniques such as convenience sampling are often used when data supports exploratory analysis (Kelley, Clark, Brown, & Sitzia, 2003) or when results will not be generalized to the larger population using inferential statistics (Banerjee & Chaudhury, 2010).

When disseminating this survey, we faced challenges recruiting instructors from a population that is not centrally documented (such as in a directory). Given the lack of a census of this population, our sample was what social scientists call a "convenience sample." This means that we recruited respondents from the population via channels that we had access to given our own professional affiliations and connections. The channels included the email listservs of professional organizations in the field, the social media sites of the National Council of Teachers of English (NCTE) college section, emails to colleagues requesting assistance distributing the survey link, and a small number of personalized emails to instructors who are particularly hard to reach, such as those at tribal colleges. To aid in recruitment, we offered a modest incentive in the form of a $5 coffee card to all who completed our survey.[3] In total, 220 respondents consented to take the survey and 154 respondents completed the survey in its entirety (a 70% completion rate). To keep our N consistent across all survey questions, we excluded the incomplete surveys from our data during the analysis phase (see Table 11.1).

In addition to proportional institutional representation, we also aimed for diversity in the faculty rank of survey respondents. We did not set formal quotas here because we knew that our recruitment techniques would necessarily be biased toward respondents who are the most professionally active. As a rough guide, however, we used data from the American Association of University Professors (AAUP) on the national percentage of all instructional staff by employment status to put our response rates by faculty rank into perspective (Curtis, 2014; see Table 11.2). The nontenure-track representation (50%), including full-time and part-time instructors and graduate students, exceeded the tenure-track representation (38%), ensuring that nontenure-track respondents had a strong voice in the aggregate survey data.

Table 11.1 Percentage of survey respondents by Carnegie classification institution

Carnegie Classification Category	Proportional Representation by Student Full Time Equivalent Enrollment (numbers available in 2014)[a]	Percentage of Survey Respondents (N = 154, complete surveys only)
Associate's colleges (predominantly 2-year institutions)	37%	38% (n = 59)
Baccalaureate colleges (largely liberal arts colleges)	7%	4% (n = 6)
Master's colleges and universities	23%	19% (n = 30)
Doctorate-granting institutions	28%	37% (n = 57)
Special focus & faith institutions (includes stand-alone law, business, and medical schools)	7%	1% (n = 1)
Tribal colleges	1%	1% (n = 1)

Note: [a] Numbers rounded up to nearest whole number. Total exceeds 100% because of rounding.

Table 11.2 Percentage of survey respondents by faculty rank

Faculty Rank	AAUP Reported Percentage of All Faculty for 2011	Percentage of Survey Respondents (N = 154, complete surveys only)
Tenure track	23.5%	38% (n =59)
Full-time, nontenure track	15.7%	23% (n = 35)
Part-time adjunct or contingent faculty	41.5%	20% (n = 31)
Graduate instructor/teaching assistant	19.3%	7% (n = 11)
Other, including full-time staff with teaching responsibility	N/A	12% (n = 18)

Note: AAUP = American Association of University Professors.

Because we were able to meet our recruitment goals within a reasonable range, we feel comfortable that our survey data generally reflects the broad experiences and views of MMPW instructors across institutional contexts and faculty ranks.

A second important issue we confronted when designing this study had to do with defining the notion of status, the pivotal term in our research question. For the purpose of this survey, we articulate *status* to include descriptive and subjective components. Descriptive components include aspects of the course that reflect its current situation in a local institutional context:

- Who teaches the course and how
- Who takes the course and why
- What kinds of materials are used to teach the course (i.e., textbooks, technology)
- How the course is situated institutionally
- What kinds of professional and institutional investments go into the course

The subjective component of status includes the assessment of instructor confidence in the effectiveness of the course:

- Student satisfaction
- Levels of student learning
- Levels of meeting student expectations
- Levels of the course meeting stated objectives
- Levels of resource availability and support for instructors
- Effectiveness of the course in meeting larger curricular goals for students and programs

Additionally, the subjective assessment portion of "status" includes instructors' perceptions of how the course can be improved (e.g., common outcomes, more opportunities for professional development, new textbooks, etc.). In sum, in investigating the status of the MMPW course, we are interested in trying to understand the many and diverse conditions under which it is taught and the subjective perceptions of those who teach it.

We designed and built our survey using the online survey tool Qualtrics. Forty-five questions on this survey supported the descriptive component, the majority of which were multiple choice (several, though, offered opportunities for write-in answers). In the descriptive arena, we asked respondents about the institutional, curricular, and faculty situation of the course as they understood it in their local context. The subjective component of the survey comprised eight questions that prompted respondents to report on their level of confidence in various aspects of the course. Respondents moved a slider along a scale from zero to 100, with zero indicating *not at all confident*, 50 indicating *sort of confident*, and 100 indicating *fully confident*. The default position of the slider was set to the middle position, 50, *sort of confident*, which we understood to reflect a position of ambivalence or uncertainty, to encourage respondents to clarify their level of confidence in one direction or the other. This turned out to be a crucial design decision for our analysis of survey data (more on this below). Finally, in the subjective section of the survey, we asked respondents to sort a list of 13 items according to whether they would improve their confidence in the course. Respondents sorted items into boxes marked *would definitely help*, *would help a little*, *would make no difference*, *already doing*, and *not relevant*. We concluded the survey by offering respondents the chance to share any information about the course and their experience of teaching it that we may have failed to solicit. To ensure that this survey would gather the data we intended and be maximally user-friendly for respondents, we user tested it during the winter and early spring of 2015 and obtained approval from the Internal Review Board at both our institutions.

In what follows, our intent is to provide descriptive evidence of patterns in the response data. Rather than present our data in a linear fashion or in the order in which it was collected in this survey, we have marshaled key findings in strategic ways so as to advance a series of arguments about the status of the MMPW course in U.S. higher education. At every turn, we have recognized the limitations of our data and resisted inferring causation for results or generalization to a broader population (which remains undocumented). Within these limitations, we believe that our data, based on a quota-based convenience sample, is adequate for initiating a conversation within the field about the implications of our study.

In this article, we begin by drawing on the descriptive data to make an argument about what we discovered to be a persistent conservatism in the teaching of the course. We then move on to the subjective data to discuss the implications of respondents' varying levels of confidence in the course, especially in regard to the preparedness of the course's instructor corps. We close by discussing the implications of our findings for key stakeholders.

Findings

In this section of our article, we share findings on the status of the MMPW course organized along three lines of inquiry that emerged as salient from the survey data:

- What pedagogical approaches shape the teaching of the MMPW course?
- How confident are instructors in the MMPW course?
- What would improve instructor confidence in the MMPW course?

What Pedagogical Approaches Shape the Teaching of the MMPW Course?

When investigating the role of pedagogy in the MMPW course, we began with our own experiences, which suggest that the pedagogical choices writing instructors make in their classrooms are complex, evolving, and multifaceted. As Tate, Tagart, Schick and Hessler (2013) write in their "Introduction" to *A Guide to Composition Pedagogies*, "rare is the teacher who does not blend the practices of many pedagogical philosophies" (p. 6). With this in mind, we set about trying to learn more about the pedagogies that shape MMPW instruction. First, we asked about the teaching materials instructors use. Next, we asked about the pedagogical approaches they draw on. Finally, hoping to learn about the ways in which instructors, departments, and institutions articulate MMPW pedagogies to themselves, students, parents, and other stakeholders, we asked respondents to share with us the title and catalogue description of the MMPW course they teach.

Although our respondents indicated that they utilize a range of teaching materials in MMPW classes (i.e., online handbooks, general and business periodicals, scholarly articles, etc.), 92% use a textbook to teach the course. This finding, as Table 11.3 shows, is consistent across most institutional types.

The finding that a majority of survey respondents utilize a textbook may not be surprising. As Warren (1996) notes in his survey of technical writing textbooks from

Table 11.3 Results for Q6: Do you or have you used a textbook to teach the MMPW course?

Institutional Type	Yes	No	Percentage Yes
Associate's college	56	3	95%
Baccalaureate college	4	2	67%
Master's college or university	27	3	90%
Doctorate granting	53	4	93%
Special focus or faith	1	0	100%
Tribal college	1	0	100%

1950 to 1970, "The textbook is the main teaching tool for instructors" (p. 155). Connors (1982), too, has shown how textbooks have historically been the primary teaching tool of MMPW instructors. In addition to learning whether or not respondents utilize a textbook, though, we were also interested in the degree of autonomy they have to choose their textbook. Our survey results show that just under two thirds (62%) of respondents who indicated that they use a textbook also indicated that they choose their own book (see Table 11.4).

Interestingly, when we looked at the question of why respondents chose a textbook, we found that greater flexibility and choice seem to be granted to instructors teaching at bachelor's (100% choose text), master's colleges and universities (87% choose text), and doctorate-granting institutions (78% choose text) than at associate's colleges (60% choose text; see Table 11.5).

The finding that almost two thirds of survey respondents choose their own textbook seems to suggest a high level of autonomy in devising pedagogies to teach their courses. At the same time, the fact that so many respondents use a textbook suggests that textbooks may exert a powerful influence over the curriculum and pedagogy of MMPW courses. This fact makes an investigation into which textbooks MMPW instructors use critical. Before we get to our findings on textbook use, though, we

Table 11.4 Results for Q76: Why did you choose this textbook?

Answer Choice	Responses	Percentage of Textbook Users (n = 142)
It is required by the program/department that all sections of this course use this textbook	32	23%
I personally chose it	88	62%
Other (write-in)	22	15%
Total	142	100%

Table 11.5 Results for Q76: Why did you choose this textbook?

Institutional Type	It Is Required by the Program or Department That All Sections of This Course Use This Textbook	I Personally Chose It	Percentage Chose Own Textbook
Associate's college	19	28	60%
Baccalaureate college	0	4	100%
Master's college or university	3	20	87%
Doctorate granting	10	35	78%
Special focus or faith	0	1	100%
Tribal college	0	0	N/A

would like to offer this qualification: We are aware that textbooks have a varying degree of influence over the curriculum and pedagogy of a writing class. Although some instructors may follow the outline of a textbook to the letter, others might dip into a book now and again to touch on certain topics or use the book behind the scenes as a resource for planning lessons. In sum, we are well aware that it is difficult to draw firm conclusions about textbook-use from survey data that indicate only textbook selection.

Having said this, the survey asked respondents to identify the textbook they use from a list of nine titles that we created based on our own anecdotal knowledge of popular titles in the field. Table 11.6 lists the nine books as well as the percentage of respondents who indicated that they use each title.

We note that though nearly one half (43%) of those who indicated that they use a textbook do not use a book on our list, nearly half (43%) use one of five well-known textbooks: *Technical Communication* (Markel, 2014), *Writing That Works* (Oliu, Brusaw, & Alred, 2016), *Essentials of Technical Communication* (Tebeaux & Dragga, 2014), *Technical Communication Strategies for Today* (Johnson-Sheehan, 2014), *Successful Writing at Work* (Kolin, 2012). Each of these five textbooks deploys what we have identified in another section of our survey as a "communications genres" approach to the teaching of MMPW. This approach walks students through a fairly stable and, as historical studies of professional writing instruction have shown (see Connors, 1982), well-established roster of professional document types (e.g., memos, letters, proposals, reports, etc.). The pervasiveness of this approach as it is formalized across the most commonly used textbooks raises the question of whether these textbooks are still in touch with research in the field of professional and technical communication and the realities of 21st-century knowledge-based communications practices. This is a larger question that can only be initiated by our analysis here and would require more in-depth analysis to pursue further.

Because we assumed that textbook choice offers only a limited window into instructors' pedagogical practice, we also asked survey respondents to report on their approach to teaching the MMPW course, giving them the chance to choose from a list

Table 11.6 Results for Q84: Which of the following is the textbook that you use NOW or have used most recently?

Textbook	Responses	Percentage of Textbook Users (n = 142)
Technical Communication (Markel, 2014)	20	14%
Writing That Works (Oliu, Brusaw, & Alred, 2016)	12	8%
Technical Communication (Lannon, 2007)	5	4%
Essentials of Technical Communication (Tebeaux & Dragga, 2014)	8	6%
Technical Communication Strategies for Today (Johnson-Sheehan, 2014)	19	13%
Strategies for Technical Communication (Lannon & Gurak, 2016)	3	2%
Business Communication (Lehman & DuFrene, 2010)	1	1%
Workplace Communication (Searles, 2013)	2	1%
Successful Writing at Work (Kolin, 2012)	11	8%
Write-in	61	43%

of common approaches gathered from our experience and the scholarly literature. Respondents first identified any/all of the approaches that they draw on when teaching MMPW courses and then were asked to identify their primary or central approach from the same list of possibilities (see Tables 11.7 and 11.8).

Table 11.7 Results for Q17: Which of the following approaches characterize how you teach your MMPW course? (Choose all that apply.)

Teaching Approach	Responses	Percentage of Respondents Choosing an Approach
Teaching communication genres (e.g., memos, letters, reports, etc.)	135	88%
Teaching professional development genres (e.g., resumes, cover letters, LinkedIn)	104	68%
Connecting students to clients for writing projects (e.g., collaboration with institutional or industry partners)	43	28%
Engaging students in service learning projects in the community (e.g., community/nonprofit partnerships)	51	33%
Teaching students about how to do their own research about writing in workplace contexts	101	66%
Engaging students in reading scholarly texts (e.g., journal articles, monographs, research reports) and experimenting with scholarly research methods (e.g., ethnography, case study, interviews, etc.)	74	48%
Exploring case studies to create contexts for writing assignments	75	49%

Table 11.8 Results for Q82: Now choose your CENTRAL or PRIMARY approach for teaching the MMPW course (choose one)

Teaching Approach	Responses	Percentage of Respondents Choosing an Approach (n = 152)
Teaching communication genres (e.g., memos, letters, reports, etc.)	78	51%
Teaching professional development genres (e.g., resumes, cover letters, LinkedIn)	7	5%
Connecting students to clients for writing projects (e.g., collaboration with institutional or industry partners)	7	5%
Engaging students in service learning projects in the community (e.g., community/ nonprofit partnerships)	13	9%
Teaching students about how to do their own research about writing in workplace contexts	23	15%
Engaging students in reading scholarly texts (e.g., journal articles, monographs, research reports) and experimenting with scholarly research methods (e.g., ethnography, case study, interviews, etc.)	13	9%
Exploring case studies to create contexts for writing assignments	11	7%

Overall, our prediction, based on the results of our textbook questions, that a genre-based approach is pervasive in the teaching of MMPW among respondents, was supported by the data. We found that the teaching communication genres approach was the most frequently chosen approach (88% of respondents) and the most frequently chosen primary or central approach (51% of respondents). Interestingly, we found that these results remained largely consistent across all institutional types. For example, we did not find significant variations in the frequencies of approaches chosen between respondents from doctorate-granting and associate's institutions.

We note that the second most common teaching approach among survey respondents is teaching professional development genres (68%), which is an approach that also focuses on explicit instruction in specific genres of writing (i.e., cover letters, resumes). Instruction in communications genres and professional development genres is a component of all five of the most commonly used MMPW textbooks, and these findings about approach are consistent across all institutional types.

Other notable findings from our questions about pedagogical approach include the finding that two thirds of respondents (66%) indicate that "teaching students how to do their own research about writing in workplace contexts" is an element of their pedagogy and that just over one half of all respondents connect their MMPW students with professional writing contexts external to the college/university (28% engage students in client-centered work and 33% engage them in service-learning projects). The relatively robust frequencies for these alternate approaches are tempered by the fact that only 14% of respondents identify one of these two approaches as their central or primary approach. Finally, we note that though 48% of respondents, almost half, indicate that they engage students in reading scholarly texts and experimenting with scholarly research methods, an approach we advance (Read & Michaud, 2015), only 9% identify this approach as their central or primary one.

A final and admittedly more complicated means of gathering data on the pedagogies of MMPW courses comes from the course titles and descriptions that instructors, programs/departments, and institutions create to communicate to themselves, students, parents, and other stakeholders the content and approach of their MMPW course. Our initial review of course titles and descriptions confirmed the complexity of using them as a means of understanding pedagogical concerns—there were simply too many variations in terms to create stable categories of analysis. However, given the emerging picture in our data about the prevalence of the formalist, genre-based approach, we limited our analysis of the course description data to a single question, asking whether course descriptions explicitly specified that students would be exposed to or taught prototypical workplace communication genres (e.g., memos, letters, proposals, and reports). We found that 69 out of the 91 descriptions we analyzed, or 76%, explicitly mentioned that communications genres would be covered in the course. We feel that this number is meaningful and that it further illustrates the extent to which teaching communications genres is an important element of our survey respondents' pedagogical approach to teaching the MMPW course.

Our findings under the banner of the question: "What pedagogical approaches shape the teaching of the MMPW course?" lead us to suggest that there are persistent norms among respondents for pedagogical approach and choice of teaching materials in MMPW instruction that are shaped by what have been known generally as current-traditional rhetorical (CTR) practices. Although we are aware that importing the disciplinarily loaded term CTR into a discussion of MMPW pedagogies is fraught with complication, we ground our use of this term in Crowley's (1998) articulation of CTR as a pedagogy that "resists changes in its rules and preserves established verbal traditions and institutional lines of authority" (p. 218). We note that all three sources of evidence we examined to investigate pedagogical considerations—choice of textbook, choice of teaching approach, and description of course content and approach—suggest the prevalence of a largely formalist, genre-driven approach to the teaching of MMPW among respondents.

How Confident Are MMPW Instructors in the MMPW Course?

Part of our motivation for undertaking this survey project was to document how respondents felt about the MMPW course. In addition, we wanted to prompt respondents to think beyond their own experiences with the course and to get an idea of their confidence in the whole enterprise of the MMPW course at their institution. We were motivated to conduct what we called the "subjective assessment" portion of the survey because of our own personal suspicions, based on experience and anecdotal evidence, that we would find a gap between the confidence that instructors of the MMPW course have in their own instantiations of the course and their confidence in the enterprise of the MMPW at their institution in general. In this section of the article we will share our findings from the subjective assessment portion of the survey and discuss what they can mean for how we understand the status of the MMPW course.

The subjective section of the survey was comprised of eight questions that each prompted the respondent to record his or her level of confidence in an aspect of the MMPW course by moving a slider to an acceptable position along a scale from *not at all confident* (0) to *fully confident* (100). The slider default was set to the middle position of *sort of confident* (50). Each of the slider prompts was a statement in which respondents rated their level of confidence. The first prompt asked respondents to record their "overall confidence in the effectiveness of the MMPW." The seven following statements addressed narrower aspects of the MMPW course. These statements fell into three categories of factors related to the course: students, instructors, and institutions. This design created internal consistency for the survey and also meant that descriptive data from the first part of the survey could be used to better understand the subjective data from the second part of the survey. Table 11.9 lists the eight slider prompts and the average slider position (on a scale from 0–100) across all respondents for each prompt, broken out by student, instructor and institutional factors.

At first glance, Table 11.9 tells a uniformly positive story. First, across all of the prompts, the average of all respondents' ratings is above 50 (*sort of confident*). What this means is that when responses from all survey respondents are considered collectively the overall picture is one of moderately high confidence. This is true across all the categories of factors, from overall confidence in the course (Q62: average 67), to confidence in whether students are readier to write in the workplace after taking the course (Student factors; Q38_2: average 78), to whether instructors have sufficient experience and training (Instructor factors; Q41_1: average 62), to whether instructors have access to sufficient professional development (Q41_2: average 55). Taken from a high level, if any average above 50 is considered confirmation of a positive level of confidence in the course, then the data suggest that confidence in the MMPW course across the three factors is positive.

Table 11.9 Subjective assessment slider prompt and average response across all respondents

Subjective Assessment Question	Average for All Respondents (N = 154)
Q62: Overall confidence in effectiveness of MMPW	67
Student factors	
Q38_1: Students feel more ready to write in workplace contexts after taking the MMPW course	78
Q38_2: The MMPW course helps students to acquire sufficient and appropriate technological skills to write in future workplace contexts	68
Q38_3: MMPW students are more ready to write in the workplace than students who have not taken the MMPW course	84
Instructor factors	
Q41_1: Instructors have sufficient experience and training	62
Q41_4: Sections are equal in rigor and presentation of standard curriculum	59
Institutional factors	
Q41_2: Instructors have access to sufficient professional development	55
Q41_3: MMPW instructors at your institution have access to sufficient technologies in the classroom to effectively teach the course	72

Note: MMPW = Multi-major Professional Writing.
Averages rounded up to the nearest whole number.

The average level of confidence is not equal, however, across all aspects of the course. There is almost a 30-point difference between the highest rated average level of confidence (whether students who have taken an MMPW course are readier to write in the workplace than students who have not taken the MMPW course [Q38_3; average 84]) and the lowest rated average level of confidence (whether instructors have access to sufficient professional development [Q41_2: average 55]). Another lower rated average level of confidence appears for whether sections of the MMPW are equal in rigor and presentation of standard curriculum (Q41_1; average 59). Even though the level of confidence in the course among respondents is on average at positive levels, the confidence levels vary depending on the aspect of the course under consideration.

RESPONDENTS REPORTED LOWER CONFIDENCE IN INSTRUCTOR FACTORS THAN STUDENT FACTORS

The average levels of confidence across all prompts for a given factor (student, instructor, institutional) make visible the variation in confidence levels by aspect of the course (see Table 11.10). The average slider response across all prompts related to students (student factors) is higher, on average, than confidence in aspects of the course that are related to instructors or institutions (instructor and institutional factors). Although average confidence levels in instructor and institutional factors are about the same (61 and 64, respectively), average confidence levels in student factors are notably higher (77).

What the overall positive levels of the averages in Table 11.10 do not account for, however, is the extent to which the responses that were below 50 (sort of confident)

Table 11.10 Average rating of confidence for all respondents summarized by factor.

Subjective Assessment Question Type	Average for All Respondents (N = 154)
Q62: Overall confidence in effectiveness of the MMPW course	67
Student factors (Q38_1-Q38_3)	77
Instructor factors (Q41_1; Q41_4)	61
Institutional factors (Q41_2; Q41_3)	64

Note: MMPW = Multi-major Professional Writing.

across all of the slider prompts affect the overall averages. As we considered the data, we found ourselves wondering what confidence levels across all three factors would look like if we broke out the responses of the 19.4% of respondents who registered a less than 50 response to question Q62 (overall confidence in the effectiveness of the MMPW course): How would the trend in lower confidence in instructor factors than student factors look within this less confident group? New insights emerged into the variation in average confidence levels between student, instructor, and institutional factors (see Table 11.11) when the data was sorted into whether respondents recorded an overall confidence level (Q62) of either above 50 or below 50.

Breaking out the respondents who recorded less than 50 for overall confidence level confirmed the trend of lower levels of confidence in instructor factors than student factors. Most notably, the disparity between the average level of confidence in student factors (60) and the average level of confidence in instructor factors (32)— a disparity of almost 30 points—is even more pronounced for the less-than-50 group than for respondents overall (difference of 16 points between student and instructor factors). In addition, this level of disparity is maintained across all three survey questions that relate to instructors: whether instructors have sufficient experience and training (Q41_1; average 31), whether sections are equal in rigor and presentation of standard curriculum (Q41_4; average 32), and whether instructors have access to sufficient professional development (Q41_2; average 34). Although the average levels of confidence in instructor factors are lower than student factors across all respondents, this trend is even more pronounced for the 19.4% of respondents who recorded an overall lower level of confidence in the course. We take this trend as confirmation that an important source of lower confidence for the less-than-50 group is in aspects of the MMPW course that relate to instructor experience, training, and effectiveness. Looking across all of the average levels of confidence in student, instructor, and institutional factors, broken out by overall confidence level in the course (see Table 11.12), the average level of confidence for instructor factors (32) for the less-than-50 group stands out as the lowest level of confidence overall.

LOWER CONFIDENCE IN INSTRUCTOR FACTORS IS NOT RELATED TO INSTITUTIONAL TYPE OR INSTRUCTOR RANK

Having noticed the trend across all respondents in the lower level of confidence in aspects of the course that relate to instructor experience, training, and effectiveness, we wondered whether respondents with a specific instructor rank or institutional

Table 11.11 Average level of confidence for student, instructor, and institutional factors by over-all confidence level (under 50 and above 50).

Subjective Assessment Question	Average for Respondents under 50 for Q62 (n = 30)	Average for Respondents over 50 for Q62 (n = 124)	Average for All Respondents (N= 154)
Q62: Overall confidence in effectiveness of MMPW	36[a]	75	67
Student factors			
Q38_1: Students feel more ready to write in workplace contexts after taking the MMPW course	58	82	78
Q38_2: The MMPW course helps students to acquire sufficient and appropriate technological skills to write in future workplace contexts	52	72	68
Q38_3: MMPW students are more ready to write in the workplace than students who have not taken the MMPW course	69	87	84
Instructor factors			
Q41_1: Instructors have sufficient experience and training	31	69	62
Q41_4: Sections are equal in rigor and presentation of standard curriculum	33	65	59
Institutional factors			
Q41_2: Instructors have access to sufficient professional development	34	60	55
Q41_3: MMPW instructors at your institution have access to sufficient technologies in the classroom to effectively teach the course	59	75	72

Note: MMPW = Multi-major Professional Writing.
[a]Averages rounded up to the nearest whole number.

Table 11.12 Average rating of confidence summarized by factor by respondent group

Subjective Assessment Question Type	Average for Respondents under 50 for Q62 (n = 30)	Average for Respondents over 50 for Q62 (n = 124)	Average for All Respondents (N = 154)
Q62: Overall confidence in effectiveness of the MMPW course	36	75	67
Student factors	60	80	77
Instructor factors	32	67	61
Institutional factors	47	68	64

Note: MMPW = Multi-major Professional Writing.

type were represented more heavily than others in the lower confidence group. Because 56% of respondents indicated that most of the MMPW course sections at their institution are taught by nontenure-track faculty (28% part-time adjunct, 27% full-time nontenure-track, 6% graduate TA) and the most frequent institutional type where most MMPW courses are taught by adjunct faculty was associate's (56%), we wondered whether respondents off the tenure track with institutional type of associate's would be represented at a higher frequency in the less-than-50 group than in respondents as a whole. This assumption would seem warranted given that, broadly speaking, instructors off the tenure track often have higher teaching loads yet less access to professional development than tenure-track colleagues, especially at institutions with fewer resources.

Notably, this assumption turned out to be wrong (see Table 11.13). When we took into consideration the institutional type and instructor rank of the lower confidence group (under 50 on Q62) we found that both indicators were in range of being proportionately represented at the same rate as among survey respondents as a whole. For example, percentage representation of institutional affiliation at associate's colleges was very similar between the under-50 group (40%) and respondents as a whole (38%). Likewise, 20% of both the under-50 group and respondents overall claimed adjunct/part-time as their instructor rank. The other categories were also similar, except for a moderate divergence between the representation of the instructor rank of tenure track in the under-50 group (47%) and respondents as a whole (38%). Overall, comparing the representation of institutional type and instructor rank between the group with overall lowest confidence in the MMPW course and survey respondents as a whole did not reveal any meaningful trends that might suggest that either one of these factors contributes more strongly than anything else to a lower level of confidence in the course.

Table 11.13 Comparing representation of institutional type and instructor rank for the overall lower confidence group to survey respondents as a whole

	Under-50 for Q62	*Percentage of Under-50 Respondents* (n = 30)	*All Respondents*	*Percentage of All Survey Respondents* (N= 154)
Institutional type				
Associate's college	12	40%	59	38%
Master's college	5	17%	30	19%
Doctoral institution	10	30%	57	37%
Baccalaureate college	1	3%	6	4%
Special focus	1	3%	1	1%
Instructor rank %				
Tenure track (TT)	14	47%	59	38%
Full-time non-TT	8	27%	35	23 %
Adjunct/part-time	6	20%	31	20%
Teaching assistant	1	3%	11	7%

INDIVIDUAL CASES OF RESPONDENTS WITH THE LOWEST CONFIDENCE LEVELS LOOK UNIQUE

Given that there was no clear trend linking institutional status and instructor rank to a lower level of confidence in the course overall, we turned to looking at other aspects of the individual situations of the 10 respondents with the lowest overall confidence levels (under-40 for Q62) to see if we could find any patterns in their situations. Among these 10 respondents, we looked at primary area of graduate training, number of MMPW sections taught per year at that institution (1–10, 11–50, more than 50), faculty rank of who teaches the most sections (tenure track [TT], full-time nontenure track [FT NTT], part-time [PT]/adjunct, teaching assistant [TA]), presence of a course coordinator (yes/no), presence of a common syllabus for the course (yes/no), and access to professional development (yes/no). Notably, even among this least confident group, we found that most combinations of answers to these questions were present across all institutional types and faculty ranks. In other words, when considered at an individual level, respondent situations look unique.

Given that the situations of the 10 respondents with the lowest amount of confidence in the course look unique, it would be possible to conclude that the overall confidence levels of these respondents are not linked to any of the elements measured by the survey, but to other things entirely (which could easily be the case). It would also be possible to conclude that the institutional situations and instructor profiles for survey respondents vary so much that each case can only be considered on its own terms. These are reasonable conclusions to draw from these 10 respondent cases; however, we do not want this ambivalence to overwhelm what we understand as a robust and meaningful trend in the subjective assessment data from this survey: that a lower level of confidence in the instructor factors related to the course (compared to confidence in the value the course has for students) is generalized in the data across all institutional types, instructor ranks, and institutional situations. In addition, though the lower level of confidence in instructor factors is more pronounced for the 19.4% of respondents who reported a lower level of confidence in the course overall, it is also true, on average, that across all survey respondents there is a lower level of confidence in the instructor corps for the MMPW course than in the confidence level for the value that it has for students. We understand this to mean that a meaningful number of MMPW course instructors across all institutional types and instructor ranks feel that the MMPW instructor corps at their institution is lacking in experience, training, and access to professional development.

WHAT DO WE KNOW ABOUT THE MMPW INSTRUCTOR CORPS THAT CAN EXPLAIN THIS LOWER CONFIDENCE?

Given the generalized sentiment that the instructor corps for the MMPW course is lacking in experience, training, and access to professional development, we turned to the descriptive data from survey respondents about levels of training, experience in industry, and access to professional development to see if there were any strong trends that might explain this sentiment.

Table 11.14 displays the frequency of responses for aspects of instructor training by institutional type, including highest degree obtained (Doctor of Philosophy [PhD],

Table 11.14 Aspect of instructor training by institutional type.

Aspect of Instructor Training	Percentage of Respondents by Institutional Type	
	Doctoral (n = 58)	*Associate's (n = 60)*
Highest degree: PhD[a]	59%	33%
Highest degree: MA/MFA/MX[a]	31%	52%
Have graduate training in PTC	28%	3%
Have graduate training in PTC or comp.	72%	35%
Have PTC industry experience	60%	57%
Have access to professional development	64%	55%

Note: MA = Master's; MFA = Master's of Fine Arts; MX = other Master's-level degrees; comp. = Composition; PTC = Professional and Technical Writing.

[a] Percentages for highest degree do not add up to 100% for institutional type because not all answer choices are represented in this table (such as "JD" or "other").

Master's [MA], Master's of Fine Arts [MFA], or any type of master's-level degree [MX]), graduate training specifically in professional and technical communication (PTC) or composition or writing studies (comp.), PTC experience in industry, and access to professional development opportunities at the home institution. In our discussion we will focus on the data from doctoral and associate's institutions because the number of respondents in each of those groups is roughly equal (n = 58 and 60, respectively) and together these two institutional types make up a large proportion of institutions of higher education overall (68%, Carnegie 2015 data [Carnegie, 2016a]).

Given this relative comparability between these two groups, Table 11.15 suggests trends in instructor training, experience, and access to professional development that are on the one hand anticipated and at the same time troubling.

Although the frequency of PhD as the highest degree earned by MMPW instructors is higher at doctoral institutions (59%) than at associate's institutions (33%), we did not find this to be unexpected, given that historically a PhD has not been a minimum requirement for full-time (tenure-track and nontenure-track) positions at 2-year colleges. Nor did we find it unexpected that the frequency of master's degrees (of any type) is in roughly equal inverse proportion to PhDs across both institutional types (31% doctorate, 52% associate's). However, when the primary focus of graduate training is taken into account, a stronger trend is noticeable. When asked to declare the primary focus of their graduate training, only 3% of respondents at associate's colleges chose PTC compared to 28% at doctorate-granting institutions. Because we don't want to suggest that only graduate training specific to PTC qualifies an instructor to teach the MMPW course, we also took into consideration the frequency with which respondents declared composition (comp.) as an area of graduate training. When the frequencies of graduate training in PTC and comp. are aggregated, the disparity between doctorate-granting and associate's institutions widens significantly: 72% of respondents from doctorate-granting institutions have graduate training in either PTC or comp., whereas this is only the case for 35% of respondents at associate's institutions—a disparity of 37 points. We find this disparity to be meaningful, especially in light of the fact that these frequencies mean that 65% of respondents at associate's colleges (and 28% at doctoral institutions) do not have graduate training

Table 11.15 Ranked item that "would help" improve confidence

Item Related to MMPW Course	Would Help (n = 154)	Would Make No Difference (n = 154)	Already Doing (n = 154)
1. Training in contemporary workplace technologies for instructors	135	9	8
2. More instructors with industry experience	106	28	15
3. More instructors with a scholarly interest in professional writing	106	23	14
4. Adding an experiential dimension to the course (internship, service learning, or client project)	97	14	33
5. Smaller class sizes	93	24	33
6. A longer MMPW course sequence (2 or more quarters or semesters)	88	44	6
7. Additional classroom technologies, including software and hardware	87	27	35
8. A set of standardized outcomes for the course from a professional organization such as WPA, ATTW, CPTSC, or other.	85	22	42
9. Resources to assess the course across sections on a regular basis	81	25	37
10. New textbooks that support contemporary workplace writing practices	69	42	39
11. A college-, department-, or program-level coordinator for the MMPW course	56	29	54
12. More sections of the course (larger course enrollments)	39	53	18
13. Changing the department location of the course (from Business to English, for example, or vice versa)	17	72	9

Note: MMPW = Multi-major Professional Writing; WPA = Writing Program Administrators; ATTW = Associated Teachers of Technical Writing; CPTSC = Council of Programs in Technical and Scientific Communication.
[a] Row numbers do not add up to 154 because the "not relevant" responses are not displayed.

in writing studies related fields. This trend suggests that a source of the lower level of confidence in the MMPW instructor corps is relevant graduate training (or lack thereof).

Notably, when it comes to experience in industry with PTC, respondents at both institutional types responded with similar frequency (60% at doctoral, 57% associate's). Respondents who answered affirmatively to having relevant experience in industry were also asked to describe that experience. In these write-in answers respondents described a full range of experiences that would be considered relevant industry experience for teaching the MMPW course: working as a professional in industries such as banking, nonprofits, construction, information technology, human resources; working as an executive secretary; being employed as a technical writer or technical editor or doing this work as an independent contractor; having careers in journalism, publishing, and other media industries.

Although we found the relatively high (compared to graduate education) frequency of industry experience among respondents to be good news overall, we do want to raise the point that one of the differences between experience in industry and graduate training in PTC or composition is the exposure to writing pedagogy and rhetorical studies. Certainly, an MMPW instructor with industry experience who is able to speak with authority about encountering writing situations in real workplaces has an edge over an instructor who has never encountered such writing situations. However, this edge is limited by the instructor's access to and openness to learning the best practices for teaching writing and the rhetorical thinking that is at the foundation of teaching for learning transformation (see, e.g., Brent, 2012). Although industry experience is valuable, it is neither sufficient nor a substitute for either graduate training or other formal training in writing pedagogy (see also Meloncon & England, 2011, for a discussion of hiring working professionals to teach TPC). As a result, a source of lower confidence in the MMPW course instructor corps could be that respondents recognize a lack in instructor training, despite a relatively high number of instructors with industry experience.

In terms of access to professional development, respondents at associate's and doctoral institutions reported having access at fairly similar rates (55% and 64%, respectively). Although more respondents from doctoral institutions reported having access to professional development than respondents at associate's institutions, the difference (9 points) is neither notable nor unexpected given historic trends in how institutions are funded and the fact that institutional investment in instructor development varies significantly by instructor rank (TT vs. adjunct, for example). Given this ambivalence, we are left with the interpretive choice of reading these frequencies as either positive (more than one half of respondents in both categories have access to professional development) or as negative (nearly one half of respondents at associate's institutions do not have access to professional development). In both cases, the survey does not provide the data or the warrant for making more than an arbitrary choice to read the data one way or the other. Either way, the interpretive choice depends on whether these frequencies exceed our individual expectations, or not, and how we see the rates of access to professional development changing over time (either increasing or decreasing). The ambivalence in the analysis of this data about professional development, however, should not prevent it from warranting arguments for greater investment in the professional development for MMPW instructors.

In sum, we argue that our data shows that across all respondents there is an overall positive level of confidence in the MMPW course. This is especially true when it comes to the confidence that MMPW instructors have that students benefit from taking this course. However, we argue that there is a meaningfully lower level of confidence in the training and experience of the instructor corps who teach the MMPW course that warrants attention from institutions and the field.

What Would Improve MMPW Instructor Confidence in the Course?

Our intention in designing and executing this survey research has always been to provide the field with a resource for discussing how the status of the MMPW course can be improved. As much as we wanted to create a snapshot in time of how this course is institutionally situated, the curricular and pedagogical norms for the course, who teaches the course, and how confident instructors are in the course, we also wanted to capture data that would point to interventions for improving the status of the course within individual institutions and, more generally, across the fields of professional and technical communication and writing studies.

In this spirit, this survey ended with an exercise that prompted respondents to sort items that might improve their confidence in the MMPW course at their institution into boxes labeled: *would definitely help, would help a little, would make no difference, already doing,* and *not relevant.* Items that might improve confidence in the course included items related to instructor training and experience, items related to changes in the institutional situation of the course, and items related to the pedagogy and instructional materials for the course.

To simplify the presentation of the data and to make positive trends clearer, we created a category called "would help" that aggregates the *would definitely help* and the *would help a little* responses for each item. We then ranked the items based on the cumulative total of the new "would help" category (see Table 11.15). Table 11.15 also includes the number of *would make no difference* and *already doing* responses for each item. As would be expected, as the number of would help responses diminish, the number of *would make no difference* and *already doing* responses rises.

This data has value as a snapshot of what kinds of interventions would improve the confidence of instructors in the MMPW course. There are several notable trends to which we would like to draw attention. First, we find it notable that the top three ranked items are all directly related to instructor experience and training: (1) Training in contemporary workplace technologies for instructors, (2) More instructors with industry experience, tied with (2) More instructors with a scholarly interest in professional writing. This trend supports the findings in the previous section about the lower level of confidence in instructor factors over student factors and the low levels of instructor training in TPC and composition, in particular at associate's institutions. From our results we can say with confidence that as a group, MMPW course instructors and coordinators have a lower level of confidence in the training and experience of the instructor corps and that as a group the most frequently chosen items that "would help" increase confidence in the course are related to instructor experience and training.

Additionally, we note that the next four most frequently chosen items that "would help" are ones that are related to curricular matters, either directly or indirectly. The number four ranked item—adding an experiential dimension to the course (internship, service learning, or client project)—is an acknowledgment of the now widely held view that connecting students with real workplace writing situations is a pedagogical best practice for the MMPW course. Smaller class sizes (five), a longer course sequence (six), and additional classroom technologies (seven) address curricular matters more indirectly yet also point to ways to enhance the curriculum: Smaller class sizes make experiential learning more possible; a longer course sequence creates room for additional content in the course, and additional classroom technologies expand the modes of learning and skill sets that students can learn in the course.

Interestingly, the items that relate to the institutional situation of the course are ranked at the bottom half of the list. This finding speaks to the fact that items such as the existence of standardized outcomes (eight), resources to assess the course (nine), and the presence of a course coordinator (11) have among the highest level of *already doing* responses (42, 37, and 54, respectively). We were surprised that we didn't find stronger negative trends in the institutional situation for the course. For example,

78% of respondents reported the existence of a common set of outcomes for the MMPW course at their institutions; almost one half of respondents reported a MMPW course coordinator (46%); and 58% of respondents reported that an assessment process for the course was in place. Aside from the definitely positive frequency of course outcomes, however, we remain unsure whether to interpret the respondent frequencies for a course coordinator and assessment process as positive or negative. Does it call out a deficit in the institutional situation of the course that around one half of survey respondents reported no MMPW course coordinator and no resources for assessing the course? This number is certainly tempered by variations in the institutional situations of the course, such as the fact that 49% of respondents reported that 10 or fewer sections of the course were taught each year, in which case a course coordinator may not be necessary. Or is the inverse true, that for a course that has often existed at the periphery of English departments these frequencies are heartening and indicative of a positive trend in the development of the MMPW course? Without data that reflects these trends over time it is hard to tell which story is more reflective of the status of the course in the field as a whole.

Conclusion

This article began with a fairly typical course description for the MMPW course. By calling attention to what, for so many stakeholders, is hidden in plain sight (i.e., the normalization of the course in U. S. higher education), our purpose in conducting this survey was to investigate the status of the MMPW course across its diverse institutional contexts—to ask, essentially, "What is going on right now with this course?"

Findings from the survey lead us to suggest that there is systemic underinvestment in the MMPW course by postsecondary institutions across all institutional types. This underinvestment has consequences that have shown up in our findings, such as the persistence of formalist, genre-based pedagogical and curricular norms that are reified in the most commonly used standard textbooks. Another consequence is that, though instructors have a high level of confidence in the value that the MMPW course has for students, they have a lower level of confidence in the instructor corps that teach the course. These findings point to underinvestment by institutions in hiring instructors with the proper qualifications to teach the course as well as underinvestment in professional development opportunities to keep instructors current in their knowledge of the field. Unsurprisingly, given this underinvestment, the most frequently chosen item for increasing confidence in the MMPW course was related to instructor training and preparedness. This finding points to an urgency to improve instructor development and strengthen hiring practices.

Our argument about underinvestment in the MMPW course is motivated by what we see as an urgency to initiate a conversation to promote change that is past due. The first step toward the conversation we are suggesting is to view the MMPW course as a shared phenomenon across institutional contexts that can be taken up by the professional bodies that serve these constituencies. One place the professional organizations with a stake in this course (e.g., National Council of Teachers of English [NCTE], Conference on College Composition and Communication [CCCC], Council of Writing Program Administrators [CWPA], Association of Teachers of Technical Writing [ATTW], Council of Programs in Technical and Scientific Communication [CPTSC], Association for Business Communication [ABC]) can begin is in the development of a shared set of standardized outcomes similar to those the Council of Writing Program Administrators (CWPA, 2014) has produced and continues to update for FYC. Although there are limits to the comparisons one can make between the MMPW course and the FYC course, the ubiquity of and nearly universally professed value in the MMPW course surely merits a similar level of attention and investment by our professional organizations.

We anticipate that such a conversation will not be easy. Few have a stake in arguing against the current status of the MMPW course: students see it as a necessity for career preparation, humanities faculty see it as a gesture toward professional training in a curricular environment that is increasingly hostile to the liberal arts, professional schools are satisfied that once their students complete an MMPW course they have had adequate exposure to training in communication skills, contingent faculty see the MMPW course as a predictable source of income, the textbook industry sees the course as a cash cow, and department and college administrators rely on MMPW course enrollments as a key source of revenue. But we argue that we, as a profession, can and should hold ourselves to a higher standard for the MMPW course.

Acknowledgments

We would like to thank Jessica Bishop-Royse at the Social Science Research Center at DePaul University for her support with survey methods and our summer research assistant, Theresa Bailey, for her work with recruitment and transcription. Additionally, we would like to thank Amy Hornat-Kaval, Rosemary Golini, and Joseph Szpila for helping us user test the survey. Finally, we would like to thank all of the survey respondents for their generosity with their time and willingness to share and we would like to thank the anonymous *TCQ* reviewers for helping to strengthen our manuscript.

Funding

This research was supported by a 2015 Research Initiative Grant from the Conference on College Composition and Communication.

Notes

1 This number is an estimate based on the number of each Carnegie type institution (doctoral, master's, baccalaureate, associate's; 2015 numbers) and the frequency with which survey respondents reported the number of sections of the MMPW course taught each year at their institution (1–10, 11–50, >50). To be conservative, calculations used the middle number in the lower ranges (5 and 30) and the lowest number (51) for the "more than 50" category. The unrounded estimated total for the overall number of sections of the MMPW course taught each year is 43,538.

2 This number is based on the National Census of Writing report (2013) conducted by Swarthmore finding that 81% of the schools in the census reported that they require first-year composition. Because a large majority of students (86%) are enrolled at institutional types that are also the most common locations for the FYC course (doctoral, master's and associate's institutions), we did not break this calculation down by institutional type because the order of magnitude of the total number would not have changed. In addition, we did not account for the percentage of students who place out of the course because of grades or testing (such as Advanced Placement (AP) testing) or first year student retention rates. We assumed that

there are an estimated 5 million college first-year students (one fourth of the total number of college students for 2015 as reported by Carnegie (2016a)) and we assumed an average of 20 students per section. Given these assumptions, the unrounded estimated total number of sections is 202,500. We assume that this number has meaning only to the order of magnitude of the closest 10,000.

3 The coffee cards were delivered electronically, thus preserving the deidentification of the survey data.

References

[This article appeared before the 7th edition of the *APA Publication Manual* was published, and the citations follow the style of the 6th edition.]

Allen, N., & Benninghoff, S. T. (2004). TPC program snapshots: Developing curricula and addressing challenges. *Technical Communication Quarterly*, *13*(2), 157–185. doi:10.1207/s15427625tcq1302_3

Banerjee, A., & Chaudhury, S. (2010). Statistics without tears: Populations and samples. *Industrial Psychiatry Journal*, *19*(1), 60–65. doi:10.4103/0972-6748.77642

Blakeslee, A. M. (2001). Bridging the workplace and the academy: Teaching professional genres through classroom- workplace collaborations. *Technical Communication Quarterly*, *10*(2), 169–192. doi:10.1207/s15427625tcq1002_4

Brent, D. (2012). Crossing boundaries: Co-op students relearning to write. *College Composition and Communication*, *63*(4), 558–592.

Carnegie Classification of Institutions of Higher Education, Center for Postsecondary Research. (2016a). *2015 update: Facts & figures*. Bloomington, IN: Indiana School of Education.

Carnegie Classification of Institutions of Higher Education, Center for Postsecondary Research. (2016b). *Standard listings*. Retrieved from http://classifications.carnegiefoundation.org

Connors, R. J. (1982). The rise of technical writing instruction in America. *Journal of Technical Writing and Communication*, *12*(4), 329–353.

Council of Writing Program Administrators. (2014). WPA Outcomes Statement for First-Year Composition (3.0), Approved July 17, 2014.

Crowley, S. (1998). *Composition in the university: Historical and polemical essays*. Pittsburgh, PA: University of Pittsburgh Press.

Curtis, J. W. (2014). *The employment status of instructional staff members in higher education*. Washington, DC: American Association of University Professors.

Dayton, D., & Bernhardt, S. A. (2004). Results of a survey of ATTW members, 2003. *Technical Communication Quarterly*, *13*(1), 13–43. doi:10.1207/S15427625TCQ1301_5

Johnson-Sheehan, R. (2014). *Technical communication strategies for today* (2nd ed.). Upper Saddle River, NJ: Pearson.

Kain, D., & Wardle, E. (2005). Building context: Using activity theory to teach about genre in multi-major professional communication courses. *Technical Communication Quarterly*, *14*(2), 113–139. doi:10.1207/s15427625tcq1402_1

Kelley, K., Clark, B., Brown, V., & Sitzia, J. (2003). Good practice in the conduct and reporting of survey research. *International Journal for Quality in Health Care*, *15*(3), 261–266. doi:10.1093/intqhc/mzg031

Kolin, P. (2012). *Successful writing at work* (10th ed.). Boston, MA: Cengage.

Lannon, J. (2007). *Technical communication* (11th ed.). New York, NY: Longman.

Lannon, J., & Gurak, L. J. (2016). *Strategies for technical communication in the workplace* (3rd ed.). New York, NY: Pearson.

Lehman, C. M., & DuFrene, D. D. (2010). *Business communication* (16th ed.). Mason, OH: South-western Cengage Learning.

Markel, M. (2014). *Technical communication* (11th ed.). New York, NY: Bedford/St. Martin's.

McKenna, B., & Thomas, G. (1997). A survey of recent technical writing textbooks. *Journal of Technical Writing and Communication, 27*(4), 441–452. doi:10.2190/CGA9-CVJY-82CX-AEFJ

Meloncon, L. (2012). Current overview of academic certificates in technical and professional communication in the United States. *Technical Communication, 59*(3), 207–222.

Meloncon, L., & England, P. (2011). The current status of contingent faculty in technical and professional communication. *College English, 73*(4), 396–408.

Meloncon, L., & Henschel, S. (2013). Current state of US undergraduate degree programs in technical and professional communication. *Technical Communication, 60*(1), 45–64.

National Census of Writing. (2013). How would you describe your first-year writing requirement? Retrieved from http://writingcensus.swarthmore.edu

Oliu, W. E., Brusaw, C. T., & Alred, G. J. (2016). *Writing that works: Communicating effectively on the job* (12th ed.). New York, NY: Bedford/St. Martin's.

Read, S., & Michaud, M. (2015). Writing about writing and the multi-major professional writing course. *College Composition and Communication, 66*(3), 427–457.

Reave, L. (2004). Technical communication instruction in engineering schools a survey of top-ranked US and Canadian programs. *Journal of Business and Technical Communication, 18*(4), 452–490. doi:10.1177/1050651904267068

Searles, G. J. (2013). *Workplace communication: The basics* (6th ed.). New York, NY: Pearson.

Spinuzzi, C. (1996). Pseudotransactionality, activity theory, and professional writing instruction. *Technical Communication Quarterly, 5*(3), 295–308. doi:10.1207/s15427625tcq0503_3

Staples, K. (1999). Technical communication from 1950–1998: Where are we now? *Technical Communication Quarterly, 8*(2), 153–164. doi:10.1080/10572259909364656

Sullivan, P. A., & Porter, J. E. (1993). Remapping curricular geography: Professional writing in/and English. *Journal of Business and Technical Communication, 7*(4), 389–422. doi:10.1177/1050651993007004001

Tate, G., Tagart, A. R., Schick, K., & Hessler, H. B. (2013). *A guide to composition pedagogies* (2nd ed.). New York, NY: Oxford University Press.

Tebeaux, E., & Dragga, S. (2014). *The essentials of technical communication* (3rd ed.). Oxford, UK: Oxford University Press.

Warren, T. L. (1996). An informal survey of technical writing textbooks: 1950–1970. *Journal of Technical Writing and Communication, 26*(2), 155–161. doi:10.2190/1QD8-PK64-X0RJ-ATWG

Wickliff, G. A. (1997). Assessing the value of client-based group projects in an introductory technical communication course. *Journal of Business and Technical Communication, 11*(2), 170–191. doi:10.1177/1050651997011002002

Yeats, D., & Thompson, I. (2010). Mapping technical and professional communication: A summary and survey of academic locations for programs. *Technical Communication Quarterly, 19*(3), 225–261. doi:10.1080/10572252.2010.481538

Appendix

Institutional Situation

Q56 What is your current or recent association with a MMPW Course? (Check all that apply.)

Q66 At how many institutions do you currently teach or coordinate a MMPW course?

Q54 Choose the type of institution where you PRIMARILY teach or coordinate with the MMPW course. (Choose one only.)

Q47 Please write in the course number and title of the MMPW course that you teach or coordinate. (Example: English 234: Introduction to Professional Writing.)

Q48 If available, please copy and paste the catalog description for the MMPW course you teach or coordinate:

Q65 In what department, program, or college is the MMPW course offered?

Q67 About how many sections of this course are offered each year?

Q35 Is there a program administrator or coordinator for the MMPW course?

Curricular Situation

Q23 Are there prerequisites for the MMPW course at your institution, other than the first-year writing sequence in writing?

Q24 Which of the following function as prerequisites for the MMPW course at your institution? (Select all that apply.)

Q27 Is the MMPW course at your institution a stand-alone or part of a larger instructional sequence?

Q28 If the MMPW course is a part of a larger instructional sequence, which of the following best characterizes the kind of sequence it's a part of?

Q25 What types of students take this course? (Check all that apply.)

Q29 To the best of your knowledge, which of the following groups of students are REQUIRED to take the MMPW course? (Select all that apply.)

Q74 To the best of your knowledge, which of the following groups of students occasionally CHOOSE to take the MMPW course, even though they may not be required to do so? (Select all that apply.)

Faculty

Q87 Which types of faculty members teach MMPW at your institution? (Check all that apply.)

Q86 Which type of faculty member teaches THE MOST sections of MMPW courses at your institution? (Choose one.)

Q49 What is your institutional/instructional status?

Q50 What is your highest degree obtained?

Q51 What is your primary field of graduate training?

Q59 Beyond teaching MMPW courses, is professional/technical writing a scholarly area of interest for you?

Q68 If no, what would you consider your PRIMARY area of interest for scholarly research or writing?

Q61 If yes, briefly describe a recent or current interest or project.

Q53 Do you bring relevant industry or other nonacademic experience to the teaching of professional writing?

Q60 If yes, please describe that experience.

Q55 How many years of experience do you have teaching ANY kind of writing?

Q57 How many years of experience do you have teaching the MMPW course?

Curriculum and Instruction: Course Materials

Q6 Do you or have you used a textbook to teach the MMPW course?

Q84 Which of the following is the textbook that you use NOW or have used most recently?

Q76 Why did you choose this textbook?

Q69 What other course materials do you use? Check all that apply:

Curriculum and Instruction: Section Standardization

Q1 Do all sections of the MMPW course at your institution share a common syllabus?

Q5 Is there a standardized set of course outcomes for MMPW instruction at your institution?

Q9 Do instructors write their own MMPW course outcomes?

Q8 Is there a college, department, or program process for assessing the success of MMPW sections meeting these outcomes?

Curriculum and Instruction: Pedagogy

Q17 Which of the following approaches characterize how you teach your MMPW course? (Choose all that apply.)

Q82 Now choose only your CENTRAL or PRIMARY approach for teaching the MMPW course? (Choose one.)

Curriculum and Instruction: Technology

Q70 Have you recently taught the MMPW course in a computer lab?

Q72 Have you recently taught an online or hybrid version of the MMPW course?

Curriculum and Instruction: Professional Development

Q20 Do you have access to professional development opportunities for faculty who teach this MMPW course?

Q22 Have you, personally, ever pursued professional development opportunities to improve your MMPW instruction?

Q75 Do you enjoy teaching and/or coordinating the MMPW course?

Q76 Please tell us why you feel this way:

Subjective Assessment

Slide the pointer to reflect your level of confidence as prompted by the question in the left-hand column.

Q62_1 Overall, how confident are you in the effectiveness of the MMPW course at your institution?

Q38_1 Students feel more ready to write in workplace contexts after taking the MMPW course.

Q38_2 The MMPW course helps students to acquire sufficient and appropriate technological skills to write in future workplace contexts.

Q38_3 MMPW students are more ready to write in the workplace than students who have not taken the MMPW course.

Q41_1 MMPW instructors at your institution have sufficient experience and training in professional writing to effectively teach this course.

Q41_2 MMPW instructors at your institution have sufficient access to professional development, including training in contemporary technologies, to effectively teach the MMPW course.

Q41_3 MMPW instructors at your institution have access to sufficient technologies in the classroom to effectively teach the course.

Q41_4 Sections of the MMPW course at your institution are equal in terms of rigor and the presentation of standard curriculum.

Q42 What would help to IMPROVE your confidence in the MMPW course at your institution?

Please drag and drop from the list of "Items" on the left to the appropriate boxes on the right. Sorting boxes are labeled as "Would Definitely Help," "Would Help A Little," "Would Make No Difference," "Already Doing," and "Not Relevant."

Items

- Adding an experiential dimension to the course (internship, service learning, or client project)
- Additional classroom technologies, including software and hardware
- A set of standardized outcomes for the course from a professional organization such as WPA, ATTW, CPTSC, or other
- Training in contemporary workplace technologies for instructors
- Smaller class sizes
- New textbooks that support contemporary workplace writing practices
- A longer MMPW course sequence (2 or more quarters or semesters)
- More instructors with industry experience
- More instructors with a scholarly interest in professional writing
- A college-, department-, or program-level coordinator for the MMPW course
- Changing the department location of the course (from Business to English, for example, or vice versa)
- Resources to assess the course across sections on a regular basis
- More sections of the course (larger course enrollments)

Commentary

Read and Michaud's article is an excellent example of a report of survey results because it avoids the problems often associated with survey result articles. It makes no attempt to generalize from the survey sample to the overall population of interest.

The subject of introductory professional communication courses is an important one. Its significance is obvious for academics, who typically teach one or more sections of such courses each term. It may also be of interest to students of technical and professional communication, all of whom are likely to take such a course as part of their bachelor's degree programs. The level of interest may not be obvious for practitioners, but a surprising number of technical and professional communicators in industry teach part time, and multi-major professional writing courses are the ones that they are most likely to teach.

Purpose and Audience

Read and Michaud provide an excellent statement of their article's purpose in the final paragraph of their Introduction.

> This survey study assumed that the MMPW course is a unit of analysis that can be studied laterally across institutional types and contexts .... The results of such a study provide a macrolevel view of who teaches the course, its institutional situation, common pedagogical practices, and perceptions of the effectiveness of the course. The value of this view is that it can motivate discussion at the level of the field's professional organizations about ways to develop and invest in the MMPW course so as to ensure that it remains relevant and responsive to changes in the 21st-century academy and economy.

Because it appears in *Technical Communication Quarterly*, the journal published by the Association of Teachers of Technical Writing, the primary audience for the article is technical and professional communication faculty. As noted at the beginning of this commentary, however, this report is likely to be of interest to some students (especially those who are preparing for a career in academe) and some industry practitioners as well.

Organization

Like the quantitative and qualitative articles that we analyzed in Chapters 9 and 10, this report uses the typical IMRAD structure.

- The *Introduction* section announces the article's purpose and reviews the literature that defines the MMPW course, establishes the importance of such courses, and explains what we already know about them from previous research.
- The *Methodology* section describes how respondents were recruited and how the survey itself was constructed.
- The *Findings* section walks the reader through the survey results regarding pedagogical approaches to the MMPW course, instructors' confidence that the course

accomplishes its purpose, and their ideas on what would improve that confidence. Most of the results are included in a series of 15 detailed tables.

- The *Conclusion* section discusses the underinvestment of educational institutions in the MMPW course and the need to pursue a serious conversation about ways to address this problem.
- At the end of the article are some notes that justify assumptions on which respondent recruitment were based, acknowledgments of colleagues who assisted the authors and a grant that supported the project, as well as citations of all the works referenced in the text.
- Finally, the *Appendix* presents the questions that comprised the survey.

There are no surprises here in terms of organization.

Survey Construction

Information Elicited by the Questions

As we saw in Chapter 6, surveys can do several things: collect data, measure self-reported behavior, and assess attitudes. The extensive questions in Read and Michaud's survey do all of these (see the article's *Appendix*).

- The *Institutional Situation* section of the survey seeks information about both the respondent and the school(s) at which the respondent teaches MMPW courses, and some basic data about the courses themselves.
- The *Curricular Situation* section seeks information about course prerequisites, the types of students who enroll in the course, and the course's role in the overall curriculum.
- The *Faculty* section asks for data about the respondent and others who teach the MMPW course at that institution.
- The *Curriculum and Instruction: Course Materials* section seeks information about textbooks and other teaching materials.
- The *Curriculum and Instruction: Section Standardization* section asks about syllabi, course outcomes, and assessment across all sections of the course.
- The *Curriculum and Instruction: Pedagogy* and the *Curriculum and Instruction: Technology* sections ask about teaching approaches, use of a computer lab, and online/hybrid versions of the course.
- The *Curriculum and Instruction: Professional Development* section asks about development opportunities provided by the institution and the respondent's satisfaction with teaching the course.
- The *Subjective Assessment* section seeks data on the respondent's level of confidence in student outcomes, faculty qualifications, and access to development opportunities and other resources.

Thus, in this survey, we see questions about individual and institutional demographics, and about self-reports of teaching "behaviors." We also see faculty attitudes toward the course, student outcomes, colleagues' qualifications, and institutional resources for the course and its teaching staff. These are precisely the types of information that surveys are intended to gather.

Design and Structure of the Questionnaire

QUESTION DESIGN

Read and Michaud have designed their 65 questions very well. (Note that the article text specifies a total of 66, but the survey itself contains 65 questions.) All of the questions successfully avoid absolute statements. They do not use absolute terms such as *always* and *never*, and all of the questions are limited to the particular experience and perceptions of the respondent. Thus, the questions are well grounded in the specific and avoid generalities. Similarly, the opinion questions avoid the problem of statements in the negative. The questions are stated in such a way that they are reasonably easy to understand and would not be confusing to the intended respondent group.

The questions also avoid bias. For example, although Question 84 requests that respondents identify the textbook used in the course from a list of nine commonly used texts, it also provides the opportunity to write in the title of another book if the respondent does not use one of the texts on the list. Finally, each question focuses on a single concept, such as identifying the respondent's institution type, the text that he or she uses in the course, and the respondent's access to professional development opportunities.

The survey also focuses on one concept per question. For example, "Beyond teaching MMPW courses, is professional/technical writing a scholarly area of interest for you?" simply asks whether the respondent reads or does research in technical and professional communication other than teaching a course in this subject. There is no problem with double-barreled questions here or elsewhere in the survey.

Because the multiple-choice responses are not provided in the list of survey questions, it isn't possible to tell whether neutral or opt-out choices are provided for the questions that aren't included in the article's tables or text. For example, for the question "To the best of your knowledge, which of the following groups of students are REQUIRED to take the MMPW course? (Select all that apply.)," we cannot determine whether there was a neutral or opt-out choice, or even whether one might be needed.

The only other fault that we can find in the survey questions is that there are so many of them. Responding carefully to a 65-item questionnaire would likely require a minimum of 20–30 minutes, especially since some questions expect the respondent to do some research to determine the number of instructors who staff the course, the number of sections offered per year, and so forth.

And as already noted, the survey questions provided in the article's *Appendix* do not include the response choices on the survey as it was distributed to potential respondents. This lapse can be partially excused because including those responses would have considerably increased the length of the *Appendix* (and thus of the article), and the responses for the most important questions are included in the text of the article or in the data tables. See, for example, the text in the *Methodology* section describing the slider scale for the survey's subjective questions, as well as Tables 11.3–11.8.

QUESTION TYPES

The survey uses a mix of question types.

- Most of the questions in the survey are closed-ended or multiple-choice. Some allow only one selection, although others allow more than one. For example,

"Are there prerequisites for the MMPW course at your institution, other than the first-year writing sequence in writing?" and "Which of the following function as prerequisites for the MMPW course at your institution? (Select all that apply.)"

- Occasionally, multiple choice questions have a write-in option if none of the choices fit the respondent's situation. For example, "Which of the following is the textbook that you use NOW or have used most recently?" Table 11.6, reporting results for this question, indicates that there was a write-in option for respondents to specify a text not on the list.
- A few questions are open-ended. For example, "Please write in the course number and title of the MMPW course that you teach or coordinate."
- The Subjective Assessment rating questions prompt the respondent to "Slide the pointer to reflect your level of confidence as prompted by the question in the left-hand column," using a scale of 0 to 100.
- The Items section asks respondents to sort a list of items into 5 categories, "Would Definitely Help," "Would Help A Little," "Would Make No Difference," "Already Doing," and "Not Relevant."

Notably, the *Subjective Assessment* and *Items* sections are not open-ended, thus not asking respondents to explain why they chose the rating or category for each one. Such open-ended questions would have provided respondents with a chance to clarify the reasoning behind their ratings and the researchers with additional data. Doing so, however, would have required respondents to spend even longer completing the survey. It would have also significantly increased the difficulty of summarizing and reporting the results because each response would have required some degree of inter-pretation by the researchers.

Because so few of the survey questions were open-ended, the demand on respond-ents' time to answer the questions and on the researchers' time to analyze the responses was significantly less than it would have been otherwise. The small number of open-ended questions made data analysis much less burdensome for the researchers. The relative simplicity of answering mostly closed-ended and multiple-choice questions may have encouraged some reluctant respondents to complete the lengthy survey. Still, only 154 of the 220 people who began the survey completed it, a 70% completion rate.

Exercise 11.1 Writing Survey Questions

Suppose that you want to learn more about the opportunities for and participation in professional development among both practitioner and academic technical communi-cators. Devise a survey consisting of five to seven questions that requests demographic information, the number and type of opportunities available, and the respondents' level of participation in those opportunities. Be sure to use a variety of question types.

RECRUITMENT OF RESPONDENTS

In their *Methodology* section, Read and Michaud tell us that they used a convenience sample because there is no central directory of instructors for multi-major professional writing courses.

[W]e recruited respondents from the population via channels that we had access to given our own professional affiliations and connections. The channels included the email listservs of professional organizations in the field, the social media sites of the National Council of Teachers of English (NCTE) college section, emails to colleagues requesting assistance distributing the survey link, and a small number of personalized emails to instructors who are particularly hard to reach, such as those at tribal colleges.

They took pains to ensure that they included a variety of faculty ranks and statuses in their recruitment effort. They also wanted to reflect

the experiences and views of MMPW instructors across the diversity of 2- and 4-year postsecondary contexts. Toward these ends, we used data published by the CCIHE [Carnegie Classification of Institutions of Higher Education] (2016b) to set quotas for respondent recruitment by institutional type from the overall population of MMPW instructors. We used CCIHE's six postsecondary institutional types for 2-year and 4-year institutions and reported proportional representation of student enrollment at each institutional type (see Table 11.1).

The authors do not provide us with the recruitment letter or overall survey instructions, so we don't know whether they explained to potential respondents the study's purpose, estimated completion time, ability to save and complete the survey later, measures to protect respondents' privacy, institutional review board approval, respondent qualifications, and the fact that participation was voluntary. These last two items are probably moot since the survey population was targeted and obviously chose to respond or not.

We also don't know the precise order in which questions were asked. We can generally assume that the questions were posed in the order reproduced in the *Appendix*, but Sarah Read shared with us that the "survey had a fairly complicated branching system that sorted questions for respondents based on their answers, so not every respondent answered the same questions" (personal communication, August 30, 2019).

The survey was created using Qualtrics, a web-based survey tool, and distributed via a web link. Read also told us that "Qualtrics assigned numbers to questions as they were added or deleted. As questions were moved around the survey, they maintained their original numbers. So, question numbers are identifiers for questions but do not necessarily have consecutive meaning" (personal communication, August 30, 2019).

A total of 220 people consented to take the survey, of whom 154 completed the questionnaire. Only the completed surveys were included in the data analysis. Respondents who completed the survey received a $5 USD coffee card as an incentive. This token gift probably also encouraged some potential respondents to complete the rather lengthy survey. (Teachers will do almost anything for a $5 coffee card!)

SURVEY TESTING

Pilot testing a survey allows researchers to evaluate the test instrument and tweak it to ensure that it elicits the results that they are looking for. There is only a brief mention of pilot testing in the article's *Methodology* section.

To ensure that this survey would gather the data we intended and be maximally user-friendly for respondents, we user tested it during the winter and early spring of 2015 ....

In the *Acknowledgments*, Read and Michaud also thank three people for "helping us user test the survey."

Their decision to pilot the survey is commendable. We don't know whether the three individuals acknowledged were the only pilot participants or whether they conducted the testing. If the former, the number of pilot participants was very small, especially considering the diversity of the participants that the authors wished to recruit. If the latter, which seems more probable, the pilot population is more likely to approximate the population recruited for the full-scale survey. The authors report no changes to the survey based on the results of the pilot.

Report of Results

Read and Michaud use frequency distribution to report the vast majority of survey responses.

- Tables 11.1 and 11.2 report the frequency of respondents by their institutions' Carnegie classifications of institution type and by faculty rank.
- Tables 11.3–11.6 specify the frequency of textbook use by institution type, reason for choosing the text overall and by institution type, and the specific text used.
- Tables 11.7 and 11.8 report the approaches used in teaching the course as well as the main approach used by each respondent.
- Table 11.13 compares (by institution type and faculty rank) the number and percentage of those who expressed overall lower confidence in the effectiveness of their MMPW courses with survey respondents as a whole.
- Table 11.14 compares the percentages of various types of instructor training for respondents at doctoral degree-granting institutions with those at associate's degree-granting institutions (highest degree earned, graduate training in professional and technical communication, graduate training in professional and technical communication or composition, industry experience, and access to professional development).

These frequency distributions provide very helpful insights about responses to what were mostly closed-ended and multiple-choice questions. Again, although these frequencies cannot be generalized to the total population of MMPW instructors, they give us good insights about where to begin the discussion of the status of the MMPW course, just as Read and Michaud intended.

The authors use averages—means—to report the results of the subjective assessment questions in Tables 11.9–11.12. These questions asked respondents to rate their assessments on a scale from 0 to 100. The responses to these questions indicated lower respondent confidence in instructor factors than in student factors. The authors also found that the lower confidence in instructor factors was not related to institution type or instructor rank. However, they do not compare means as we discussed in

Chapter 4. Instead, they simply report the mean responses to subjective assessment questions.

Finally, Table 11.15 reports raw number counts for "Would Help," "Would Make No Difference," and "Already Doing" responses in sorting the 13 *Items* at the end of the survey.

Margin of Error and Confidence Intervals

Read and Michaud do not address margin of error or confidence interval to infer that their results are representative of the population of interest—that is, all those who teach MMPW courses. This is a wise decision on the authors' part. As they acknowledge in the next-to-last paragraph of their *Methodology* section,

> At every turn, we have recognized the limitations of our data and resisted inferring causation for results or generalization to a broader population (which remains undocumented). Within these limitations, we believe that our data, based on a quota-based convenience sample, is adequate for initiating a conversation within the field about the implications of our study.

The type of survey that the authors have undertaken is quite different from a political poll. In such polls the pollsters carefully define the sample population that they wish to recruit so that it reflects the larger population in terms of demographics, political affiliation, and geography. They then draw truly random samples to fit those definitions. As a result, political pollsters can report margin of error and confidence interval scientifically, something not possible with a convenience sample.

The authors didn't know the number of people who received an invitation to complete the survey via listservs, social media, or personal contact, so it was not possible for them to calculate a response rate. Neither did they know the total number of people who qualify as instructors of MMPW courses because, as they note, there is no central directory of these instructors.

So although Read and Michaud use frequency distributions in reporting their results, they make no claim that their results are representative of the larger population. They present their study as exploratory research only. By acknowledging from the beginning that they are "initiating a conversation" rather than "inferring causation for results or generalization to a broader population," Read and Michaud have wisely focused our attention on the exploratory nature of their research. Because most surveys in technical and professional communication are exploratory in nature and rely on convenience samples, other researchers would be wise to emulate Read and Michaud's example here.

Exercise 11.2 Analyzing Population and Sample

Suppose that you want to repeat Read and Michaud's research with a much smaller population—let's say the people who have taught the multi-major professional writing courses at five large state universities in your region of the US (or five universities in your country if you live outside the US) during the past academic year.

1. How would you initially determine the size of the population of interest?
2. How would you identify and recruit participants for your sample?

3. Why would your survey allow you to determine response rate, margin of error, and level of confidence in your results?

Is It RAD?

As we saw in Chapter 2, RAD research as defined by Haswell (2005) must be replicable, aggregable, and data-supported. In other words, the methodology must be sufficiently defined that others can repeat the study. The results from the study must be reported in sufficient detail that they can be aggregated or combined with the results of repeated studies to build a body of data that can be compared and further built upon. And the conclusions of the study must be supported by those data, not simply reflect the impressions or gut feelings of the researchers.

It is clear from our analysis that Read and Michaud's article meets these requirements. Their methodology is described in enough detail to allow another researcher to replicate the survey, though inclusion of the responses to each question would make that task easier and more precise. Similarly, the data that they report can be aggregated with results from further studies. And finally, the conclusions drawn in the article are entirely based on the data reported. As a result, we can conclude with confidence that this is a RAD article.

Summary

This chapter has expanded on the concepts and methods presented in Chapter 6 by examining an article-length report of the results of a survey on multi-major professional communication courses. Following the full text of the article, we have explored how its authors went about writing it, examining its purpose, audience, and organization.

We also looked carefully at the design of the survey questions and the selection of the survey population. Finally, we noted that the authors did not discuss the survey report in terms of response rate, margin of error, and confidence interval because they were not able to determine the response rate and do not claim that their sample is representative of the larger population.

For Further Study

Read one of the following articles reporting the results of a survey.

Brumberger, E. (2007). Visual communication in the workplace: A survey of practice. *Technical Communication Quarterly, 16*(4), 369–395.
Kreth, M. L., & Bowen, E. (2017). A descriptive survey of technical editors. *IEEE Transactions on Professional Communication, 60*(3), 238–255.

When you have completed a careful reading of the article, analyze it using the Commentary in this chapter as a model, as well as the information about survey construction and analysis in Chapter 6.

Answer Key

Exercise 11.1

The answer to this exercise will be unique for each person who prepares it, so there is no key to this exercise.

Exercise 11.2

The answer to this exercise will be unique for each person who prepares it, so there is no key to this exercise.

References

Haswell, R. H. (2005). NCTE/CCCC's recent war on scholarship. *Written Communication*, 22(2), 198–223.

Read, S., & Michaud, M. (2018). Hidden in plain sight: Findings from a survey on the multi-major professional writing course. *Technical Communication Quarterly*, 27(3), 227–248.

12 Analyzing a Report on the Results of a Usability Study

Introduction

In Chapter 7, we explored usability research in terms of constructing and implementing a study, and analyzing the data to draw a reliable conclusion. We explored the kinds of questions that a usability study can assess. We also considered various study methodologies: usability tests, heuristic evaluations, interviews, and questionnaires.

In this chapter, we will analyze a specific report about a usability test to see how its authors have approached the task. The chapter contains the full text of Colton J. Turner, Barbara S. Chaparro, and Jibo He's "Texting while walking: Is it possible with a smartwatch?" which appeared in the February 2018 issue of the *Journal of Usability Studies*, as well as a detailed commentary about the article. The article is a follow-up to a 2016 article by the same authors (see the *References* at the end of the article), a fact that illustrates that many research projects—including this present study by Turner and colleagues—may lead to further work. Indeed, the conclusion of this article calls for additional research to further develop our knowledge about the usability of texting while walking.

Learning Objectives

After you have read this chapter, you should be able to:

- Analyze a report on the results of a usability study
- Apply the results of your analysis to designing, conducting, and reporting on a usability study of your own

The Article's Context

As we noted in Chapter 7, usability tests collect data from users as they interact in an authentic way with the product or process being studied. We discussed the need to plan carefully to ensure a systematic and consistent collection of data, just as you would for any rigorous study. We also examined how to observe participants interacting with a product or process, and capture as much of the data regarding the interaction as possible.

Further, we considered analyzing the results of the study using descriptive and inferential statistics, the advantages of small vs. large studies, and top-down vs. bottom-up analysis. We also discussed diagramming, categorizing, and summarizing data to make sense of them. Finally, we examined how to find meaning in and report those data.

The article that we examine in this chapter reports on a very good example of a laboratory usability experiment. The authors want to discover how quickly and accurately novice and expert participants can type on a full QWERTY smartwatch keyboard while standing and walking, using two input methods, tracing and tapping. The two posture conditions that they tested are more realistic for a smartwatch product than previous tests in which participants were seated. Turner and his colleagues also wanted to determine which input method their test participants preferred.

We should note that although the authors and we ourselves refer to the test described in the article as an experiment, it is actually quasi-experimental. Unlike a true experiment, there is no hypothesis stated that the authors wish to validate or refute, there is no control group or condition, and the authors are testing multiple independent variables. Nevertheless, we will use the terms *experiment* and *experimental* because this test has a distinctly different character from most of the usability tests our readers are familiar with, as we discuss in our consideration of the article's *Methods* section.

The article is important because there are relatively few examples of full-scale usability test reports outside of usability textbooks because most such tests are conducted for corporate clients, even in university usability labs, and the results are therefore confidential. Moreover, the article shows that typing speed and accuracy were significantly better than most people would be likely to predict for participants who were standing or walking while typing brief messages on a remarkably small keyboard. Because this is a preliminary study with 32 student users, the findings cannot be generalized, but the results are certainly interesting and (we think) unexpected.

Texting While Walking: Is It Possible With a Smartwatch?

Colton J. Turner, Barbara S. Chaparro, and Jibo He

[This article was originally published in 2018 in the *Journal of Usability Studies 13* (2), 94–118. Reprinted with permission of the *Journal* and the User Experience Professionals Association.]

Introduction

In a society that appears to be "always on," personal computers that offer a unique and convenient contribution to consumer lives are of great value. The smartwatch is the latest in a line of personal computers that aim to be the next great step in convenient technology. Smartwatches are billed with the promise of bringing the power of the smartphone to the convenient location of the wrist. However, nearly all smartwatches are lacking one crucial feature—typing capabilities—which is a primary function of smartphones. Predefined responses (e.g., "In a meeting," "Call you back later," and "Hello!") and voice input are the typical solutions to this issue, yet these methods lack the versatility and customization that typing on a keyboard allows. If smartwatches are to be the next level of convenient technology, then an efficient method of keyboard typing is essential.

Alternative Typing and Interaction Methods for Smartwatches

When smartwatches made their debut, early thoughts of including a keyboard for typing purposes faced much skepticism for three main reasons. First, the size of the smartwatch screen and resulting size of the keyboard was thought to be too small for effective use (Arefin Shimon et al., 2016; Hong, Heo, Isokoski, & Lee, 2015). Second, users' fingers were assumed to be too large in relation to the keyboard to accurately hit the small keys, also known as the "fat finger issue" or "fat finger problem" (Arefin Shimon et al., 2016; Kim, Sohn, Pak, & Lee, 2006; Oney, Harrison, Ogan, & Wiese, 2013; Siek, Rogers, & Connelly, 2005). Third, the users' input finger was thought to be too large in relation to the size of screen and could occlude the users' view of the screen (Arefin Shimon et al., 2016; Funk, Sahami, Henze, & Schmidt, 2014). In response to these issues alternative forms of input, not limited specifically to typing, for small screen devices were developed. These include gesture recognition systems, wristband input, and skin-based input among others.

The development of gesture recognition systems is an attempt to circumvent the limited screen space of small screen devices, including smartwatches, by expanding interaction to the mid-air space around the watch. Examples of gesture recognition systems include HoverFlow (Kratz & Rohs, 2009), MagiWrite (Ketabdar, Roshandel, & Yüksel, 2010), Gesture Watch (Kim, He, Lyons, & Starner, 2007), zSense (Withana, Peiris, Samarasekara, & Nanayakkara, 2015), WristFlex (Dementyev & Paradiso, 2014), Transture (Han, Ahn, & Lee, 2015), Abracadabra (Harrison & Hudson, 2009), and mid-air gestural input (Katsuragawa, Wallace, & Lank, 2016; Song et al., 2014). Gesture recognition systems paired with finger rings/discs have also been explored, such as eRing (Wilhelm, Krakowczyk, Trollmann, & Albayrak, 2015) and Magic Ring (Jing, Cheng, Zhou, Wang, & Huang, 2013). Darbar, Sen, Dash, and Samanta (2016) introduced a sensor-based mechanism paired with a magnetic disk on the index finger for text input on smartwatches; the authors of the study found that users were able to input four words per minute (WPM).

Attempts have also been made to use the wristband and bezel of smartwatches as a means of input. Designs for wristband input include BandSense (Ahn, Hwang, Yoon, Gim, & Ryu, 2015), Watchit (Perrault, Lecolinet, Eagan, & Guiard, 2013), and CircularSelection (Plaumann, Müller, & Rukzio, 2016). Funk et al. (2014) evaluated a touch-sensitive wristband and found users were able to type three WPM using an on-band linear keyboard and four WPM using the on-band multi-tap keyboard layout. Modified bezel designs use the watch bezel as an interactive input method. TiltType (Partridge, Chatterjee, Sazawal, Borriello, & Want, 2002), 2D panning and twist with binary tilt and click (Xiao, Laput, & Harrison, 2014), WatchMI (Yeo, Lee, Bianchi & Quigley, 2016), EdgeTouch (Oakley & Lee, 2014), WatchOut (Zhang, Yang, Southern, Starner, & Abowd, 2016), and B2B-Swipe (Kubo, Shizuki, & Tanaka, 2016) are all examples of modified bezel designs. Kerber, Kiefer, and Löchtefeld (2016) compared the input techniques of a digital crown, a rotating bezel, and touch input on a one-dimensional selection task using a smartwatch. Kerber et al. (2016) found both the touch input and digital crown were rated as more usable than the rotating bezel. Other bezel designs extend input to the side of the smartwatch, such as PressTact (Darbar, Sen, & Samanta, 2016).

Additional input methods include the use of the back of the device for interaction (Baudisch & Chu, 2009), a smartwatch camera (WatchMe; Van Vlaenderen, Brulmans, Vermeulen, & Schöning, 2015), a non-smartwatch camera-based keyboard (CamK; Yin et al., 2016), thumb slide movement of the watch hand (ThumbSlide; Aoyama, Shizuki, & Tanaka, 2016), blowing air (Blowatch; Chen, 2015), a non-voice acoustic input (Whoosh; Reyes et al., 2016), lightful interaction (Yoon, Park, & Lee, 2016), multi-screened bracelets (Facet; Lyons, Nguyen, Ashbrook, & White, 2012), a finger-mounted fine-tip stylus (NanoStylus; Xia, Grossman, & Fitzmaurice, 2015), single-tap interaction with different areas of finger pads (TouchSense; Huang et al., 2014), and gaze interaction (Akkil et al., 2015).

Even the skin of the user has been utilized as an input area by SkinWatch (Ogata & Imai, 2015), Skin Buttons (Laput, Xiao, Chen, Hudson, & Harrison, 2014), iSkin (Weigel et al., 2015), and Skinput (Harrison, Tan, & Morris, 2010). Knibbe et al. (2014) combined gesture and skin input for a bimanual gesture input system.

A mobile typing method must meet three requirements to be acceptable to the mass consumer market (Zhai & Kristensson, 2012). First, the input method must be fast, allowing users to type quickly. Second, it should be intuitive for new users to efficiently use the entry method. Third, the input method should support increasing efficiency through practice in use. Based on these requirements, it is doubtful the alternative input methods discussed in this section may ever be adopted by the mass consumer market for typing on smartwatches as they all fail at least one of these requirements.

Alternative Keyboards for Smartwatches

Despite the original skepticism regarding the feasibility of typing on a smartwatch, several studies have shown keyboard-based typing is feasible and more effective than alternative input methods. In recent years, numerous keyboards have either been designed or adapted for use on smartwatches. In a review of the current existing smartwatch keyboards, Arif and Mazalek (2016) provided a summary table and detailed descriptions and illustrations of these keyboards. In this study, we updated the summary table presented in Arif and Mazalek's (2016) article with the latest research findings, and we added columns for participant mobility (seated, standing, walking) and subjective measures (see Appendix). As alternative keyboards continue to be developed, it is important to know how participant mobility affects performance, perceived workload, user satisfaction, and intent to use. As shown in the Appendix, few studies report detailed subjective ratings for alternative input methods, and the studies focus primarily on performance metrics. The studies that do report subjective ratings tend to be limited to preference ratings and non-standardized questionnaires. We believe solely relying on performance as a measure of keyboard quality is shortsighted; if users do not like the keyboard or use it, typing speed is irrelevant.

Despite all the research done with both non-QWERTY and QWERTY alternative keyboards, they have primarily failed at mass adoption largely due to their steep learning curves (Bi & Zhai, 2016). In addition, several of these alternative keyboards demonstrate very low text entry speeds. According to Arif and Mazalek (2016), most techniques using predictive technology achieved about 20 WPM, and for non-predictive, the range was from 4 to 22 WPM. The fastest typing speeds observed on a smartwatch have been accomplished with the use of a sentence- based decoder: Velocitap (41 WPM; Vertanen, Memmi, Emge, Reyal, & Kristensson, 2015) and trace input (24 WPM and 37 WPM; Gordon, Ouyang, & Zhai, 2016; Turner, Chaparro, & He, 2016, respectively). According to the guidelines of Zhai and Kristensson (2012), an existing smartphone keyboard that users are already familiar with may be best suited for use on a smartwatch, especially if the end goal is mass adoption.

In this paper, keyboards designed specifically for use on small screen devices, or those requiring an extra interaction outside of typing (i.e., zooming or panning), are referred to as *alternative keyboards*. In addition, traditional point-and-tap input is referred to as *tap*, and gesture input is referred to as *trace*. Trace is unique over point-and-tap in that it requires the user to use one continuous on-screen finger motion to type a word (Figure 12.1).

Walking and Typing

The majority of the aforementioned keyboards have only been studied with the users in a seated position. Yet, typing on a smartphone is rarely static; in fact, typing while standing or walking is quite common (Yatani & Truong, 2009). For example, Turner et al. (2016) observed some of the highest reported typing speeds on a smartwatch in a seated position. The authors went on to state in their limitations and future studies the importance of observing typing performance and garnering user feedback in mobile scenarios to more accurately represent normal user typing behavior. They also mentioned that additional research is needed to determine how users perform with smartwatch keyboards in mobile conditions. Based on the findings of Schildbach and Rukzio (2010), alternative keyboards that utilize zoom or panning functions may not be effective for text input while walking; however, little research has been done to determine what keyboard may be better suited. Not only do smartwatches need an effective typing method, but the method must be versatile and forgiving enough for mobile typing.

Figure 12.1 Examples of standard point-and-tap input (left) and trace input (right)

SMARTPHONE

Walking and typing on a smartphone is a complex, yet common, task that requires the coordination of multiple cognitive and physical resources. To achieve an accurate text message when walking and typing, visual-motor coordination, finger movements, and cognitive attention must integrate to compensate for hand and body oscillations experienced during walking (Agostini, Fermo, Massazza, & Knaflitz, 2015; Bergstrom-Lehtovirta, Oulasvirta, & Brewster, 2011). Walking while using a smartphone has been shown to negatively affect text legibility (Mustonen, Olkkonen, & Hakkinen, 2004), reading comprehension (Barnard, Yi, Jacko, & Sears, 2007; Schildbach & Rukzio, 2010), working memory (Lamberg & Muratori, 2012), target selection (Kane, Wobbrock, & Smith, 2008), and increase mental workload and stress (Vadas, Patel, Lyons, Starner, & Jacko, 2006). In addition, walking while typing affects user walking behavior, such as walking speed, gait pattern, and situational awareness (Agostini et al., 2015; Bergstrom-Lehtovirta et al., 2011; Hatfield & Murphy, 2007; Lamberg & Muratori, 2012; Licence, Smith, McGuigan & Earnest, 2015; Lopresti-Goodman, Rivera & Dressel, 2012; Plummer, Apple, Dowd, & Keith, 2015; Schabrun, van den Hoorn, Moorcroft, Greenland, & Hodges, 2014). Bergstrom-Lehtovirta et al. (2011) showed the preferred walking speed of participants dropped from 2.4 mph while undistracted to 1.8 mph while interacting with a touchscreen device. In addition, accuracy for the target selection task significantly decreased when walking only 20–40% of their preferred walking speed. The decrease in walking speed observed by Bergstrom-Lehtovirta et al. (2011) is not surprising as typing has been shown to affect walking more than either talking or reading (Lamberg & Muratori, 2012; Schabrun et al., 2014).

Most relevant to our review is walking's impact on typing. Several studies have shown typing on a smartphone declines both in speed and accuracy with walking compared to sitting or standing (Clawson, Starner, Kohlsdorf, Quigley, & Gilliland, 2014; Conradi, Busch, & Alexander, 2015; Mizobuchi, Chignell, & Newton, 2005; Nicolau & Jorge, 2012; Schildbach & Rukzio, 2010; Yatani & Truong, 2009). Attempts have been made to improve typing performance when walking such as the exploration of the following technologies: walking user interfaces (WUIs; Kane et al., 2008), games to improve the typing and walking experience (Rudchenko, Paek, & Badger, 2011), use of a smartphone accelerometer to increase accuracy (Goel, Findlater & Wobbrock, 2012), and feedback of user surroundings (Arif, Iltisberger & Stuerzlinger, 2011). Other research has studied optimal key size for mobile text input, recommendations range from 3 to 14 mm depending on the device used (Conradi et al., 2015; Mizobuchi et al., 2005; Parhi, Karlson, & Bederson, 2006). Lin, Goldman, Price, Sears, and Jacko (2007), using Fitts' Law, stated target size should be dynamic and change for the user's mobility: 4.2 mm in diameter when seated, 5.3 mm when walking on a treadmill, and 6.4 mm when walking an obstacle course. Optimal key size for use on a smartwatch is thought to be 5.7 x 5.7mm to 7 x 7mm (Dunlop, Komninos, & Durga, 2014; Shao et al., 2016).

SMARTWATCH

It would appear that walking and typing is quite difficult and demanding for a user, yet it is a common user behavior on a smartphone. Only two studies have evaluated typing on a smartwatch in mobile scenarios: Hong, Heo, Isokoski, and Lee (2016) and Darbar, Dash, and Samanta (2016). This is most likely due to the fact that most smartwatches do not currently include a keyboard for typing.

Hong et al. (2016) compared user performance with SplitBoard, Zoomboard, and a standard QWERTY keyboard using a Samsung Gear 1 smartwatch with the auto-correct feature disabled. Participants in the study completed the study tasks while standing or walking on a treadmill. Participants were allowed to set their own walking speed, walking 2.4 mph on average. Performance decreased for all three keyboards from the standing to walking condition: SplitBoard (15 WPM to 13 WPM), ZoomBoard (10 WPM to 9 WPM), and the standard QWERTY (13 WPM to 12.5 WPM). Declines in performance when walking are to be expected, as seen in the literature on walking and typing on a smartphone. However, declines observed by Hong et al. (2016) on a smartwatch were quite small, refuting the idea that key size has to be increased to avoid degraded performance while mobile (Lin et al., 2007).

Darbar et al. (2016) compared their ETAO keyboard prototype to SplitBoard, Zoomboard, and a standard QWERTY keyboard using a LG W100 Watch without an auto-correct feature. Study participants' performance was compared when they used the different keyboards while sitting or walking in the lab. As with Hong et al. (2016), performance worsened with all keyboards from the sitting to walking condition: ETAO (12 WPM to 9 WPM), SplitBoard (12 WPM to 9 WPM), ZoomBoard (9 WPM to 8 WPM), and standard QWERTY (7 WPM to 5 WPM).

Experience

In addition to mobility, prior experience with text input methods may influence typing performance on a smartwatch. Reyal, Zhai and Kristensson (2015) found that novice trace users were able to increase trace typing speed on a smartphone from 26 WPM to 34 WPM over a 10-day period. Relatively little research has been reported on how prior typing experience affects typing performance on a smartwatch. Kim et al. (2006) found entry speeds increased 18% over 5 days with no difference in error rate when using the One-Key Keyboard. Gupta and Balakrishnan (2016) demonstrated user performance increased over a 10-day span with both the DualKey QWERTY and DualKey SWEQTY keyboards. Turner et al. (2016) showed that self-reported experts with trace input on a smartphone typed 6 WPM faster, when tracing on a smartwatch, than novice trace users. This finding provides evidence that experience with trace input may carry over to smartwatch performance.

Purpose

This study is a follow-up to Turner et al. (2016). In the current study, we investigated participants' typing performance and subjective user ratings while they performed the study tasks on a full QWERTY smartwatch keyboard while standing or walking. This study aims to answer four research questions:

- What impact does mobility (standing vs. walking) have on typing performance using trace and tap input?
- Which input method (trace vs. tap) results in better typing performance when walking?

- Does prior experience with trace input on a smartphone influence typing performance on the smartwatch?
- Which input method (trace vs. tap) results in better subjective ratings when walking?

Methods

The metrics gathered in this study mirror those used in Turner et al. (2016) with the exception that participants typed while standing and while walking rather than sitting. Mobility (standing vs. walking), tracing experience (novice vs. expert on a smartphone), and text input method (trace vs. tap) were the independent variables. Typing performance (WPM), accuracy (word error rate [WER]), and subjective measures of performance were the dependent variables. Multiple hand dimensions were also measured to assess if there was a relationship between hand and finger size and typing performance.

Participants

Thirty-two college-age participants (20 female, 12 male), ranging from 18–34 years of age ($M = 22.53$, $SD = 4.42$), participated in this study for course credit. Participants were recruited based on their expertise with trace on a smartphone (all had experience with tap). None had experience typing on a smartwatch. Participants self-reported their experience level with trace on a 1–7 scale (1 = no experience; 7 = expert). Novices were categorized by a 1 or 2 ($M = 1.25$, $SD = 0.44$) and experts by a 6 or 7 ($M = 6.38$, $SD = 0.5$) rating. Participants were not made aware of the expertise criteria prior to their self-evaluation. Those who identified as a 3–5 rating were dismissed from the study and given partial course credit. Two participants were dismissed for not meeting the study criteria. Sixteen novices (11 female, 5 male) and 16 experts (9 female, 7 male) participated; all typed on the smartwatch using their index finger.

All participants were fluent English speakers, had normal or corrected-to-normal vision, and did not have any physical limitations to their hands that would prevent them from being able to type on a smartwatch. All participants were experienced with sending and receiving text messages on their touchscreen smartphone.

Materials

A Samsung Galaxy Gear 1 (display size of 1.63 inches) with the Swype word-gesture keyboard (version 1.6.5.23769) was used in this study. The keyboard measured 17.5 mm wide x 30 mm high, and each key 4 mm x 3 mm. All 35 keys on the keyboard were fully functioning and the autocorrect feature was enabled.

A subset of phrases were randomly selected from a list of 500 composed by Mac-Kenzie and Soukoreff (2003). Ten practice phrases and 15 experimental phrases were randomly chosen for each condition; there was no overlap between the practice and experimental phrases. The following are some example phrases: "time to go shopping," "a great disturbance in the force," and "all good boys deserve fudge." Novices and experts of the same participant number received the same phrases (i.e., p1 novice received the same phrases as p1 expert, but a different set than p2, p3 ... novice and expert). The phrases contained lowercase letters only (no numbers, symbols, punctuation, or uppercase letters). Phrases ranged from 16 to 42 characters for all conditions. A JAS Trackmaster (model number: TX425C) treadmill was used to simulate walking conditions (Hong et al., 2016). Participants were allowed to choose their walking speed but instructed to select a comfortable speed they could maintain for the entirety of the walking conditions (Hong et al., 2016). Walking speeds ranged from 1.5 to 2.5 mph (M = 2.04, SD = 0.30). Figure 12.2 shows the experimental setup.

Procedure

After providing consent, participants were given a brief demographic survey assessing smartphone texting behavior and text input method usage. Based on their experience with trace input on a smartphone, participants were placed in either the novice or expert group. Participants were then introduced to the first condition, either tap or trace and walking or standing, and given a brief tutorial by the experimenter. Next they were given 10 practice phrases to type before the experimental trials began. For the walking conditions, participants started off at a speed of 1.0 mph and were allowed to increase walking speed after each practice phrase until a comfortable speed was selected. The order of input method and mobility condition was partially counterbalanced across all participants to prevent participants from doing two consecutive walking conditions.

For the experimental trials, 15 phrases were presented one at a time on a computer screen in front of the participants. They were instructed to read each phrase aloud to ensure that they understood the phrase and to verbally indicate when they started and stopped typing (Arif et al., 2011; MacKenzie & Read, 2007). Time was recorded by a researcher using a digital stop watch. Participants were instructed to type the phrases as quickly and accurately as possible. They were allowed to correct mistakes but not

Figure 12.2 Experimental setup

required to do so. Phrases were saved as a text file on the watch and later scored manually by an experimenter.

Once participants had completed the 15 phrases of the condition, they completed a perceived usability survey and a mental workload assessment. After finishing the mental workload assessment, participants were introduced to the second condition and the steps were repeated.

After all four conditions were completed, participants were asked to rate the four conditions on perceived performance and preference scales and an intent to use scale. Finally, the participants' typing hand and finger dimensions were measured. They were then debriefed and thanked for their time.

Design

A 2 x 2 x 2 mixed design was used for this study. The independent variables were input method, mobility, and experience. Input method (trace vs. tap) and mobility (standing vs. walking) were the within-subjects factors. Experience (novice vs. expert) was the between- subjects factor. Dependent variables included typing speed, typing accuracy, subjective perceptions of usability, workload, performance, and intent to use.

Performance

Performance was measured by typing speed, words per minute (WPM), and typing accuracy as reflected by the word error rate (WER). Typing speed was calculated using WPM = 12 * (T − 1)/S where T is the number of transcribed characters, S is the number of seconds, and one word is assumed to be 5 characters (MacKenzie & Tanaka-Ishii, 2010). Typing accuracy, or WER, was calculated using the number of word errors per phrase divided by the total number of words per phrase.

Typing accuracy was investigated by the WER and the type of errors: substitution, insertion, and omission error rate. Substitution errors occurred when a word was transcribed other than what was intended. Insertion errors occurred when an extra word was transcribed. Omission errors occurred when an intended word was omitted from the transcription.

Subjective Measures

The subjective measures were determined by measuring the workload, perceived usability, perceived performance and preference, and intent to use.

SUBJECTIVE WORKLOAD

The raw NASA Task Load Index (NASA TLX—R; Hart & Staveland, 1988) was used to measure participants' perceived workload and performance after each condition. Participants provided ratings on a 21-point scale for perceived mental, physical, and temporal effort; performance; overall effort; and frustration. A higher score indicates a more demanding experience or worse perceived performance.

PERCEIVED USABILITY

An adapted System Usability Scale (SUS) was used to measure participants' perceived usability of each input method with the mobility condition. The SUS is an industry-standard 10-item questionnaire with 5 response options (Strongly Disagree to Strongly Agree) that is summarized as a single score between 0–100 (Brooke, 2013). Higher scores indicate higher perceived usability. The scale was adapted by replacing "system" with "input method."

PERCEIVED PERFORMANCE AND PREFERENCE

Perceived accuracy, perceived speed, and overall preference with each input method and mobility condition were measured using a 50-point scale with higher scores reflecting more preferred or better in terms of accuracy or speed.

INTENT TO USE

Participants rated the likelihood they would use each input method with each mobility condition on a 0–10 scale with a 10 being very likely.

Anthropometric Measurements

A sliding digital caliper was used to measure the typing hand of each participant. Hand measurements included the length and width of hand, and length, width, and circumference of the index finger and thumb in millimeters. Thumb dimensions were later excluded from analysis because no participants used their thumb to type.

Results

All dependent measures were analyzed using a 2 x 2 x 2 mixed model ANOVA. Partial eta squared (η_p^2) was used to estimate effect size for all ANOVA tests. Analyses of simple main effects were conducted to follow-up on all significant interactions. Bonferroni correction was used to control for family-wise Type I error across multiple comparisons.

Typing Speed

Significant main effects of input method and mobility were found for typing speed (WPM), with participants typing faster with trace ($M = 35.33$, $SD = 9.01$) than tap ($M = 29.88$, $SD = 6.86$) and when standing ($M = 32.25$, $SD = 7.76$) than walking ($M = 29.88$, $SD = 8.21$): $F(1, 30) = 77.42$, $p < 0.001$, $\eta_p^2 = 0.72$; $F(1, 30) = 19.69$, $p < 0.001$, $\eta_p^2 = 0.40$, respectively. No other main effects or interactions were found for WPM, $p > 0.05$. Figure 12.3 shows typing speed by input method, mobility, and experience.

Typing Accuracy

Significant main effects of input method and mobility were found for typing accuracy (WER), with participants typing more accurately with trace ($M = 0.10$, $SD = 0.07$)

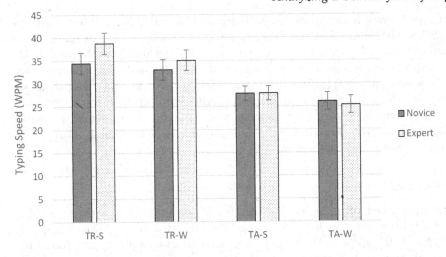

Figure 12.3 Typing speed. Error bars represent ±1 standard error. TR = Trace, TA = Tap, S = Stand, W = Walk

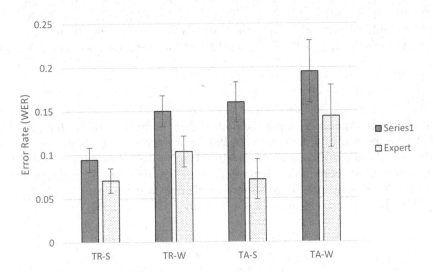

Figure 12.4 Typing accuracy. Error bars represent ± 1 standard error. TR = Trace, TA = Tap, S = Stand, W = Walk

than tap ($M = 0.14$, $SD = 0.12$) and when standing ($M = 0.10$, $SD = 0.08$) than walking ($M = 0.15$, $SD = 0.11$): $F(1, 30) = 5.82$, $p = 0.02$, $\eta_p^2 = 0.16$; $F(1, 30) = 20.57$, $p < 0.001$, $\eta_p^2 = 0.41$, respectively.

A significant main effect of mobility was found for substitution error rate with participants typing fewer substitution errors when standing ($M = 0.07$, $SD = 0.06$) than walking ($M = 0.11$, $SD = 0.08$); $F(1, 30) = 14.65$, $p = 0.001$, $\eta_p^2 = 0.33$.

Significant main effects of experience, input method, and mobility were found for insertion error rate, with experts ($M = 0.02$, $SD = 0.03$) committing fewer insertion errors than novices ($M = 0.04$, $SD = 0.05$); $F(1, 30) = 4.86$, $p = 0.04$, $\eta_p^2 = 0.14$.

Participants typed fewer insertion errors with trace ($M = 0.01$, $SD = 0.01$) than tap ($M = 0.05$, $SD = 0.05$) and when standing ($M = 0.02$, $SD = 0.04$) than walking ($M = 0.03$, $SD = 0.05$): $F(1, 30) = 20.45$, $p < 0.001$, $\eta_p^2 = 0.41$; $F(1,30) = 6.66$, $p = 0.02$, $\eta_p^2 = 0.18$, respectively. A significant interaction of input method and experience for insertion errors was found; $F(1,30) = 4.52$, $p = 0.04$, $\eta_p^2 = 0.13$. Follow-up analysis revealed novice participants made more insertion errors with tap than trace, and novices made more insertion errors than experts, $p < 0.05$. No other significant main effects or interactions were found, $p > 0.05$. Figure 12.4 shows typing accuracy by input method, mobility, and experience.

Subjective Workload

A significant main effect of input method was found for frustration with trace ($M = 6.73$, $SD = 4.79$) being rated as less frustrating than tap ($M = 10.30$, $SD = 5.25$); $F(1, 30) = 16.38$, $p < 0.001$, $\eta_p^2 = 0.35$. A significant main effect of mobility was found for mental, physical, temporal, performance, effort, and frustration with standing being rated as less demanding than walking: $F(1, 30) = 31.30$, $p < 0.001$, $\eta_p^2 = 0.51$; $F(1, 30) = 54.49$, $p < 0.001$, $\eta_p^2 = 0.65$; $F(1,30) = 19.09$, $p < 0.001$, $\eta_p^2 = 0.39$; $F(1,30) = 11.24$, $p = 0.002$, $\eta_p^2 = 0.27$; $F(1,30) = 21.40$, $p < 0.001$, $\eta_p^2 = 0.42$; $F(1,30) = 22.00$, $p < 0.001$, $\eta_p^2 = 0.42$, respectively. No other main effects or interactions were found for subjective workload, $p > 0.05$. Figure 12.5 shows perceived workload by mobility.

Perceived Usability

Significant main effects of experience, input method, and mobility were found for perceived usability, with experts ($M = 75.39$, $SD = 15.98$) reporting higher scores than novices ($M = 57.50$, $SD = 17.68$); $F(1,30) = 22.43$, $p < 0.001$, $\eta_p^2 = 0.43$. Trace ($M = 74.10$, $SD = 16.80$) was perceived as more usable than tap ($M = 58.79$, $SD = 18.15$), and standing ($M = 67.77$, $SD = 19.48$) was perceived as more usable than walking ($M = 65.12$, $SD = 18.64$): $F(1, 30) = 22.53$, $p < 0.001$, $\eta_p^2 = 0.43$; $F(1, 30) = 5.25$, $p = 0.03$, $\eta_p^2 = 0.15$.

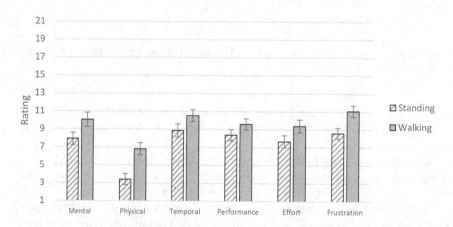

Figure 12.5 Perceived workload by mobility (1 = least). Error bars represent ± 1 standard error

No other main effects or interactions were found for perceived usability, $p > 0.05$. Figure 12.6 shows perceived usability score by input method, mobility, and experience.

Perceived Accuracy, Speed, and Preference

Significant main effects of experience, input method, and mobility were found for perceived accuracy, with experts ($M = 37.45$, $SD = 7.63$) having higher perceived accuracy ratings than novices ($M = 29.22$, $SD = 11.50$); $F(1, 30) = 11.57$, $p = 0.002$, $\eta_p^2 = 0.28$. Trace ($M = 36.42$, $SD = 9.65$) was perceived as more accurate overall than tap ($M = 30.25$, $SD = 10.61$) and standing ($M = 35.75$, $SD = 10.34$) more than walking ($M = 30.92$, $SD = 10.31$): $F(1, 30) = 11.90$, $p = 0.002$, $\eta_p^2 = 0.28$; $F(1,30) = 24.65$, $p < 0.001$, $\eta_p^2 = 0.45$. No other main effects or interactions were found for perceived accuracy, $p > 0.008$.

Significant main effects of input method and mobility were found for perceived speed, with trace ($M = 40.08$, $SD = 6.57$) having higher perceived speed ratings than tap ($M = 29.38$, $SD = 9.12$) and standing ($M = 36.89$, $SD = 8.84$) higher than walking ($M = 32.56$, $SD = 9.84$): $F(1,30) = 45.77$, $p < 0.001$, $\eta_p^2 = 0.60$; $F(1,30) = 19.48$, $p < 0.001$, $\eta_p^2 = 0.39$. No other main effects or interactions were found for perceived speed, $p > 0.008$.

Significant main effects of experience, input method, and mobility were found for overall preference, with experts ($M = 34.70$, $SD = 12.29$) having higher overall preference ratings than novices ($M = 30.02$, $SD = 11.80$); $F(1, 30) = 4.32$, $p = 0.05$, $\eta_p^2 = 0.13$. Trace ($M = 38.59$, $SD = 9.01$) was preferred more overall than tap ($M = 26.13$, $SD = 11.89$) and standing ($M = 34.52$, $SD = 11.53$) more than walking ($M = 30.20$, $SD = 12.62$): $F(1, 30) = 27.13$, $p < 0.001$, $\eta_p^2 = 0.48$; $F(1, 30) = 11.51$, $p = 0.002$, $\eta_p^2 = 0.28$. No other main effects or interactions were found for overall preference, $p > 0.008$. Figure 12.7 shows overall preference by input method, mobility, and experience.

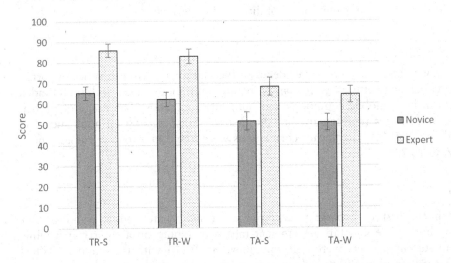

Figure 12.6 Perceived usability score (100 = highest). Error bars represent ± 1 standard error. TR = Trace, TA = Tap, S = Stand, W = Walk

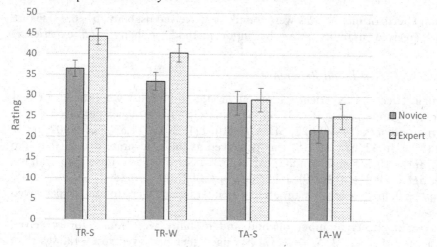

Figure 12.7 Overall preference (50 = best). Error bars represent ± 1 standard error. TR = Trace, TA = Tap, S = Stand, W = Walk

Intent to Use

Significant main effects of input method and mobility were found for intent to use, with trace (*M* = 7.95, *SD* = 2.48) having a higher intent to use rating than tap (*M* = 4.30, *SD* = 2.88) and standing (*M* = 6.67, *SD* = 2.95) higher than walking (*M* = 5.58, *SD* = 3.45): $F(1,30)$ = 30.29, $p < 0.001$, η_p^2 = 0.50; $F(1,30)$ = 22.33, $p < 0.001$, η_p^2 = 0.43, respectively. A significant interaction of input method and mobility also was found; $F(1,30)$ = 9.02, p = 0.01, η_p^2 = 0.23. Follow-up analysis revealed participants rated standing and walking with trace higher on intent to use than standing and walking with tap, and they rated tap standing higher than tap walking, $p < 0.05$. No other main effects or interactions were found, $p > 0.05$. Figure 12.8 shows intent to use by input method, mobility, and experience.

Hand Measurements

To determine whether there was any evidence of the "fat finger" issue, a series of correlations were conducted between performance, hand width, index finger width, and index finger length in both mobile conditions and both input methods. The range of participants' hand widths was representative of the first to 75th percentile of adult men and women (White, 1980). No significant correlations were found, $p > 0.05$ (*r* values ranged from –0.27. to +0.23).

Discussion

This study is the first to explore trace input on a smartwatch while walking. Our results show both trace and tap are efficient means of typing on a smartwatch while walking and standing. Users were able to achieve 35 WPM with trace and 30 WPM with tap, regardless of mobility or experience. These observed trace and tap typing speeds are among the fastest observed on a smartwatch even though users were

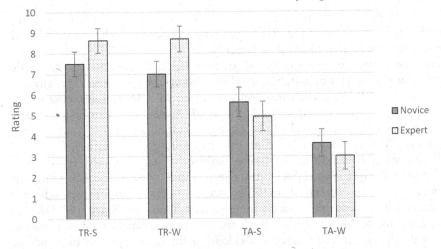

Figure 12.8 Intent to use (10 = best). Error bars represent ± 1 standard error. TR = Trace, TA = Tap, S = Stand, W = Walk

standing or walking (see Appendix). The observed superiority of trace over tap is consistent with previous findings (Gordon et al., 2016; Turner et al., 2016). Surprisingly, experience with trace input on a smartphone had no significant impact on entry speed, an effect previously observed by Turner et al. (2016). It is possible the lack of difference between experts and novices is due to the increased variability in performance in the walking condition (participants were seated for Turner et al., 2016). Regardless, this suggests that users completely unfamiliar to trace input are able to quickly reach the performance level of trace experts when typing on a smartwatch after very little practice. In addition, users typed 32 WPM when standing and 30 WPM when walking; these speeds are nearly three times faster than other reported typing speeds on alternative smartwatch keyboards in stationary and mobile scenarios (Darbar et al., 2016; Hong et al., 2016).

Prior experience with trace on a smartphone did not seem to have an effect on accuracy. Users typed more accurately with trace (10% WER) than tap (14% WER) and more accurately when standing (10% WER) than when walking (15% WER). These word error rates are consistent with other observed error rates on smartwatches (see Appendix). The increased error rate from standing to walking is consistent with the literature (Bergstrom-Lehtovirta et al., 2011; Darbar et al., 2016; Hong et al., 2016). It is possible the higher error between trace and tap when walking is because tap input requires users to lift their finger before and after each keystroke. When walking, this task is even more difficult due to the constant motion of the body with each step. In contrast, trace requires the user to use one continuous motion to type, so the finger is always in contact with the screen. This is likely the reason why tap was rated as more frustrating than trace; however, future research should examine the biomechanics of each interaction method to investigate further.

Performance with both input methods and mobility conditions remained high despite key sizes, 4 mm x 3 mm, being significantly smaller than the recommended key size range for use with a smartwatch, 5.7 to 7 mm (Dunlop et al., 2014; Shao et al., 2016). In addition, no evidence of the "fat finger" issue or screen occlusion was found as performance was not related to hand or finger size.

We believe the observed superiority of trace over the alternative keyboards shown in the Appendix is attributable to three factors. First, users are already familiar with the QWERTY keyboard layout, resulting in a shorter learning curve than alternative keyboard layouts. Second, the small screen size required less distance for the user's finger to travel while typing, resulting in faster input. Third, the keyboard used in this study included an effective auto-correct feature.

Results of this study also add to the limited subjective data typically reported in smartwatch typing studies. Subjectively, trace was rated more favorably than tap across all measures. The perceived usability scores of both input methods fell within the marginally acceptable to acceptable ranges: *Good* for trace and *OK* for tap (Bangor, Kortum, & Miller, 2009). With the exception of perceived frustration, users reported no difference in perceived workload, a finding supported by Sonaike, Bewaji, Ritchey, and Peres (2016). Users also perceived their performance as better when using trace and indicated they would prefer to use trace over tap if given the choice when typing on a smartwatch. Standing was consistently rated more favorably than walking, as expected. Thumb size has been shown to be correlated with user satisfaction when typing on smartphone keyboards (Balakrishnan & Yeow, 2008). Interestingly, no significant correlations between hand size, or finger size, and any of the subjective measures gathered were found in our study. One potential reason for this is that in Balakrishnan and Yeow (2008) all participants typed using a 3 x 4 key keypad and not a full QWERTY keyboard. It is likely the 3 x 4 key keyboard layout yielded more cumbersome typing behavior for users with larger thumbs.

Conclusion

This study expands upon the limited research on smartwatch typing and is the first to explore trace input on a smartwatch while walking. We demonstrated both tap and trace are efficient methods of typing on a smartwatch QWERTY keyboard in a mobile scenario. Users completely naïve to typing on a smartwatch were able to achieve high typing speeds with little practice. Trace input appears to be especially well suited for typing on a smartwatch as users were able to type 30–35 WPM depending on the mobility condition, regardless of prior experience with trace. In addition, users subjectively rated trace easier to use, preferred it over tap, and suggested they would use it in the future. Pulvirent (2015) notes, "To make smartwatches a long-term device and not simply a quick hit, manufacturers and developers are going to need to make them relevant and necessary for daily activities" (para. 6). We believe the addition of a familiar, easy-to-use keyboard that yields accurate typing is both relevant and necessary. Smartwatch manufacturers should include QWERTY keyboards with trace input as a standard feature.

Limitations

While this study investigated more realistic smartwatch usage than sitting at a desk, it was still conducted as a controlled study in a laboratory setting.

A treadmill was used to simulate normal walking behavior so we could investigate typing performance in a steady, walking condition. Treadmills have been used to simulate walking environments in other studies; one benefit of treadmill use is that the participant must maintain a steady walking pace. Walking in more natural environments, while more ecologically valid, results in inconsistent gait, as well as starting and stopping. More research should be done to examine typing performance in such environments where distractions are more likely to occur. In addition, this study evaluated college-aged individuals who are most likely to be interested in using smartwatch technology. It is unknown how the results from this study would transfer to older age groups less familiar with the technology. Future research should include a wider range of ages, experience, and education levels to test the generalizability of these results.

Recommendations

The following are recommendations for the development of smartwatch keyboard technology and for future smartwatch studies:

- Smartwatch manufacturers should incorporate a trace based QWERTY keyboard in all smartwatch designs.
- Developers of novel keyboards should emphasize the importance of gathering subjective measures to inform design improvements from the user's point of view.
- Future studies should compare trace input against alternative keyboards, such as WatchWriter and Swipeboard, in different mobile scenarios on different smartwatch designs to see if the superiority of trace generalizes to different mobile scenarios and smartwatch designs.
- Future studies should seek to explore different user age groups, experience, and education levels.

Tips for Usability Practitioners

The following are suggestions for usability practitioners:

- Familiarity with keyboard layout may help new users learn new typing techniques quickly.
- Subjective ratings in addition to performance ratings should be collected when studying mobile device text input to provide the maximal insight to user satisfaction and acceptance.
- Treadmills can be used in lieu of more natural walking tasks to provide a controlled simulation of walking.
- Partial counterbalancing of experimental conditions should be used to minimize participant fatigue in mobile conditions.

Acknowledgments

We would like to thank Dr. Jeremy Patterson and Morgan Colling, in the Human Performance Lab at Wichita State University, for the use of their treadmill equipment and their contributions to this work.

References

[This article appeared before the 7th edition of the *APA Publication Manual* was published, and the citations follow the style of the 6th edition.]

Agostini, V., Fermo, F. L., Massazza, G., & Knaflitz, M. (2015). Does texting while walking really affect gait in young adults? *Journal of Neuroengineering and Rehabilitation, 12*(1), 1.

Ahn, Y., Hwang, S., Yoon, H., Gim, J., & Ryu, J.-H. (2015). *BandSense: Pressure-sensitive multitouch interaction on a wristband.* Paper presented at the Proceedings of the 33rd Annual ACM Conference Extended Abstracts on Human Factors in Computing Systems.

Akkil, D., Kangas, J., Rantala, J., Isokoski, P., Spakov, O., & Raisamo, R. (2015). *Glance awareness and gaze interaction in smartwatches.* Paper presented at the Proceedings of the 33rd Annual ACM Conference Extended Abstracts on Human Factors in Computing Systems.

Aoyama, S., Shizuki, B., & Tanaka, J. (2016). *ThumbSlide: An interaction technique for smartwatches using a thumb slide movement.* Paper presented at the Proceedings of the 2016 CHI Conference Extended Abstracts on Human Factors in Computing Systems.

Arefin Shimon, S. S., Lutton, C., Xu, Z., Morrison-Smith, S., Boucher, C., & Ruiz, J. (2016). *Exploring non-touchscreen gestures for smartwatches.* Paper presented at the Proceedings of the 2016 CHI Conference on Human Factors in Computing Systems.

Arif, A. S., Iltisberger, B., & Stuerzlinger, W. (2011). *Extending mobile user ambient awareness for nomadic text entry.* Paper presented at the Proceedings of the 23rd Australian Computer-Human Interaction Conference.

Arif, A. S., & Mazalek, A. (2016). A survey of text entry techniques for smartwatches. In M. Kurosu (Ed.), *Human-computer interaction. Interaction platforms and techniques: 18th international conference, HCI international, proceedings, part II-Volume 9732* (pp. 255–267). New York, NY: Springer-Verlag New York, Inc.

Balakrishnan, V., & Yeow, P. H. P. (2008). A study of the effect of thumb sizes on mobile phone texting satisfaction. *Journal of Usability Studies, 3*(3), 118–128.

Bangor, A., Kortum, P., & Miller, J. (2009). Determining what individual SUS scores mean: Adding an adjective rating scale. *Journal of Usability Studies, 4*(3), 114–123.

Barnard, L., Yi, J. S., Jacko, J. A., & Sears, A. (2007). Capturing the effects of context on human performance in mobile computing systems. *Personal and Ubiquitous Computing, 11*(2), 81–96.

Baudisch, P., & Chu, G. (2009). *Back-of-device interaction allows creating very small touch devices.* Paper presented at the Proceedings of the SIGCHI Conference on Human Factors in Computing Systems.

Bergstrom-Lehtovirta, J., Oulasvirta, A., & Brewster, S. (2011). *The effects of walking speed on target acquisition on a touchscreen interface.* Paper presented at the Proceedings of the 13th International Conference on Human Computer Interaction with Mobile Devices and Services.

Bi, X., & Zhai, S. (2016). *IJQwerty: What difference does one key change make? Gesture typing keyboard optimization bounded by one key position change from QWERTY.* Paper presented at the Proceedings of the 2016 CHI Conference on Human Factors in Computing Systems.

Brooke, J. (2013). SUS: A retrospective. *Journal of Usability Studies, 8*(2), 29–40.

Cha, J.-M., Choi, E., & Lim, J. (2015). Virtual sliding QWERTY: A new text entry method for smartwatches using tap-n-drag. *Applied Ergonomics*, *51*, 263–272.

Chaparro, B. S., He, J., Turner, C., & Turner, K. (2015). Is touch-based text input practical for a smartwatch? In C. Stephanidis (Ed.), *HCI international 2015—Posters' extended abstracts. HCI 2015. Communications in computer and information science* (Vol. 529, pp. 3–8). Springer.

Chen, W.-H. (2015). *Blowatch: Blowable and hands-free interaction for smartwatches*. Paper presented at the Proceedings of the 33rd Annual ACM Conference Extended Abstracts on Human Factors in Computing Systems.

Chen, X. A., Grossman, T., & Fitzmaurice, G. (2014). *Swipeboard: A text entry technique for ultra-small interfaces that supports novice to expert transitions*. Paper presented at the Proceedings of the 27th Annual ACM Symposium on User Interface Software and Technology.

Clawson, J., Starner, T., Kohlsdorf, D., Quigley, D. P., & Gilliland, S. (2014). *Texting while walking: An evaluation of mini-qwerty text input while on-the-go*. Paper presented at the Proceedings of the 16th International Conference on Human-Computer Interaction with Mobile Devices & Services.

Conradi, J., Busch, O., & Alexander, T. (2015). Optimal touch button size for the use of mobile devices while walking. *Procedia Manufacturing*, *3*, 387–394.

Darbar, R., Dash, P., & Samanta, D. (2016). ETAO keyboard: Text input technique on smartwatches. *Procedia Computer Science*, *84*, 137–141.

Darbar, R., Sen, P. K., Dash, P., & Samanta, D. (2016). Using hall effect sensors for 3D space text entry on smartwatches. *Procedia Computer Science*, *84*, 79–85.

Darbar, R., Sen, P. K., & Samanta, D. (2016). *PressTact: Side pressure-based input for smartwatch interaction*. Paper presented at the Proceedings of the 2016 CHI Conference Extended Abstracts on Human Factors in Computing Systems.

Dementyev, A., & Paradiso, J. A. (2014). *WristFlex: Low-power gesture input with wrist-worn pressure sensors*. Paper presented at the Proceedings of the 27th Annual ACM Symposium on User Interface Software and Technology.

Dunlop, M. D., Komninos, A., & Durga, N. (2014). *Towards high quality text entry on smartwatches*. Paper presented at the CHI'14 Extended Abstracts on Human Factors in Computing Systems.

Funk, M., Sahami, A., Henze, N., & Schmidt, A. (2014). *Using a touch-sensitive wristband for text entry on smart watches*. Paper presented at the CHI'14 Extended Abstracts on Human Factors in Computing Systems.

Goel, M., Findlater, L., & Wobbrock, J. (2012). *WalkType: Using accelerometer data to accommodate situational impairments in mobile touch screen text entry*. Paper presented at the Proceedings of the SIGCHI Conference on Human Factors in Computing Systems.

Gordon, M., Ouyang, T., & Zhai, S. (2016). *WatchWriter: Tap and gesture typing on a smartwatch miniature keyboard with statistical decoding*. Paper presented at the Proceedings of the 2016 CHI Conference on Human Factors in Computing Systems.

Gupta, A., & Balakrishnan, R. (2016). *DualKey: Miniature screen text entry via finger identification*. Paper presented at the Proceedings of the 2016 CHI Conference on Human Factors in Computing Systems.

Han, J., Ahn, S., & Lee, G. (2015). *Transture: Continuing a touch gesture on a small screen into the air*. Paper presented at the Proceedings of the 33rd Annual ACM Conference Extended Abstracts on Human Factors in Computing Systems.

Harrison, C., & Hudson, S. E. (2009). *Abracadabra: Wireless, high-precision, and unpowered finger input for very small mobile devices*. Paper presented at the Proceedings of the 22nd Annual ACM Symposium on User Interface Software and Technology.

Harrison, C., Tan, D., & Morris, D. (2010). *Skinput: Appropriating the body as an input surface*. Paper presented at the Proceedings of the SIGCHI Conference on Human Factors in Computing Systems.

Hart, S. G., & Staveland, L. E. (1988). Development of NASA-TLX (Task Load Index): Results of empirical and theoretical research. *Advances in Psychology*, *52*, 139–183.

Hatfield, J., & Murphy, S. (2007). The effects of mobile phone use on pedestrian crossing behaviour at signalised and unsignalised intersections. *Accident Analysis & Prevention, 39*(1), 197–205.

Hong, J., Heo, S., Isokoski, P., & Lee, G. (2015). *SplitBoard: A simple split soft keyboard for wristwatch-sized touch screens.* Paper presented at the Proceedings of the 33rd Annual ACM Conference on Human Factors in Computing Systems.

Hong, J., Heo, S., Isokoski, P., & Lee, G. (2016). Comparison of three QWERTY keyboards for a smartwatch. *Interacting with Computers, 28*(6), 811–825.

Hsiu, M.-C., Huang, D.-Y., Chen, C. A., Lin, Y.-C., Hung, Y.-P., Yang, D.-N., & Chen, M. (2016). *ForceBoard: Using force as input technique on size-limited soft keyboard.* Paper presented at the Proceedings of the 18th International Conference on Human-Computer Interaction with Mobile Devices and Services Adjunct.

Huang, D.-Y., Tsai, M.-C., Tung, Y.-C., Tsai, M.-L., Yeh, Y.-T., Chan, L., … Chen, M. Y. (2014). *TouchSense: Expanding touchscreen input vocabulary using different areas of users' finger pads.* Paper presented at the Proceedings of the SIGCHI Conference on Human Factors in Computing Systems.

Jing, L., Cheng, Z., Zhou, Y., Wang, J., & Huang, T. (2013). *Magic ring: A self-contained gesture input device on finger.* Paper presented at the Proceedings of the 12th International Conference on Mobile and Ubiquitous Multimedia.

Kane, S. K., Wobbrock, J. O., & Smith, I. E. (2008). *Getting off the treadmill: Evaluating walking user interfaces for mobile devices in public spaces.* Paper presented at the Proceedings of the 10th International Conference on Human Computer Interaction with Mobile Devices and Services.

Katsuragawa, K., Wallace, J. R., & Lank, E. (2016). *Gestural text input using a smartwatch.* Paper presented at the Proceedings of the International Working Conference on Advanced Visual Interfaces.

Kerber, F., Kiefer, T., & Löchtefeld, M. (2016). *Investigating interaction techniques for state-of-the-art smartwatches.* Paper presented at the Proceedings of the 2016 CHI Conference Extended Abstracts on Human Factors in Computing Systems.

Ketabdar, H., Roshandel, M., & Yüksel, K. A. (2010). *MagiWrite: Towards touchless digit entry using 3D space around mobile devices.* Paper presented at the Proceedings of the 12th International Conference on Human Computer Interaction with Mobile Devices and Services.

Kim, J., He, J., Lyons, K., & Starner, T. (2007). *The gesture watch: A wireless contact-free gesture based wrist interface.* Paper presented at the 2007 11th IEEE International Symposium on Wearable Computers.

Kim, S., Sohn, M., Pak, J., & Lee, W. (2006). *One-key keyboard: A very small QWERTY keyboard supporting text entry for wearable computing.* Paper presented at the Proceedings of the 18th Australia Conference on Computer-Human Interaction: Design: Activities, Artefacts and Environments.

Knibbe, J., Martinez Plasencia, D., Bainbridge, C., Chan, C.-K., Wu, J., Cable, T., … Coyle, D. (2014). *Extending interaction for smart watches: Enabling bimanual around device control.* Paper presented at the CHI'14 Extended Abstracts on Human Factors in Computing Systems.

Komninos, A., & Dunlop, M. (2014). Text input on a smart watch. *Pervasive Computing, IEEE, 13*(4), 50–58.

Kratz, S., & Rohs, M. (2009). *HoverFlow: Expanding the design space of around-device interaction.* Paper presented at the Proceedings of the 11th International Conference on Human-Computer Interaction with Mobile Devices and Services.

Kubo, Y., Shizuki, B., & Tanaka, J. (2016). *B2B-Swipe: Swipe gesture for rectangular smartwatches from a bezel to a bezel.* Paper presented at the Proceedings of the 2016 CHI Conference on Human Factors in Computing Systems.

Lamberg, E. M., & Muratori, L. M. (2012). Cell phones change the way we walk. *Gait & Posture, 35*(4), 688–690.

Laput, G., Xiao, R., Chen, X. A., Hudson, S. E., & Harrison, C. (2014). *Skin buttons: Cheap, small, low-powered and clickable fixed-icon laser projectors*. Paper presented at the Proceedings of the 27th Annual ACM Symposium on User Interface Software and Technology.

Leiva, L. A., Sahami, A., Catalá, A., Henze, N., & Schmidt, A. (2015). *Text entry on tiny QWERTY soft keyboards*. Paper presented at the Proceedings of the 33rd Annual ACM Conference on Human Factors in Computing Systems.

Licence, S., Smith, R., McGuigan, M. P., & Earnest, C. P. (2015). Gait pattern alterations during walking, texting and walking and texting during cognitively distractive tasks while negotiating common pedestrian obstacles. *PLOS ONE, 10*(7), e0133281. https://doi.org/10.1371/journal.pone.0133281

Lin, M., Goldman, R., Price, K. J., Sears, A., & Jacko, J. (2007). How do people tap when walking? An empirical investigation of nomadic data entry. *International Journal of Human-Computer Studies, 65*(9), 759–769.

Lopresti-Goodman, S. M., Rivera, A., & Dressel, C. (2012). Practicing safe text: The impact of texting on walking behavior. *Applied Cognitive Psychology, 26*(4), 644–648.

Lyons, K., Nguyen, D., Ashbrook, D., & White, S. (2012). *Facet: A multi-segment wrist worn system*. Paper presented at the Proceedings of the 25th Annual ACM Symposium on User Interface Software and Technology.

MacKenzie, I. S., & Read, J. C. (2007). *Using paper mockups for evaluating soft keyboard layouts*. Paper presented at the Proceedings of the 2007 Conference of the Center for Advanced Studies on Collaborative Research.

MacKenzie, I. S., & Soukoreff, R. W. (2003). *Phrase sets for evaluating text entry techniques*. Paper presented at the CHI'03 Extended Abstracts on Human Factors in Computing Systems.

MacKenzie, I. S., & Tanaka-Ishii, K. (2010). *Text entry systems: Mobility, accessibility, universality*. Morgan Kaufmann.

Mizobuchi, S., Chignell, M., & Newton, D. (2005). *Mobile text entry: Relationship between walking speed and text input task difficulty*. Paper presented at the Proceedings of the 7th International Conference on Human Computer Interaction with Mobile Devices & Services.

Mottelson, A., Larsen, C., Lyderik, M., Strohmeier, P., & Knibbe, J. (2016). *Invisiboard: Maximizing display and input space with a full screen text entry method for smartwatches*. Paper presented at the Proceedings of the 18th International Conference on Human-Computer Interaction with Mobile Devices and Services.

Mustonen, T., Olkkonen, M., & Hakkinen, J. (2004). *Examining mobile phone text legibility while walking*. Paper presented at the CHI'04 Extended Abstracts on Human Factors in Computing Systems.

Nicolau, H., & Jorge, J. (2012). *Touch typing using thumbs: Understanding the effect of mobility and hand posture*. Paper presented at the Proceedings of the SIGCHI Conference on Human Factors in Computing Systems.

Oakley, I., & Lee, D. (2014). *Interaction on the edge: Offset sensing for small devices*. Paper presented at the Proceedings of the SIGCHI Conference on Human Factors in Computing Systems.

Ogata, M., & Imai, M. (2015). *SkinWatch: Skin gesture interaction for smart watch*. Paper presented at the Proceedings of the 6th Augmented Human International Conference.

Oney, S., Harrison, C., Ogan, A., & Wiese, J. (2013). *ZoomBoard: A diminutive qwerty soft keyboard using iterative zooming for ultra-small devices*. Paper presented at the Proceedings of the SIGCHI Conference on Human Factors in Computing Systems.

Parhi, P., Karlson, A. K., & Bederson, B. B. (2006). *Target size study for one-handed thumb use on small touchscreen devices*. Paper presented at the Proceedings of the 8th Conference on Human-Computer Interaction with Mobile Devices and Services.

Partridge, K., Chatterjee, S., Sazawal, V., Borriello, G., & Want, R. (2002). *TiltType: Accelerometer-supported text entry for very small devices*. Paper presented at the Proceedings of the 15th Annual ACM Symposium on User Interface Software and Technology.

Perrault, S. T., Lecolinet, E., Eagan, J., & Guiard, Y. (2013). *Watchit: Simple gestures and eyes-free interaction for wristwatches and bracelets.* Paper presented at the Proceedings of the SIGCHI Conference on Human Factors in Computing Systems.

Plaumann, K., Müller, M., & Rukzio, E. (2016). *CircularSelection: Optimizing list selection for smartwatches.* Paper presented at the Proceedings of the 2016 ACM International Symposium on Wearable Computers.

Plummer, P., Apple, S., Dowd, C., & Keith, E. (2015). Texting and walking: Effect of environmental setting and task prioritization on dual-task interference in healthy young adults. *Gait & Posture, 41*(1), 46–51.

Poirier, F., & Belatar, M. (2016). *UniWatch: A soft keyboard for text entry on smartwatches using 3 keys.* Paper presented at the International Conference on Human-Computer Interaction.

Pulvirent, S. (2015, January 8). Does anyone really want a smartwatch? Retrieved from www.bloomberg.com/news/articles/2015-01-08/does-anyone-really-want-a-smart-watch

Reyal, S., Zhai, S., & Kristensson, P. O. (2015). *Performance and user experience of touchscreen and gesture keyboards in a lab setting and in the wild.* Paper presented at the Proceedings of the 33rd Annual ACM Conference on Human Factors in Computing Systems.

Reyes, G., Zhang, D., Ghosh, S., Shah, P., Wu, J., Parnami, A., … Edwards, W. K. (2016). *Whoosh: Non-voice acoustics for low-cost, hands-free, and rapid input on smartwatches.* Paper presented at the Proceedings of the 2016 ACM International Symposium on Wearable Computers.

Rudchenko, D., Paek, T., & Badger, E. (2011). *Text text revolution: A game that improves text entry on mobile touchscreen keyboards.* Paper presented at the International Conference on Pervasive Computing.

Schabrun, S. M., van den Hoorn, W., Moorcroft, A., Greenland, C., & Hodges, P. W. (2014). Texting and walking: Strategies for postural control and implications for safety. *PLOS ONE, 9*(2), e91489. https://doi.org/10.1371/journal.pone.0091489

Schildbach, B. & Rukzio, E. (2010). *Investigating selection and reading performance on a mobile phone while walking.* Paper presented at the Proceedings of the 12th International Conference on Human Computer Interaction with Mobile Devices and Services.

Shao, Y.-F., Chang-Ogimoto, M., Pointner, R., Lin, Y.-C., Wu, C.-T., & Chen, M. (2016). *SwipeKey: A swipe-based keyboard design for smartwatches.* Paper presented at the Proceedings of the 18th International Conference on Human-Computer Interaction with Mobile Devices and Services.

Shibata, T., Afergan, D., Kong, D., Yuksel, B. F., MacKenzie, S., & Jacob, R. J. (2016). *Text entry for ultra-small touchscreens using a fixed cursor and movable keyboard.* Paper presented at the Proceedings of the 2016 CHI Conference Extended Abstracts on Human Factors in Computing Systems.

Siek, K. A., Rogers, Y., & Connelly, K. H. (2005). Fat finger worries: How older and younger users physically interact with PDAs. In M. F. Costabile & F. Paternò (Eds.), *Human- Computer Interaction—INTERACT 2005. Lecture Notes in Computer Science* (Vol. 3585, pp. 267–280). Berlin, Germany: Springer.

Sonaike, I. A., Bewaji, T. A., Ritchey, P., & Peres, S. C. (2016). *The ergonomic impact of Swype.* Paper presented at the Proceedings of the Human Factors and Ergonomics Society Annual Meeting.

Song, J., Sörös, G., Pece, F., Fanello, S. R., Izadi, S., Keskin, C., & Hilliges, O. (2014). *In-air gestures around unmodified mobile devices.* Paper presented at the Proceedings of the 27th Annual ACM Symposium on User Interface Software and Technology.

Turner, C. J., Chaparro, B. S., & He, J. (2016). Text input on a smartwatch QWERTY keyboard: Tap vs. trace. *International Journal of Human–Computer Interaction, 1–8.* doi:10.1080/10447318.2016.1223265

Vadas, K., Patel, N., Lyons, K., Starner, T., & Jacko, J. (2006). *Reading on-the-go: A comparison of audio and hand-held displays.* Paper presented at the Proceedings of the 8th Conference on Human-Computer Interaction with Mobile Devices and Services.

Van Vlaenderen, W., Brulmans, J., Vermeulen, J., & Schöning, J. (2015). *Watchme: A novel input method combining a smartwatch and bimanual interaction.* Paper presented at the Proceedings of the 33rd Annual ACM Conference Extended Abstracts on Human Factors in Computing Systems.

Vertanen, K., Memmi, H., Emge, J., Reyal, S., & Kristensson, P. O. (2015). *VelociTap: Investigating fast mobile text entry using sentence-based decoding of touchscreen keyboard input.* Paper presented at the Proceedings of the 33rd Annual ACM Conference on Human Factors in Computing Systems.

Weigel, M., Lu, T., Bailly, G., Oulasvirta, A., Majidi, C., & Steimle, J. (2015). *Iskin: Flexible, stretchable and visually customizable on-body touch sensors for mobile computing.* Paper presented at the Proceedings of the 33rd Annual ACM Conference on Human Factors in Computing Systems.

White, R. M. (1980). *Comparative anthropometry of the hand* (No. NATICK/CEMEL-229). Army Natick Research and Development Labs Ma Clothing Equipment and Materials Engineering Lab.

Wilhelm, M., Krakowczyk, D., Trollmann, F., & Albayrak, S. (2015). *eRing: Multiple finger gesture recognition with one ring using an electric field.* Paper presented at the Proceedings of the 2nd International Workshop on Sensor-based Activity Recognition and Interaction.

Withana, A., Peiris, R., Samarasekara, N., & Nanayakkara, S. (2015). *zSense: Enabling shallow depth gesture recognition for greater input expressivity on smart wearables.* Paper presented at the Proceedings of the 33rd Annual ACM Conference on Human Factors in Computing Systems.

Xia, H., Grossman, T., & Fitzmaurice, G. (2015). *NanoStylus: Enhancing input on ultra-small displays with a finger-mounted stylus.* Paper presented at the Proceedings of the 28th Annual ACM Symposium on User Interface Software & Technology.

Xiao, R., Laput, G., & Harrison, C. (2014). *Expanding the input expressivity of smartwatches with mechanical pan, twist, tilt and click.* Paper presented at the Proceedings of the SIGCHI Conference on Human Factors in Computing Systems.

Yatani, K. & Truong, K. N. (2009). An evaluation of stylus-based text entry methods on handheld devices studied in different user mobility states. *Pervasive and Mobile Computing, 5*(5), 496–508.

Yeo, H.-S., Lee, J., Bianchi, A., & Quigley, A. (2016). *WatchMI: Pressure touch, twist and pan gesture input on unmodified smartwatches.* Paper presented at the Proceedings of the 18th International Conference on Human-Computer Interaction with Mobile Devices and Services.

Yin, Y., Li, Q., Xie, L., Yi, S., Novak, E., & Lu, S. (2016). *CamK: A camera-based keyboard for small mobile devices.* Paper presented at the IEEE INFOCOM 2016-The 35th Annual IEEE International Conference on Computer Communications.

Yoon, H., Park, S.-H., & Lee, K.-T. (2016). Lightful user interaction on smart wearables. *Personal and Ubiquitous Computing*, 1–12.

Zhai, S., & Kristensson, P. O. (2012). The word-gesture keyboard: Reimagining keyboard interaction. *Communications of the ACM, 55*(9), 91–101.

Zhang, C., Yang, J., Southern, C., Starner, T. E., & Abowd, G. D. (2016). *WatchOut: Extending interactions on a smartwatch with inertial sensing.* Paper presented at the Proceedings of the 2016 ACM International Symposium on Wearable Computers.

Appendix
Observed Typing Performance and Subjective Measures on Smartwatch Sized Keyboards

Keyboard	Reference	Participants	Participant Mobility	Entry Speed (WPM)	Error Rate %	Subjective Measures
Callout	Leiva, Sahami, Catalá, Henze, & Schmidt, 2015	20	Seated	4.3[1] 7.1[1] 8.3[1]	2.6[CER] 0.8[CER] 0.7[CER]	NASA-TLX, SUS
DriftBoard	Shibata et al., 2016	10	*	9.7[2]	0.6[ER]	-
DualKey[QWERTY]	Gupta & Balakrishnan, 2016	10	*	19.6[3]	5.3[TER]	Q & A
DualKey[SWEQTY]	Gupta & Balakrishnan, 2016	8	*	7.1[1]	0.8[CER]	Q & A
ETAO	Darbar, Dash, & Samanta, 2016	10	Seated Walking	8.3[1] 9.4[3]	0.7[CER] 7.1[TER]	Perceived Learning Time
Fleksy	Chaparro, He, Turner, & Turner, 2015	18	Seated	20.3[3]	16.0[TER]	NASA-TLX, SUS, Perceived Performance & Preference
ForceBoard	Hsiu et al., 2016	12	*	12.4[1]	9..2[TER]	User Preference
Invisiboard	Mottelson, Larsen, Lyderik, Strohmeier, & Knibbe, 2016	12	*	9.5[2]	3.2[MWD]	-
Optimized Alphabetic Layout (OAL)[4]	Komninos & Dunlop, 2014	20	*	8.1[3]	-	NASA-TLX, Qualitative Feedback
QWERTY-like Keypad (QLKP)	Hong et al., 2015	12	*	9.2[3]	4.3[TER]	Questionnaire, Preference Ratings
SlideBoard	Hong et al., 2015	12	*	12.1[3]	7.9[TER]	Questionnaire, Preference Ratings
SplitBoard	Hong, Heo, Isokoski, & Lee, 2016	12	Seated	14.8[3]	9.0[TER]	Questionnaire, Preference Ratings
		18	Seated	10.5[3]	14.0[TER]	
				11.5[3]	11.0[TER]	
			Standing	15.0[3]	8.0[TER]	
				14.5[3]	7.0[TER]	
		12	Walking	13.0[3]	12.0[TER]	
	Hong et al., 2015	24	*	14.8[3]	7.5[TER]	Questionnaire, Preference Ratings
	Hsiu et al., 2016	12	*	11.9[3]	10.1[TER]	User Preference
	Darbar, Dash, & Samanta, 2016	10	Seated Walking	12.2[3] 9.3[3]	10.5[TER] 12.8[TER]	Perceived Learning Time
Standard QWERTY	Hong et al., 2016	12		13.7[3]	21.0[TER]	Questionnaire, Preference Ratings
		18	Seated	10.0[3]	28.0[TER]	
				12.0[3]	20.0[TER]	
				14.5[3]	20.0[TER]	
		12	Standing	13.0[3]	23.0[TER]	
			Walking	13.0[3]	23.0[TER]	
	Hong et al., 2015	12	*	12.9[3]	21.4[TER]	Questionnaire, Preference Ratings
		10	Seated	7.1[3]	22.1[TER]	Perceived Learning Time

Keyboard	Reference	Participants	Participant Mobility	Entry Speed (WPM)	Error Rate %	Subjective Measures
	Darbar, Dash, & Samanta, 2016		Walking	5.2³	28.5 TER	
Swipeboard Alphabetical	Shao et al., 2016	12	*	7.3³	9.0 CER	Questionnaire, Interview
Swipeboard QWERTY	Chen, Grossman, & Fitzmaurice, 2014	8	*	19.6²	17.5 TER	-
	Shao et al., 2016	12	*	7.2³	10.0 CER	Questionnaire, Interview, Preference
SwipeKey4	Shao et al., 2016	12	*	11.0³	4.4 CER	Questionnaire, Interview, Preference
SwipeKey5	Shao et al., 2016	12	*	10.9³	7.4 CER	Questionnaire, Interview, Preference
Swype Tap	Turner, Chaparro, & He, 2016	16 / 16	Seated	27.0³ / 26.0³	8.0 TER / 5.0 TER	NASA-TLX, SUS, Intent to Use, Perceived Performance & Preference
Swype Trace	Chaparro et al., 2015	18	Seated	29.3³	9.0 TER	NASA-TLX, SUS, Perceived Performance & Preference
	Turner et al., 2016	16 / 16		31.0³ / 37.0³	6.0 TER / 5.0 TER	NASA-TLX, SUS, Intent to Use, Perceived Performance & Preference
UniWatch⁴	Poirier & Belatar, 2016	5	Seated	9.8³	-	-
Virtual Sliding QWERTY (VSQ)	Cha, Choi, & Lim, 2015	20	Seated	10.8² / 11.7² / 11.3² / 10.6² / 10.0²	- / - / - / - / -	Preference, Ease of Use
WatchWriter Gesture	Gordon, Ouyang, & Zhai, 2016	18	Seated	24.0³	3.7 CER	-
WatchWriter Tap	Gordon et al., 2016	18	Seated	22.0³	1.5 CER	-
ZoomBoard	Oney et al., 2013	6	*	9.3²	-	Qualitative Survey
	Chen et al., 2014	8	*	17.1²	19.6 TER	-
	Mottelson et al., 2016	12	*	9.3¹	2.1 MWD	-
	Hong et al., 2015	12	*	9.2³	7.1 TER	Questionnaire, Preference Ratings
	Leiva et al., 2015	20	Seated	6.0² / 7.8² / 8.2²	1.1 CER / 1.2 CER / 1.4 CER	NASA-TLX, SUS
	Hong et al., 2016	12		9.8³	7.0 TER	Questionnaire, Preference Ratings
		18	Seated	8.0³ / 9.0³ / 9.0³	10.0 TER / 6.0 TER / 7.0 TER	
		12	Standing / Walking	9.0³ / 8.5³	5.0 TER / 8.0 TER	
	Hsiu et al., 2016	12	*	9.5³	6.1 TER	User Preference
	Darbar, Dash, et al., 2016	10	Seated / Walking	8.7³ / 8.0³	8.6 TER / 9.8 TER	Perceived Learning Time
ZShift	Leiva et al., 2015	20	Seated	5.4¹ / 7.2¹ / 9.1¹	1.3 CER / 1.3 CER / 0.9 CER	NASA-TLX, SUS

* Mobility not specifically stated, [1] Observed on a smartphone, [2] Observed on a tablet, [3] Observed on smartwatch,
[4] Did not use phrase set from MacKenzie and Soukoreff (2003)

Commentary

We selected this article by Turner and colleagues for several reasons. First, it is a fine example of an article reporting the results of usability testing. Despite some minor flaws, there are no major problems with the design or execution of the experiment, nor with the analysis or reporting of the results.

Second, although the article presents very detailed quantitative performance data, the authors make very clear that "solely relying on performance as a measure of keyboard quality is shortsighted; if users do not like the keyboard or use it, typing speed is irrelevant." Thus, in addition to reporting quantitative results, Turner and colleagues also measure user preference. Unlike many usability research reports, however, these qualitative data are reported not in terms of "preference ratings and non-standardized questionnaires," but rigorous investigation of user preference using industry-standard tools.

Finally, we chose this article because it deals with an increasingly popular wearable technology among all kinds of users, from youth and young professionals to retirees. The size of the device's display also poses major usability challenges. Finally, the fact that the manufacturers of the most popular and highly-rated smartphones, Apple and Samsung, have entered the smartwatch market over the past six years is ample evidence that the product should rank highly as a subject for usability professionals.

Purpose and Audience

The purpose of this article is to explore the authors' four research questions about user performance and subjective preference stated in the last paragraph of the *Introduction* section.

This study aims to answer four research questions:

- What impact does mobility (standing vs. walking) have on typing performance using trace and tap input?
- Which input method (trace vs. tap) results in better typing performance when walking?
- Does prior experience with trace input on a smartphone influence typing performance on the smartwatch?
- Which input method (trace vs. tap) results in better subjective ratings when walking?

Because the article was published in the *Journal of Usability Studies*, the journal of the User Experience Professionals Association, we can assume that its primary audience is usability and user experience practitioners, with a secondary audience of teachers and researchers in those fields. Because of the article's topic, smartwatch manufacturers and keyboard developers are likely also potential audiences. Some of these practitioners and academics are technical communicators, but note that two of the article's authors are psychologists, and the third is an engineering psychologist. All three authors are heavily involved in studying human factors. We should remember, then, that usability and user experience are highly interdisciplinary, and thus, only a minority of the article's intended audience is likely to have an extensive background in technical and professional communication.

The study is exploratory, and the results cannot be generalized beyond the study's scope. It was conducted with a population of university students, addressed only a single product, and utilized a treadmill to simulate walking rather than a more realistic walking environment. Nevertheless, technical communication practitioners will be interested in the article because it is so tightly focused in terms of task and deals with major issues posed by this "bleeding edge" technology.

Organization

Like the other three articles we have studied, the organization of this one is a variation on the IMRAD format for research reports: Introduction, Methods, Results, and Discussion.

- *Introduction:* This section reviews the literature, providing the context for the experiment that the authors designed and conducted. The authors demonstrate a thorough knowledge of current technologies and alternatives. Their literature review considers typing and other input methods for smartwatches, various keyboards designed for smartwatches, the challenge of typing while walking, and the influence of experience with various typing methods on user performance, all significant areas of inquiry. The *Introduction* ends with the research questions that the authors want to investigate.
- *Methods:* In the *Methods* section, the authors describe recruitment of participants; the watch and treadmill used, and the phrases to be input; the experimental procedure; the details of the study's design, including independent and dependent variables; and the ways in which performance, subjective preference, and anthropometric measures were recorded.
- *Results:* The *Results* section reports the typing speed; typing accuracy; participants' perceptions of workload and the device's usability; participants' perceived accuracy, speed, and input method preferences; participants' intent to use the input methods in the future; and hand measurements.
- *Discussion:* The *Discussion* section reviews the main findings of user performance and preference.
- *Conclusion:* The *Conclusion* section characterizes the study's significance, acknowledges its limitations, and makes recommendations for smartwatch keyboard technology and future research.
- *Tips for Usability Practitioners:* This section offers four practical suggestions for developers of keyboard technology and for future research.
- *Acknowledgments:* The authors thank the individuals and lab that provided equipment for the study and contributed to the research.
- *References:* This list of sources provides a valuable starting point for those who wish to extend the work done by Turner and colleagues.
- *Appendix:* Information in this lengthy table includes data about previous studies of smartwatch keyboard typing performance and subjective measures.

As we have observed in Chapters 9–11, the IMRAD format is typically used by authors of quantitative and qualitative research studies in medicine, the physical sciences, engineering, and the social sciences. Using the IMRAD structure adds to the article's ethos or credibility by observing the conventions for presenting quantitative and qualitative research results in a form that readers will understand and expect.

Methods

In Chapter 7, we devoted significant attention to usability testing as a method for conducting a usability study. What we find in "Texting while walking," however, seems rather different from what we described there. The number of participants is significantly larger (32 rather than the five recommended by Nielsen and Landauer (1993)). There is no mention of asking participants to think aloud while interacting with the product. Data are not coded. In fact, what Turner, Chaparro, and He describe sounds more like an experiment than what many technical communicators think of as a usability test. In fact, the *Methods* section uses the word "experimental" numerous times to refer to the test.

Of course, the usability study conducted by de Jong and colleagues that we examined in Chapter 9 was also described in experimental terms. We should also keep in mind that the usability field grew out of the experimental work done by human factors engineers and psychologists since World War II. That work continues today, but it is not as common in the usability testing done by technical communicators as the "discount" usability studies that we talked about in Chapter 7. Finally, we should recall that the authors of this article have backgrounds in psychology, so the experimental nature of their test is not surprising.

The article's *Methods* section provides us with significant information about the following:

- Participants in the study
- The product tested and equipment used
- The experimental procedure
- The experimental design
- The ways that performance was measured
- The ways that subjective ratings were assessed
- The way that participants' hands, index fingers, and thumbs were measured

Participants

The authors recruited 32 college students, ranging in age from 18 to 34, who participated in the study for course credit. The major criterion for selection was prior experience with trace typing on a smartphone. In our opinion, relying on this convenience sample is an experimental flaw that could have been avoided. Because smartwatches are used by a wide variety of people of different ages, from teenagers to seniors, it is unfortunate that the participants had a mean age of only 22.53. We also learn that the participants all "had normal or corrected-to-normal vision and did not have any physical limitations to their hands that would prevent them from being able to type on a smartwatch." Relying on young people, whom we would expect to be more physically agile, have better vision, and be at least minimally more familiar with the technology tested, is certainly problematic.

As we noted, experience with trace typing on a smartphone was the only major criterion for selection. This prior experience requirement is understandable in terms of experimental feasibility because gaining adequate proficiency with this technique might have required time that the authors could not spare.

Finally, only novices and experts at trace typing were included in the participant group. Excluding those in the middle range between novices and experts simplified the experiment by requiring testing with only two expertise groups instead of three.

Because the experiment was conducted in Wichita, KS, a metropolitan area of more than 600,000, it should have been relatively easy to recruit participants from a wider range of ages, with a wider range of visual acuity, and a greater range of physical abilities. Opening up recruitment would introduce some potential problems, the greatest of which would be the inconvenience and the time required to recruit a more diverse group of participants and the need to pay them—a small-dollar value gift card, for example. The only compensation offered to the student participants in Turner and colleagues' experiment was course credit.

Materials

In the *Materials* subsection, Turner and colleagues provide us with information about materials that would be needed to repeat their experiment. These data include the model of watch used, the source of the corpus of phrases given to participants, the number of phrases for both practice and experimental input, the range of number of characters in those phrases, the model of treadmill used, and the range of treadmill speeds tested.

Procedure and Experimental Design

The *Procedure* subsection summarizes the experiment's test plan. We learn that the participants first gave consent for their participation and then completed a brief demographic survey that identified their behavior and experience when texting on a smartphone. We will note that although the authors sought participants' consent at the beginning of each person's test, there is no mention that the authors submitted their research plan to the university's institutional review board. Some journals such as the *IEEE Transactions on Professional Communication* require that this issue be specifically addressed, but apparently the *Journal of Usability Research* does not.

Next, the participants received a short tutorial. The subject of the tutorial is not specified but was presumably an overview of the first condition to be tested. The participants were then given 10 practice phrases to type using the input method and mobility condition for the first condition. Then the test began for the first condition.

This process was repeated for all four conditions tested: tapping while standing, tracing while walking, tracing while standing, and tapping while walking. For each condition, participants typed 15 phrases. Performance of each condition was timed, and after each condition, participants responded to questionnaires on perceived usability and mental workload for that condition.

Participants were allowed to correct their typing but were not required to do so. Typing was timed, and the texts were downloaded to be scored for accuracy.

The *Design* subsection tells us that the experiment used a 2 x 2 x 2 mixed design: two input methods (trace and tap), two mobility conditions (standing and walking), and two experience levels (novice and expert). Input method and mobility condition were the independent, within-subjects variables. (That is, each participant was asked to perform both input methods under both mobility conditions.) Experience was the independent, between-subjects variable. (That is, both experts and novices were asked to perform both input methods under both mobility conditions.) Typing speed, accuracy, and subjective assessments (user perceptions of usability, workload, performance, and intent to use) were the dependent variables.

Performance

The *Performance* subsection describes how typing speed and accuracy were measured. It also categorizes the types of errors identified. These errors were factored into the typing accuracy calculations.

Subjective Measures

This subsection explains how subjective workload, perceived usability, perceived performance and preference, and intent to use were determined.

- Subjective workload was measured by having each participant complete the NASA Task Load Index (TLX) after each of the four conditions. The US National Aeronautics and Space Administration web site describes this instrument as "the gold standard for measuring subjective workload across a wide range of applications" (https://humansystems.arc.nasa.gov/groups/tlx/index.php). The TLX's connection to NASA is important because the agency has more than 60 years of experience in human factors research by studying human responses to work under extreme conditions to ensure the safety of crews and vehicles.
- Perceived usability was determined using an adapted version of the industry standard System Usability Scale (SUS) to measure each participant's perception of usability after each condition. As we noted in Chapter 7, because it is easy to use, reliable, and valid, it has become a standard for testing many types of products.
- Perceived performance and preference were measured using a 50-point scale for each condition. Unfortunately, the article does not tell us any more about this instrument.
- Intent to use reflected the participant's rating of how likely they were to use each input method on a 10-point scale for each mobility condition.

Remember that in the introduction to this Commentary, we noted that the authors insisted that "solely relying on performance as a measure of keyboard quality is short-sighted." As a result, in addition to reporting quantitative performance results, they also measured user preference. Unlike many usability research reports, however, these subjective data were not reported in terms of "preference ratings and non-standardized questionnaires," but rigorous investigation of user preference. The use of the NASA Task Load Index and the industry standard SUS are suitably rigorous methods of assessing these user perceptions and preferences. Although the 50-point scale used to measure perceived performance and user preference sounds as though it is likely to be rigorous, we cannot guarantee that it is. Similarly, the 10-point scale used to estimate the user's intent to use the input method is also probably a reliable tool, but we don't know enough about it to establish that reliability.

Anthropometric Measurements

Each participant's typing hand, index finger, and thumb were measured.

Exercise 12.1 Developing Research Questions and Drafting the Methods Section

Choose another technology product type that is currently popular, such as activity trackers or camera drones. Develop at least two research questions that address potential user problems or difficulties with a specific product selected from the chosen product type. Then draft a *Methods* section that includes brief descriptions of test participants, materials, test procedure, and user performance and perception measures similar to those in the article by Turner, Chaparro, and He.

Exercise 12.2 Recruiting Diverse Participants

Assume that you want to replicate the research that Turner and colleagues report in this article using a group of participants that more accurately represents the diverse population of potential users. How would you recruit participants for your study to make them more representative of the potential user population? What other changes would you need to make to your research design to accommodate this more diverse pool of test participants?

Results

Because we did a thorough analysis of the quantitative results in the de Jong, Yang, and Karreman article in Chapter 9, we will simply summarize the findings in the Turner and colleagues article. The following results are provided:

- *Typing speed:* Tracing and standing result in significantly greater speed than tapping and walking.
- *Typing accuracy:* Tracing and standing result in significantly greater accuracy than tapping and walking. There were significantly fewer substitution errors with standing than with walking, and significantly fewer insertion errors for experts than for novices, for tracing than for tapping, and for standing than for walking. Novices made more insertion errors with tapping than with tracing, and made more insertion errors than experts.
- *Subjective workload:* Tracing was perceived as significantly less frustrating than tapping. Standing was perceived as significantly less demanding than walking.
- *Perceived usability:* Usability was perceived as significantly higher by experts than by novices. Tracing was perceived as more usable than tapping, and standing was perceived as more usable than walking.
- *Perceived accuracy, speed, and preference:* Experts perceived higher accuracy than novices, tracing was perceived to be more accurate than tapping, and standing was perceived to be more accurate than walking. Tracing was perceived to be

faster than tapping, and standing was perceived to be faster than walking. Experts had higher preference ratings than novices for tracing and standing than for tapping and walking.

- *Intent to use:* Tracing yielded a significantly greater intent-to-use score than tapping, and standing yielded a significantly greater intent-to-use score than walking. Tracing while both standing and walking resulted in a significantly higher intent-to-use score than tapping.
- *Hand measurements:* There was no evidence of a "fat finger" issue for any of the participants.

Because each test in the experiment involved within-subjects and between-subjects independent variables, the authors first used a mixed model ANOVA to analyze the measurements. Then they used partial eta squared (η_p^2) to estimate the size of the effects of the ANOVA tests. Significance of the results was measured at p less than or equal to 0.001 for 20 of the effects reported, and between 0.002 and 0.05 for another 11 effects reported. Note that these p values are at a significantly higher degree of significance than the $p = 0.1$ level that is often used in technical communication research.

The results reported in the text are illustrated by corresponding bar charts for all of the results except for the hand measurements.

Validity

We need to examine whether these results meet the standards of rigor for inferences based on measurements: the validity of the measurement and the reliability of the inference.

INTERNAL VALIDITY

You'll remember that in Chapter 4 we stated that internal validity in a quantitative study addresses whether the researchers measured the concept that they wanted to study. The concepts that Turner and his co-authors wanted to study were the differences in usability for experts and novices, for tracing rather than for tapping input, and for standing rather than walking. The description of the experiment and the results reported indicate that the experiment was carefully designed and that the concepts that the authors wanted to explore were validly operationalized using accepted techniques reported in the literature review. So it is clear that Turner, Chaparro, and He measured the concepts that they intended to measure.

EXTERNAL VALIDITY

External validity addresses whether what the researchers measured in the test environment reflects what would be found in the real world. We manage external validity by carefully designing the test environment conditions to match the conditions in the general environment as much as possible and by ensuring that the participants resemble the general population as much as possible.

As with de Jong and colleagues' experiments, because the authors here do not generalize their findings to the general population, the question of external validity is not really an issue. The *Conclusion* section indicates the limitations of the study as to generalizability to other user populations or to normal walking rather than walking on a treadmill.

Turner and colleagues managed to improve the "real world" environment reported in earlier experiments covered in their literature review by asking participants to type while standing or walking rather than while seated. However, the walking was performed on a treadmill rather than in a natural walking environment, so the test environment did not entirely match "real world" conditions.

We looked earlier at the sample population for this study, and we remember from Chapter 4 the importance of random selection of subjects. That is, the selection and assignment of test participants should be independent and equal. Selection is independent if selecting a participant as a member of one group does not influence the selection of another participant. Selection is equal if every member of a population has an equal chance of being selected. Because the test population for this experiment was a sample of convenience, its members were not selected independently. Students volunteered to participate in the experiment to gain course credit. Their assignment to the novice or expert groups, however, was dependent on their degree of expertise with trace input, and potential participants in the middle range of expertise were excluded from the experiment. Therefore, selection was not independent.

Reliability

In Chapter 4, we said that a study's reliability involves the likelihood that the results would be the same if the experiment were repeated, either with a different sample or by different researchers. The measure of the study's reliability is the statistical significance of the results. In other words, how confident can we be that the differences in performance for the input and mobility conditions and for the experience conditions are a result of the intervention introduced (the combinations of independent variables) rather than to coincidence.

As we stated earlier in Chapter 4 and Chapter 9, we judge the reliability of conclusions about a set of data based on two principles.

- The smaller the variance in the data, the more reliable the inference.
- The bigger the sample size, the more reliable the inference.

When we assess the variance in data, we want to know the probability that the results were caused by the intervention (that is, the manipulation of the independent variables) and not by sampling error or coincidence. In Chapter 4 we noted that most research in technical and professional communication is sufficiently rigorous when results have a p value of 0.1 or lower; in other words, there is no more than 1 chance in 10 that the results were caused by coincidence or sampling error. For more conservative situations where harm could be done by accepting a false claim, a p value of less than 0.05 or even 0.01 might be more appropriate; in other words, there are no more than 5 chances in 100 or 1 chance in 100 that the difference is not statistically significant.

As we noted earlier, the significance of the results in the Turner, Chaparro, and He article was measured at p less than or equal to 0.001 for 20 of the effects reported, and between 0.002 and 0.05 for another 11 effects reported. Thus, the findings in this article are strongly statistically significant, despite the fact that the authors cannot generalize the results beyond the study's participants.

As for the number of participants, the 32 who performed the tests reported here comprise a much larger group than the 5 recommended by Nielsen and Landauer. On the other hand, this is a significantly smaller sample than those reported in de Jong and colleagues' two studies (69 and 83, respectively). Nevertheless, because Turner and colleagues make no attempt to generalize their results, the number of participants here is not an issue.

Discussion and Conclusion

Turner, Chaparro, and He summarize the results of their experiment in the *Discussion* section.

- "Users were able to achieve 35 WPM with trace and 30 WPM with tap, regardless of mobility or experience."
- "... experience with trace input on a smartphone had no significant impact on entry speed ...."
- "Prior experience with trace on a smartphone did not seem to have an effect on accuracy. Users typed more accurately with trace (10% WER [word error rate]) than tap (14% WER) and more accurately when standing (10% WER) than when walking (15% WER)."
- "Performance with both input methods and mobility conditions remained high despite key sizes ...."
- "... trace was rated more favorably than tap across all measures. The perceived usability scores of both input methods fell within the marginally acceptable to acceptable ranges: *Good* for trace and *OK* for tap ...."
- "Standing was consistently rated more favorably than walking ...."

The *Conclusion* section puts the results of this study in the context of the previous literature. It also considers the limitations of treadmill walking versus walking in a more natural environment, as well as the problem with the convenience sample used. Finally, the authors make four recommendations for smartwatch manufacturers, keyboard developers, and future studies.

Is It RAD?

In Chapter 2, we saw that RAD research as defined by Haswell (2005) must be replicable, aggregable, and data-supported. That is, the methodology must be defined well enough that others can repeat the study. The results of the study must be detailed enough that they can be aggregated or combined with the results of other studies to build a body of data that can be compared and further built upon. And the conclusions of the study must be supported by those data, not simply be the impressions or gut feelings of the researchers.

It is clear from our analysis that Turner, Chaparro, and He's article meets these requirements. Their methodology is described in enough detail to allow the study to be replicated. The data are reported in great detail, enabling those data to be aggregated with data collected in repetitions of their experiments. And finally, the conclusions drawn in the article are based entirely on the data reported. As a result, we can conclude with confidence that this is a RAD article.

Summary

This chapter has considered the challenges of using smartwatch keyboards while standing and walking, and has expanded on the concepts presented in Chapter 7. Following the full text of the sample article, we have explored its purpose and audience, its structure, the study design, the results, and the conclusions drawn from them.

We have seen that the results of the usability experiment conducted by Turner and his co-authors meet the requirements of internal validity because the study operationalized the concept that the authors wanted to examine. It does not entirely meet the requirements of external validity because the treadmill test environment and the sample population studied are not sufficiently like the "real world" environment.

Finally, the inferences that the authors draw from their results are reliable because of the very low probability of error, but only within the experiment's constraints (convenience sample of students and walking on a treadmill).

For Further Study

Select an article reporting the results of a usability test that you collected as part of the literature review or annotated bibliography that you assembled for your own research project in Exercise 3.1. Alternately, find another usability test report on the web. Perform the same kind of analysis of that article as the *Commentary* section of this chapter, examining the article's purpose and audience, organization, methods or study design, report of results, analysis or discussion, and conclusion.

1. What similarities do you see between the approach used by Turner, Chaparro, and He and the article that you have selected?
2. How are the two articles different?
3. What might be reasons for those differences?

Answer Key

Exercise 12.1

The answer to this exercise will be unique for each person who prepares it, so there is no key to this exercise.

Exercise 12.2

Each person who responds to this exercise will answer differently, but they may include the following:

- Recruiting participants at independent computer shops, community colleges offering lifelong-learning computer courses, or public libraries
- Offering participants a coffee card or other gift card as compensation
- Providing brief tutorials to those unfamiliar with the swiping method of typing on a smartphone
- Changing the walking speed to accommodate less agile participants

References

Haswell, R. H. (2005). NCTE/CCCC's recent war on scholarship. *Written Communication, 22* (2), 198–223.

National Aeronautics and Space Administration. (2019, August 15). *NASA TLX Task Load Index*. https://humansystems.arc.nasa.gov/groups/tlx/index.php

Nielsen, J., & Landauer, T. K. (1993). A mathematical model of the finding of usability problems. In *Proceedings of the INTERACT '93 and CHI '93 Conference on Human Factors in Computing Systems* (pp. 206–213). ACM.

Turner, C. J., Chaparro, B. S., & He, J. (2018). Texting while walking: Is it possible with a smartwatch? *Journal of Usability Studies, 13*(2), 94–118.

Appendix
Citation Styles

Introduction

Any time you quote, paraphrase, or summarize information from any source, you must cite the source. This Appendix introduces the two citation styles most commonly used in technical communication, APA and IEEE style. We introduce those styles, provide examples of citations for the most commonly used types of reference citations, explain how to select between the two styles, and briefly discuss citation management applications.

Learning Objectives

After you have read this appendix, you should be able to:

- Decide which citation style to use for a particular assignment or journal submission
- Build a reference list in accordance with the *APA Publication Manual* and the *IEEE Editorial Style Manual*
- Cite references within the text of a manuscript in accordance with the *APA Publication Manual* and the *IEEE Editorial Style Manual*
- Decide whether a citation management tool would be useful for you.

Citation Styles Used in Technical Communication

There are many types of citation styles used by various journals and book publishers, but the two used in the major technical communication journals are published by the American Psychological Association (APA) and the Institute of Electrical and Electronics Engineers (IEEE).

The most common citation style used in the social and behavioral sciences is commonly called APA style. The *Publication Manual of the American Psychological Association*, 7th edition, published in 2020, provides the details about APA style as well as examples for various kinds of sources. APA is the style choice of four of the five major journals in technical communication: *The Journal of Business and Technical Communication* (published by SAGE), the *Journal of Technical Writing and Communication* (also published by SAGE), *Technical Communication* (published by the Society for Technical Communication), and *Technical Communication Quarterly* (published by Taylor & Francis for the Association of Teachers of Technical Communication).

The exception is the *IEEE Transactions on Professional Communication* (published by the IEEE Professional Communication Society), which of course uses the *IEEE Editorial Style Manual* published by the Institute of Electrical and Electronics Engineers and now in its 9th version. Although IEEE style is superficially similar to APA style in some ways, the differences cause confusion for both students and scholars in the field who publish in the *IEEE Transactions* as well as in the other four journals.

If you are a student, be sure that you use the citation style prescribed by your instructor for the assignment that you are preparing. If you are writing a manuscript to submit to a journal, consult the instructions for authors (usually published on the journal's website or printed in the journal's front matter or on one of its covers for print journals). For a manuscript that you intend to submit to a book publisher, send an inquiry to the editor asking which style the publisher prefers.

When you started writing research papers in high school, if not earlier, you (and your teachers) certainly hated one aspect of such assignments more than any other: learning (and teaching) whatever citation style was prescribed by the handbook used in the course. These details can be a pain because they seem so trivial and arbitrary.

So why do we bother with footnotes, endnotes, or in-text citations? Why the hassle with bibliographies or lists of references? We include them in reports of our research to acknowledge the owners of intellectual property on which we've drawn and to help those who want to further explore the background of the work that we've done—perhaps to learn more about our choice of research methodology or to use our secondary research as a jumping-off point for their own studies.

And why do we format these reference lists and citations using a standard style? Well, it's essentially a matter of professional courtesy. We want to ensure that those who read our work are able to find the sources on which we've relied so that they can read them as they pursue their own research. And by making sure that our reference lists and citations conform to a particular style, we avoid accidentally omitting information about those sources that our own readers will need to locate them.

We noted in Chapter 3 that some library databases (such as EBSCO's Academic Search Complete) can generate citations in APA style, ProQuest Research Library can generate IEEE-style citations, and some desktop citation management software (such as Zotero) can generate citations in both APA and IEEE style. However, you should always check these automated citations against the format in the appropriate style manual because the versions that they supply do not always correspond exactly to the formats specified by the respective style manual.

APA Style

The *Publication Manual of the American Psychological Association* was originally published as a seven-page article in the APA's *Psychological Bulletin* in 1929. Now in its 7th edition and more than 400 pages long, the APA *Manual* is used by more than 1000 journals and book publishers, as well as in academic papers in the behavioral and social sciences. Because technical communication research often uses the research methods of the social sciences (particularly for qualitative research), APA style is a particularly apt choice for assignments and publications in our field.

APA maintains a blog that provides examples of reference and citation styles as well as updates to the *Manual* at https://blog.apastyle.org. This blog ensures that the

Manual keeps up with changes in technology and with the conventions that journals and bibliographers use.

> **Note:** If you have been accustomed to using the 6th edition of the APA manual, you will see that although the 7th is similar, there are some differences—perhaps most notably, the omission of place of publication for books. (Note that the lists of References in Chapters 9, 11, and 12 do not conform to the 7th edition of the APA manual because they were published before it was released.) So even if you think that you know APA style, the appearance of the 7th edition means that you need a refresher course!

The following are examples of the more common types of reference citations provided in APA style. If you are not able to find an example that corresponds to a work that you need to cite, consult the APA *Publication Manual* or the APA blog. Additional instructions and examples are also provided at websites such as the Purdue Online Writing Lab (see https://owl.english.purdue.edu/owl/section/2/10/).

Book with One Author

Willerton, R. (2015). *Plain language and ethical action: A dialogic approach to technical content in the twenty-first century*. Routledge.

> **Note:** For any book or portion of a book, add the Digital Object Identifier (DOI), if available, after the publisher's name.

Book with Two or More Authors

Hayhoe, G. F., & Brewer, P. E. (2021). *A research primer for technical communication: Methods, exemplars, and analyses* (2nd ed.). Taylor & Francis.

> **Note:** If applicable, provide edition information in parentheses following the title.

Editor, Translator, or Compiler

Zachry, M., & Thralls, C. (Eds.). (2007). *Communicative practices in workplaces and the professions: Cultural perspectives on the regulation of discourse and organizations*. Baywood.

> **Note:** Use (Ed.) for a single editor.

Essay in an Edited Collection

Henry, J. (2013). How can technical communicators fit into contemporary organizations? In J. Johnson-Eilola & S. A. Selber (Eds.), *Solving problems in technical communication* (pp. 75–97). University of Chicago Press.

Chapter or Portion of an Edited or Republished Volume

Aristotle. (1984). The rhetoric, Book I, *The rhetoric and the poetics of Aristotle* (W. R. Roberts & I. Bywater, Trans.; E. P. J. Corbett, Ed.), (pp. 19–89). Random House.

Preface, Foreword, Introduction, and Similar Parts of a Book

Redish, J. C. (2002). Foreword. In B. Mirel & R. Spilka (Eds.), *Reshaping technical communication* (pp. vii–xi). Erlbaum.

Electronic Book

Markel, M. (2012). *Technical communication* (10th ed.) http://bit.ly/TechComm10th PDFFree

Journal Article

Zhou, S., Jeong, H., and Green, P. A. (2017). How consistent are the best-known readability equations in estimating the readability of design standards? *IEEE Transactions on Professional Communication, 60*(1), 97–111.

> **Note:** To cite the online version of the article, add the DOI or URL to the end of the citation. For example: https://doi.org/10.1109/TPC.2016.2635720

Magazine Article

Breker, M. (2017, September). Aim and THEN fire: The business value of content strategy. *Intercom, 64*(8), 6–8.

> **Note:** To cite the online version of the article, add the DOI or URL to the end of the citation. For example: www.stc.org/intercom/download/2017/

Newspaper Article

Rutenberg, J. (2017, October 15). As the world tweets: Social media chiefs remain tight-lipped. *New York Times.* www.nytimes.com/2017/10/15/business/social-media-transparency.html

> **Note:** Most references to newspaper articles these days cite an online version. To cite the print edition of the article, specify the section and page number(s) instead of the URL following the newspaper title. In the following example, note that the date of the print edition in which the article appeared is a day later than its online posting.

Rutenberg, J. (2017, October 16). As the world tweets: Social media chiefs remain tight-lipped. *New York Times,* B1.

Published Thesis or Dissertation

Andersen, R. (2009). The diffusion of content management technologies in technical communication work groups: A qualitative study on the activity of technology transfer (Publication No. 3363411) [Doctoral dissertation, University of Wisconsin-Milwaukee]. ProQuest Dissertations Publishing.

Unpublished Thesis or Dissertation

Homburg, R. 2017. *The influence of company-produced and user-generated instructional videos on perceived credibility and usability* [Unpublished master's thesis]. University of Twente. https://essay.utwente.nl/72074/1/Homburg_MA_BMS.pdf

> **Note:** If the thesis or dissertation is unpublished but there is a URL, cite it as in the example above.

Conference Presentation

> **Note:** A paper included in the published proceedings of a meeting is treated as an essay in an edited collection.

Price, B. (2015, July). *Peeling back the layers of the information onion: Using complex layered visuals for knowledge management* [Conference presentation]. IEEE International Professional Communication Conference 2015, Limerick, Ireland.

> **Note:** The bracketed description following the title can be adjusted to reflect the type of presentation: Workshop, Paper presentation, etc.

Personal Communications

A personal communication such as a telephone conversation, letter, or e-mail message is not included in the list of references. Instead, it is cited in the text, as follows:

> J. Jones observed that research methods are very difficult to teach (personal communication, October 31, 2017).
> or
> Research methods are very difficult to teach (J. Jones, personal communication, October 31, 2017).

An Entire Web Site

An entire website is not included in the list of references. Instead, it is cited in the text, as follows:

> We consulted the US Government Services Association Section 508 website to identify requirements for web accessibility (www.section508.gov).

An Article, Document, or Short Work from a Web Site

US Government Services Administration. (2020, January). *Create accessible products.* www.section508.gov/create/documents

IEEE Style

The 2018 version of the *IEEE Reference Guide* is published in PDF form at https://ieeeauthorcenter.ieee.org/wp-content/uploads/IEEE-Reference-Guide.pdf. It provides guidance for preparing manuscripts for journals published by IEEE. A PDF containing only the instructions for citing references within the text of a manuscript and for preparing lists of references is available in *How to Cite References: IEEE Documentation Style*, which can be found at https://ieee-dataport.org/sites/default/files/analysis/27/IEEE%20Citation%20Guidelines.pdf

The major differences between APA and IEEE styles is that IEEE uses abbreviations for journal titles, numbers the entries consecutively in the list of references in the order that they are cited in the text, and uses the numbers in the text of the manuscript instead of parenthetical author/date citations.

The following are examples of the more common types of reference citations provided in IEEE style. In cases where common IEEE style does not provide a sample citation, these examples have been supplemented by the usage of the *IEEE Transactions on Professional Communication*. If you are not able to find an example that corresponds to a work you need to cite, consult the examples given in the *IEEE Editorial Style Manual* (https://www. ieee.org/content/dam/ieee-org/ieee/web/org/conferences/style_references_manual.pdf).

Book with One Author

[1] R.Willerton, *Plain Language and Ethical Action: A Dialogic Approach to Technical Content in the Twenty-first Century*. New York: Routledge, 2015.

Book with Two or More Authors

[2] G. F. Hayhoe and P. E. Brewer, *A Research Primer for Technical Communication: Methods, Exemplars, and Analyses*, 2nd ed. New York: Taylor & Francis, 2021.

Editor, Translator, or Compiler

[3] M. Zachry and C. Thralls, Eds., *Communicative Practices in Workplaces and the Professions: Cultural Perspectives on the Regulation of Discourse and Organizations*. Amityville, NY: Baywood Publishing, 2007.

Note: For cities that are not generally well known, IEEE style includes the state/ province and country in which the city is located.

Essay in an Edited Collection

[4] J. Henry, "How can technical communicators fit into contemporary organizations?" in *Solving Problems in Technical Communication*, J. Johnson-Eilola and S. A. Selber, Eds. Chicago, IL: University of Chicago Press, 2013, pp. 75–97.

Chapter or Portion of an Edited Volume

[5] Aristotle. "The rhetoric, Book I," in *The Rhetoric and the Poetics of Aristotle*, W. R. Roberts and I. Bywater, Trans., Edward P. J. Corbett, Intro. New York: Random House, 1984, pp. 19–89.

Preface, Foreword, Introduction, and Similar Parts of a Book

[6] J. C. Redish, "Foreword," in *Reshaping Technical Communication*, B. Mirel and R. Spilka, Eds. Mahwah, NJ: Lawrence Erlbaum Associates, 2002, pp. vii–xii.

Electronic Book

[7] M. Markel, *Technical Communication*, 10th ed. Boston, MA: Bedford/St. Martin's, 2012. [Online]. Available: http://bit.ly/TechComm10thPDFFree

Journal Article

[8] S. Zhou, H. Jeong, and P. A. Green, "How consistent are the best-known readability equations in estimating the readability of design standards?" *IEEE Trans. Prof. Commun.*, vol. 60, no. 1, pp. 97–111, Mar. 2017.

   Note: If the article is available online, add the DOI or URL. Example: [Online]. Available: http://ieeexplore.ieee.org/stamp/stamp.jsp?arnumber=7839917

Magazine Article

[9] M. Breker, "Aim and THEN fire: The business value of content strategy," *Intercom*, pp. 6–8, Sept. 2017.

Newspaper Article

[10] J. Rutenberg, "As the world tweets: Social media chiefs remain tight-lipped," *New York Times*, p. B1, Oct. 15, 2017. [Online]. Available: www.nytimes.com/2017/10/15/business/social-media-transparency.html.

Published Thesis or Dissertation

[11] Andersen, R. The diffusion of content management technologies in technical communication work groups: A qualitative study on the activity of technology transfer, Ph.D. dissertation, Dept. of Engl., Univ. of Wisconsin-Milwaukee, Milwaukee, WI, 2009. ProQuest Dissertations Publishing, Publication No. 3363411.

Unpublished Thesis or Dissertation

[12] Homburg, R., "The influence of company-produced and user-generated instructional videos on perceived credibility and usability." M.S. thesis, Dept. of Comm. Studies, Univ. of Twente, Enschede, the Netherlands, 2017.

   Note: If the thesis or dissertation is unpublished but there is a URL, cite it.

Unpublished Presentation at a Meeting or Conference

[13] B. Price, "Peeling back the layers of the information onion: Using complex layered visuals for knowledge management," presented at the IEEE Int. Prof. Commun. Conf., Limerick, Ireland, 2015.

   Note: A paper included in the published proceedings of a meeting is treated as an essay in an edited collection.

Personal Communications

A personal communication such as a telephone conversation, letter, or e-mail message is not included in the list of references. Instead, it is cited in the text, as follows:

J. Jones observed that research methods are also very difficult to teach (personal communication, October 31, 2017).

or

Research methods are also very difficult to teach (J. Jones, personal communication, October 31, 2017).

> **Note:** Personal communications are not typically used in engineering publications, and neither the *IEEE Editorial Style Manual* nor *How to Cite References: IEEE Documentation Style* contains a citation style for this type of source. The examples given here follow the style used by the *IEEE Transactions on Professional Communication*.

An Entire Web Site

[14] US Government Services Administration, Section 508.gov. [Online]. Available: www.section508.gov

An Article, Document, or Short Work from a Web Site

[15] US Government Services Administration. (2020, Jan.). Create accessible products. [Online]. Available: www.section508.gov/create/documents

Notable Differences between APA and IEEE Reference Styles

As you have probably noticed, there are a few similarities between the APA and IEEE styles for reference list entries, but there are also many differences. Let's consider the most significant of those differences in an effort to better understand the two systems.

- *Authors' names:* APA reverses all author names, surname first for multiple author works, and uses an ampersand (*&*) before the last author's surname for multiple-author works. IEEE gives authors' names in the usual order, and spells out *and* before the last author's name. (Both use only initials for authors' given and middle names.)
- *Publisher location:* APA omits place of publication, but IEEE supplies it before the publisher name.
- *Publication dates:* APA gives the publication date after the names of the author(s), places parentheses around the publication date, and follows the parenthetical date with a period. IEEE style provides the date at the end of the entry for most types of sources.
- *Publisher names:* APA shortens publisher names; IEEE uses full publisher names.
- *Journal titles:* APA provides full journal titles; IEEE uses abbreviations for most journal titles. See the list of abbreviations used by the Web of Science (https://images.webofknowledge.com/images/help/WOS/A_abrvjt.html) and convert those title abbreviations from all caps to caps and lower case with periods after each abbreviated word.
- *Journal volumes/issue numbers:* APA italicizes the volume numbers of journals; IEEE does not.

- *Article DOIs and URLs:* APA includes the DOI for journal articles when available, or the article or the journal's home page URL when the DOI is not available but the article is available online. IEEE uses the URL for articles available online.
- *Newspaper articles:* APA provides section and page references for newspaper articles; IEEE omits this information.

Exercise A-1 Preparing Reference Citations

Using APA or IEEE style (choose the form specified by the publisher, journal, or course for which you are preparing a submission), assemble a reference list that includes the appropriate citations for the following works. All of the information you need for the citations is contained within this exercise.

1. A 1990 dissertation for the doctor of philosophy degree written by Kim Sydow Campbell for the Department of English at Louisiana State University in Baton Rouge, LA. The title is "Theoretical and Pedagogical Applications of Discourse Analysis to Professional Writing."
2. Article on pages 21 through 23 in the January 2001 issue of *Intercom* magazine entitled "Technical Communicators: Designing the User Experience." The author is Lori Fisher.
3. E-mail message from Robin Willis to you dated March 8, 2006.
4. Article from the web (URL: www.nngroup.com/articles/concise-scannable-and-object ive-how-to-write-for-the-web/) entitled "Concise, SCANNABLE, and Objective: How to Write for the Web." The authors are John Morkes and Jakob Nielsen. It was published on January 1, 1997.
5. Conference paper entitled "Interact to Produce Better Technical Communicators: Academia and Industry" published in 1995. It was written by Thea Teich, Janice C. Redish, and Kenneth T. Rainey, and appeared on pages 57–60 in the *Proceedings of the STC 42nd Annual Conference*. The publisher is the Society for Technical Communication, located in Arlington, Virginia.
6. Article in the *New York Times* by Eduardo Porter. The title is "Send Jobs to India? Some Find It's Not Always Best." It was published on April 28, 2004, and it appeared in section A on page 1.
7. Usability Body of Knowledge web site of the User Experience Professionals' Association (URL: www.usabilitybok.org/). No publication date is provided.
8. Book entitled *Designing Visual Language: Strategies for Professional Communicators*. It was published by Allyn and Bacon, which is located in Needham Heights, Massachusetts. It was written by Charles Kostelnick and David D. Roberts, and published in 1998.
9. Article (URL: http://kairos.technorhetoric.net/7.3/binder2.html?coverweb/fishman/ index.html) in the journal *Kairos* by T. Fishman. The title is "As It Was in the Beginning: Distance Education and Technology Past, Present, and Future," and it was published in volume 7, number 3, in 2002. There are no page numbers.
10. Article in the third issue of *Technical Communication*'s volume 47 in 2000. The author is Judith Ramey. The title is "Guidelines for Web Data Collection: Understanding and Interacting with Your Users." It appeared on pages 397 through 410. The URL is www.ingentaconnect.com/contentone/stc/tc/2000/00000047/ 00000003/art00010. No DOI is available.

Citing Your References in the Text

As we noted in the introduction to this chapter, any time you quote, paraphrase, or summarize information from any source, you must cite the source. APA style uses parenthetical author-date citations rather than footnotes or endnotes. IEEE numbers each source consecutively in the order that it is cited, and uses bracketed citation numbers in the text instead of parenthetical author-date citations. Here are some examples.

APA In-Text Citations

Simple Parenthetical Citation for Paraphrase

Plain language helps readers find, understand, and use information (Redish, 2000).

Parenthetical Citation for Paraphrase with Author's Name Incorporated into the Text

According to Redish (2000), plain language helps readers find, understand, and use information.

Parenthetical Citation for a Direct Quotation

According to Redish, plain-language advocates seek to create communications that enable people to "find what they need, understand what they find, and use what they understand appropriately" (2000, p. 163).

IEEE In-Text Citations

Simple Reference Number Citation for Paraphrase

Plain language helps readers find, understand, and use information [1].

Reference Number Citation for Paraphrase with Author's Name Incorporated into the Text

According to Redish [1], plain language helps readers find, understand, and use information.

Reference Number Citation for a Direct Quotation

According to Redish, plain-language advocates seek to create communications that enable people to "find what they need, understand what they find, and use what they understand appropriately" [1, p. 163].

Avoiding Plagiarism

As we noted in Chapter 3, you must provide a citation any time you summarize, quote, or paraphrase one of your sources. Using another person's intellectual property without attribution is known as plagiarism, and that is an offense that you want to avoid at any cost!

How do you know when to cite a source? The easiest rule of thumb is to think back to the point before you began your secondary research on the topic. What did you know about the subject then? Any information that doesn't fall within your knowledge before beginning your research on the topic likely requires a citation.

An exception to this rule is what is called "common knowledge." For example, you don't need to provide a citation if you explain the difference between serif and sans serif typefaces unless you are using a direct quotation from one of your sources. This distinction qualifies as common knowledge for those in the field of technical communication—it's something that anyone who has studied technical communication knows.

The problem is that common knowledge is not always common. For example, unless you are a wizard at English-metric conversions, you probably don't know that an 8.5 x 11-inch page size is 21.6 x 27.9 cm, but this equivalence also qualifies as common knowledge because it is a simple matter of converting from one established unit of measure to another.

Using a Citation Manager

Some writers find that a citation management tool like Endnote or Zotero helps them not only to format citations for items in their lists of references, but also to find and organize sources, and collaborate on secondary research.

Endnote X9 (www.endnote.com) is a software tool with a wide array of capabilities. It helps you format your citations, find full-text versions of articles, search online for books and articles similar to those that you have already located, and build a library of sources that you have identified. In addition, if you are working on a team research project, Endnote has collaboration features that allow you and your team to share references. But this power comes at a significant price. Students may be able to buy Endnote at their campus bookstore for about $115 USD, but the professional version costs $250 USD to download or $300 USD for a shipped copy.

If you aren't a graduate student working on a doctoral dissertation or a university faculty member who does secondary research frequently as part of your job, you will probably find the cost of Endnote X9 prohibitive, especially because many university libraries subscribe to Endnote Basic, a subset of the X9 product that allows you to find sources, store and share them, and create citations. These capabilities will be sufficient for most users.

Another frequently used product, Zotero (www.zotero.org), is a free open-source tool that handles many of the same chores as Endnote X9 at no cost. Zotero works with your web browser, Microsoft Word, LibreOffice, and Google Docs, and supports both APA and IEEE citation formats. It also has collaboration features if you work on team research projects.

In addition to Endnote and Zotero, a number of other citation managers are available. These include the following:

- Mendeley (www.mendeley.com/)
- RefWorks (www.refworks.com/content/products/content.asp)
- Papers (www.papersapp.com)

Answer Key

Exercise A-1

APA Style

Campbell, K. S. (1990). *Theoretical and pedagogical applications of discourse analysis to professional writing* [Unpublished doctoral dissertation]. Louisiana State University.

Fisher, L. (2001, January). Technical communicators: Designing the user experience. *Intercom*, 21–23.

Fishman, T. (2002). As it was in the beginning: Distance education and technology past, present, and future. *Kairos*, 7(3). http://kairos.technorhetoric.net/7.3/binder2.html?coverweb/fishman/index.html

Kostelnick, C., & Roberts, D. D. (1998). *Designing visual language: Strategies for professional communicators*. Allyn and Bacon.

Morkes, J., & Nielsen, J. (1997, January 1). Concise, scannable, and objective: How to write for the web. *Nielsen Norman Group*. www.nngroup.com/articles/concise-scannable-and-objective-how-to-write-for-the-web/

Porter, E. (2004, April 28). Send jobs to India? Some find it's not always best. *New York Times*, A1.

Ramey, J. 2000. Guidelines for web data collection: Understanding and interacting with your users. *Technical Communication*, 47(3), 397–410. www.ingentaconnect.com/contentone/stc/tc/2000/00000047/00000003/art00010

Teich, T., Redish, J. C., & Rainey, K. T. (1995). Interact to produce better technical communicators: Academia and industry. In *Proceedings of the STC 42nd Annual Conference* (pp. 57–60). Society for Technical Communication.

Note: The items in the list above are arranged alphabetically by (first) author's last name, rather than in the order they are given in the numbered list in the exercise. Also, because it is a personal communication, item 3 is not included in the list of references. Similarly, because it is an entire website, item 7 is excluded as well. Both would simply be cited parenthetically in the text.

IEEE Style (Using the Exercise Numbers as Citation Numbers)

[1] K. S. Campbell, "Theoretical and pedagogical applications of discourse analysis to professional writing." Ph.D. dissertation, Dept. of English, Louisiana State University, Baton Rouge, LA, 1990.

[2] L. Fisher, "Technical communicators: Designing the user experience," *Intercom*, pp. 21–23, Jan. 2001.

Note: Because it is a personal communication, item 3 is not included in the list of references. It would simply be cited parenthetically in the text.

[4] J. Morkes and J. Nielsen, Concise, scannable, and objective: How to write for the web. [Online]. Available: www.nngroup.com/articles/concise-scannable-and-objective-how-to-write-for-the-web/. [Accessed: March 12, 2020].

[5] T. Teich, J. C. Redish, and K. T. Rainey, "Interact to produce better technical communicators: Academia and industry," in *Proceedings of the STC 42nd Annual Conference*. Arlington, VA: Society for Technical Communication, 1995, pp. 57–60.

[6] E. Porter, "Send jobs to India? Some find it's not always best," *New York Times*, Apr. 28, 2004. [Online]. Available: www.nytimes.com/2004/04/28/business/send-jobs-to-india-us-companies-say-it-s-not-always-best.html

[7] User Experience Professionals Association. (n.d.). Usability Body of Knowledge. [Online]. Available: http://usabilitybok.org

[8] C. Kostelnick and D. D. Roberts, *Designing Visual Language: Strategies for Professional Communicators*, Needham Heights, MA: Allyn and Bacon, 1998.

[9] T. Fishman, "As it was in the beginning: Distance education and technology past, present, and future," *Kairos*, vol. 7, no. 3, 2002. [Online]. Available: http://kairos.technorhetoric.net/7.3/binder2.html?coverweb/fishman/index.html

[10] J. Ramey, "Guidelines for web data collection: Understanding and interacting with your users," *Tech. Commun.*, vol. 47, no. 3, pp. 397–410, 2000.

References

American Psychological Association. (2020). *Publication manual of the American Psychological Association* (7th ed.). APA.

Institute of Electrical and Electronics Engineers. (2016). *How to cite references: IEEE documentation style.* https://ieee-dataport.org/sites/default/files/analysis/27/IEEE%20Citation%20Guidelines.pdf

Institute of Electrical and Electronics Engineers. (2018). *IEEE reference guide.* http://journals.ieeeauthorcenter.ieee.org/wp-content/uploads/sites/7/IEEE-Reference-Guide.pdf

Institute of Electrical and Electronics Engineers. (2020). IEEE editorial style manual for authors. https://www.ieee.org/content/dam/ieee-org/ieee/web/org/conferences/style_references_manual.pdf

Index

Printed in the United States
by Baker & Taylor Publisher Services